Progress in Mathematics
Volume 10[illegible]

Series Editors
J. Oesterlé
A. Weinstein

Progress in Mathematics
Volume 103

Series Editors
J. Oesterlé
A. Weinstein

Bernd Anger
Claude Portenier

Radon Integrals

An abstract approach to integration and Riesz representation through function cones

Birkhäuser
Boston · Basel · Berlin

Bernd Anger
Mathematisches Institut
Universität Erlangen-Nürnberg
8520 Erlangen
Germany

Claude Portenier
Fachbereich Mathematik
Universität Marburg
3550 Marburg
Germany

Printed on acid-free paper.

ISBN 0-8176-3630-7
ISBN 3-7643-3630-7

Camera-ready text prepared by the Authors.
Printed and bound by Quinn Woodbine, Woodbine, N.J.
Printed in the U.S.A.

9 8 7 6 5 4 3 2 1

PREFACE

In topological measure theory, Radon measures are the most important objects. In the context of locally compact spaces, there are two equivalent canonical definitions. As a set function, a Radon measure is an inner compact regular Borel measure, finite on compact sets. As a functional, it is simply a positive linear form, defined on the vector lattice of continuous real-valued functions with compact support.

During the last few decades, in particular because of the developments of modern probability theory and mathematical physics, attention has been focussed on measures on general topological spaces which are no longer locally compact, e.g. spaces of continuous functions or Schwartz distributions.

For a *Radon measure on an arbitrary Hausdorff space*, essentially three equivalent definitions have been proposed :

As a set function, it was defined by L. Schwartz as an inner compact regular Borel measure which is locally bounded. G. Choquet considered it as a strongly additive right continuous content on the lattice of compact subsets. Following P.A. Meyer, N. Bourbaki defined a Radon measure as a locally uniformly bounded family of compatible positive linear forms, each defined on the vector lattice of continuous functions on some compact subset.

Compared with the simplicity of the functional analytic description in the locally compact case, it seems that the "linear functional aspect" of Radon measures has been lost in the general situation. It is our aim to show how to define "Radon integrals" as certain linear functionals, and then how to develop a theory of integration in a functional analytic spirit. Obviously, the vector lattice of continuous functions with compact support is too small, possibly even degenerate, to serve as the domain of a Radon integral. As a substitute, we consider the function cone $\mathcal{I}(X)$, i.e. the positively homogeneous and additive set of all lower semicontinuous functions on a Hausdorff space X which take values in $\tilde{\mathbb{R}} := \mathbb{R} \cup \{+\infty\}$ and are positive outside a suitable compact set. The integral on $\mathcal{I}(X)$ with respect to a Radon measure is an increasing linear $\tilde{\mathbb{R}}$-valued functional, i.e. it is positively homogeneous and additive. Among all these functionals, Radon integrals are characterized by the following regularity property, which reflects the inner regularity of Radon measures :

For every function $s \in \mathcal{S}(X)$, the integral of s can be approximated from below by the integrals of minorants of s in $-\mathcal{S}(X)$.

This leads us to define *Radon integrals* on X as regular linear functionals on $\mathcal{S}(X)$. The concept of integrability is introduced by the coincidence and finiteness of the upper and lower integrals, defined by approximation from above with functions in $\mathcal{S}(X)$, respectively from below with functions in $-\mathcal{S}(X)$. Apart from the asymmetry of approximation, this is a kind of *abstract Riemann*, i.e. finitely additive, integration theory.

One of the advantages of our simple description of Radon integrals is its extendability to functionals on cones of semicontinuous sections in a non-trivial line bundle, which may be used to treat the concept of conical measures. Note that in this context, there is no adequate set-theoretical notion of measure. However, this generalization would go beyond the limits of our exposition.

On the other hand, since a large part of the theory of integration depends only on the regularity property, the consideration of regular linear functionals on *arbitrary function cones* needs no additional tools, but enriches the theory considerably. It allows us to treat Radon integrals as a fundamental example and applies simultaneously to Radon measures in the sense of Choquet, to abstract set-theoretical aspects of integration with respect to contents on lattices of sets, to Loomis' abstract Riemann integration for positive linear forms on vector lattices of real-valued functions, as well as to the Daniell and Bourbaki integration theories. Furthermore, we have in mind future applications to potential theory and convexity, in particular Choquet representation theory. Typically, the function cones to be considered there are min-stable but not lattice cones.

We present a unified functional analytic approach to integration, in an abstract Riemann spirit, based on the following two fundamental objects :

a min-stable *cone* $\mathcal{S}$ of $\widetilde{\mathbb{R}}$-valued functions on an arbitrary set X,
a *regular* linear $\widetilde{\mathbb{R}}$-valued functional on $\mathcal{S}$.

Note that a positive linear form μ on a vector lattice of functions is always regular, whereas on a cone of positive functions the only regular linear functional is the trivial one. It is therefore just the asymmetry of function *cones* that enables us to treat regularity. On the other hand, for Daniell and Bourbaki integration, even if $\mathcal{S}$ is a vector lattice, one has to consider the function cones $\mathcal{S}^{\sigma}$ and $\mathcal{S}_{\phi}$ of all upper envelopes of sequences, respectively families of functions in $\mathcal{S}$. Using the relevant convergence properties, one first extends μ to a regular linear functional on these cones and then applies our abstract Riemann theory to the extensions.

Even for *Riesz representation theorems*, concentration on the abstract Riemann theory achieves clarity, convergence properties playing no role at all. In our abstract setting, we study the representability of a linear functional τ on a function cone $\mathcal{T}$ as an abstract Riemann integral with respect to another linear functional μ on a given function cone $\mathcal{S}$. Since the functions in $\mathcal{T}$ may be $\mathcal{S}$-unbounded, whereas integrable functions with respect to μ are always $\mathcal{S}$-bounded, we have to develop a suitable theory of *essential integration*.

The cornerstone of our approach to integration and essential integration is the concept of an *upper functional*. This is an abstract version of an upper integral, possibly without convergence properties. The general theory of integration based on this concept is developed in the first chapter.

For Riesz representation theorems, which together with regularity and Radon integrals are the main topics of the second chapter, there are two more indispensable fundamental notions to be mentioned. The first one, *tightness*, controls the representability as well as the manner of representation. The second notion, a certain kind of *measurability* with respect to the cone $\mathcal{S}$, is needed since the functions in $\mathcal{T}$ have to be measurable for the initially unknown representing functional.

As a fundamental result, we prove that every tight regular linear functional defined on a sufficiently rich lattice cone of lower semicontinuous functions on a Hausdorff space is represented by a unique Radon integral. This is a fairly general representation theorem with Radon integrals.

As mentioned before, we also treat *set-theoretical aspects* of integration for contents, defined on lattices of sets. We incorporate this abstract measure theory by first proving a correspondence between regular contents and regular linear functionals defined on a suitable cone of step functions. In particular, we infer that the set-theoretical counterparts of Radon integrals are *Radon measures*, i.e. finite contents on the lattice of compact subsets, regular with respect to the lattice of open subsets. Secondly, we introduce an adequate set-theoretical concept of *measurability* , more general than that of Loomis. We show its equivalence to the functional analytic concept of M.H. Stone which is defined by the property that truncation by integrable functions leads to integrable functions.

In order to get representation by contents, we show more generally that measurability with respect to the cone of step functions is equivalent to lattice-measurability, a concept which for a δ-lattice coincides with semicontinuity.

It was not our aim to treat all aspects of integration theory. But we hope that our functional analytic approach to integration, and in particular to Radon integrals, will lead to fruitful further developments.

We would like to express our deep gratitude to Prof. R. B. Burckel for his kind and helpful assistance in revising the English version of the manuscript. We are very much indepted to Prof. H. Bauer and Prof. J. Lembcke for encouraging conversations and stimulating discussions on the subject, as well as to our students for suggesting valuable improvements. We also thank Frau J. Happel for typing previous versions of the manuscript with great patience.

Erlangen and Marburg
November 1991

Bernd Anger
Claude Portenier

CONTENTS

INTRODUCTION

The text consists of three chapters dealing with the general theory of integration, the functional analytic aspects including Riesz representation theorems and Radon integrals, and the set-theoretical aspects, including contents and measures on lattices of sets, e.g. Radon measures.

The following description of the sections is intended as a guide for the reader as well as a brief source of definitions and results.

§ 0. We collect the special notations and terminology used, whereas for the general ones we refer to Bourbaki.

§ 1. We discuss *linear* and *(strongly) sublinear* functionals μ on *(lattice) cones* $\mathcal{S}$ of functions defined on some set X, and their associated *upper* and *lower functionals*, defined on $\bar{\mathbb{R}}^X$ by

$$\mu^*(f) := \inf_{s\in\mathcal{S},\, s\geq f} \mu(s) \quad \text{and} \quad \mu_*(f) := -\mu^*(-f).$$

At the end of this section, we present all fundamental examples to be discussed in the sequel. In particular, we give a simple construction of the Lebesgue integral in $\mathbb{R}^n$, using a particular cone of lower semicontinuous functions (Examples 1.10.2 and 8.2).

§ 2. We develop a general theory of integration for *abstract upper functionals*. These are defined as functionals ν on $\bar{\mathbb{R}}^X$ which are *determined* by a lattice cone $\mathcal{S}$ of $\tilde{\mathbb{R}}$-valued functions, i.e. satisfy

$$\nu(f) = \inf_{s\in\mathcal{S},\, s\geq f} \nu(s)$$

for all functions f, and which are strongly sublinear on this cone. Roughly speaking, this means that

$$\nu(f+g) \leq \nu(\min(f,g)) + \nu(\max(f,g)) \leq \nu(f) + \nu(g)$$

holds when addition is restricted to functional-values in $-\tilde{\mathbb{R}}$. A function f is integrable w.r.t. ν if

$$\nu(f) = \nu_*(f) := -\nu(-f) \in \mathbb{R}.$$

The interplay between the canonical cone structure of $\tilde{\mathbb{R}}$ and the order structure of $\bar{\mathbb{R}}$ is reflected in the above determination property. This is of central importance and makes the theory rich enough to allow a reasonable *calculus of null sets*, thereby avoiding infinite values. At the end of this section, we discuss the *operations* sum, multiplication with densities, images and inverse images of upper functionals.

§ 3. We treat essential integration, which comprises *abstract improper Riemann integrals* as well as *essential Radon integrals*. It is based on the *essential upper functional* $\nu^\bullet$ defined by

$$\nu^\bullet(f) := \inf_{u \in \mathcal{I}_-} \sup_{v \in \mathcal{I}_-} \nu(\mathrm{med}(f,-v,u)) ,$$

where $\mathcal{I}_-$ denotes the set of all negative functions integrable with respect to ν. The most important concept in this context is that of *almost coinitiality* with respect to ν of a set $\mathcal{T}$ to a set $\mathcal{S}$ of functions, i.e. the condition

$$\nu(s) = \inf_{t \in \mathcal{T}} \nu(\max(s,t))$$

for all $s \in \mathcal{S}$.

§ 4. We discuss *measurability* in the sense of Stone for functions f, defined as integrability of the truncated functions $\mathrm{med}(f,-v,u)$ for all negative integrable functions u and v. This enables us to formulate integrability criteria. For a min-stable function cone $\mathcal{S}$, we introduce the concept of *$\mathcal{S}$-measurability*, using one-sided relatively uniform approximation, and show in Proposition 4.11 that functions of this type are universally measurable.

§ 5. We study *upper integrals*, i.e. upper functionals ν having the *Daniell property*

$$\nu(\sup f_n) = \sup \nu(f_n)$$

for every increasing sequence (f_n) of functions with $\nu(f_n) > -\infty$. For upper integrals, the usual convergence theorems hold.

In the second chapter, the general theory is applied to upper functionals derived from linear functionals μ on a min-stable function cone $\mathcal{S}$.

§ 6. We define *regularity* by the condition

$$\mu(s) = \mu_*(s) = \sup_{t \in \mathcal{S},\ -t \le s} - \mu(t)$$

for all $s \in \mathcal{S}$, which essentially means that all functions $s \in \mathcal{S}$ with finite value $\mu(s)$ are integrable. For positive, respectively negative functions in $\mathcal{S}$, this can be interpreted as inner, respectively outer regularity. However, if all functions in $\mathcal{S}$

are positive, respectively negative, then regularity is impossible unless $\mu = 0$. The treatment of non-regular functionals is reduced to that of regular ones by consideration of the canonical linear form $\tilde{\mu}$ which is induced by μ on the function space $\mathcal{S}_- - \mathcal{S}_-$ generated by the negative functions in $\mathcal{S}$. The integration theory for μ is defined by the upper functional μ^* in the regular case, respectively by $\mu^\times := \tilde{\mu}^*$ in the non-regular case. The associated essential upper functionals, both denoted by $\mu^\bullet$, coincide in the regular case, and therefore so do the concepts of measurability with respect to μ and $\tilde{\mu}$.

Regularity is described in Theorem 6.9 by *semiregularity*, which means that

$$\mu(t) = \mu(s) + \mu_*(t - s)$$

holds for all $s \in \mathcal{S}_-$ and $t \in \mathcal{S}$ with $s \leq t$, and by a boundedness condition which only depends on the abundance of positive functions in $\mathcal{S}$.

The section ends with the study of operations on linear functionals. It turns out that semiregularity and a determination condition conserve measurability.

§ 7. We prove *Riesz representation theorems* in an abstract setting. Conditions are given for an increasing linear functional τ on a function cone $\mathcal{T}$ to be represented by a suitable linear functional μ on a given function cone $\mathcal{S}$. This means that

$$\tau(t) = \mu_\bullet(t) = \mu^\bullet(t)$$

holds for all $t \in \mathcal{T}$. The *$\mathcal{S}$-tightness* of τ, defined by

$$\tau(t) = \inf_{s\in\mathcal{S}_-} \tau_\times(\max(t,s))$$

for all $t \in \mathcal{T}_-$, is a necessary condition for τ to be representable by a functional on $\mathcal{S}$. It is also sufficient if $\mathcal{T}$ is a min-stable cone of $\mathcal{S}$-measurable functions and $\tau_\bullet$ is linear on $\mathcal{S}_-$ (Corollary 7.4). Our most general representation theorems 7.7 and 7.8 make use of a Hahn-Banach-Andeneas theorem for cones, to be proved in § 17 of the appendix.

§ 8. Due to their importance, *Radon integrals* are treated separately in this section. They have the Bourbaki property, hence a strong convergence theorem for lower semicontinuous functions holds. We include a direct proof of the fundamental representation by Radon integrals (Theorem 8.4), mentioned in the preface, without using all the machinery of § 7, and give applications to weighted cones of lower semicontinuous functions. In particular, for every linearly separating Stonian vector lattice $\mathcal{T}$ of continuous functions on a Hausdorff space, there exists a bijection between the tight positive linear forms on $\mathcal{T}$ and those Radon integrals which essentially integrate all functions in $\mathcal{T}$ (Theorem 8.8).

Regularity is interpreted set-theoretically, and Lusin's measurability theorem is proved as well as the Theorem of Egoroff. This finally enables us to discuss operations on Radon integrals.

§ 9. In this last section of the second chapter, we discuss *Daniell integration theory* . According to our general philosophy, this is reduced to abstract Riemann integration with respect to an integral on a min-stable function cone which in addition is stable with respect to upper envelopes of increasing sequences. If $\mathcal{S}^\sigma$ denotes the smallest such cone containing $\mathcal{S}$ and μ is an *integral* on $\mathcal{S}$, i.e. a linear functional which, in the regular case, has the Daniell property, we first have to extend μ to an integral μ^σ on $\mathcal{S}^\sigma$. Then we can apply the general theory of the first chapter and of § 6, including the results for upper integrals achieved in § 5, to this extension. As in § 6, the treatment of non-regular integrals μ , indispensable for extension problems, is reduced to that of regular ones considering the associated regular linear form $\tilde{\mu}$.

A large part of the section is devoted to the comparison of the different integration theories. The most important result (Theorem 9.10) states that essential integration in the sense of Daniell for a semiregular integral μ coincides with that in the abstract Riemann sense for μ^σ .

In the third chapter, we show how abstract measure theory is incorporated into the functional analytic framework developed so far.

§ 10. We study the concept of *measurability with respect to a lattice* $\mathfrak{K}$ of sets and prove its equivalence with the functional analytic concept of $\mathcal{E}_-(\mathfrak{K})$-measurability, where $\mathcal{E}_-(\mathfrak{K})$ denotes the cone generated by all negative indicator functions -1_K with $K \in \mathfrak{K}$ (Theorem 10.10). In order to handle regularity properties, we introduce the concept of compatible lattices $\mathfrak{K}$ and $\mathfrak{G}$ by

$$K \setminus G \in \mathfrak{K} \quad \text{and} \quad G \setminus K \in \mathfrak{G}$$

for all $K \in \mathfrak{K}$ and $G \in \mathfrak{G}$. We also introduce the cone $\mathcal{E}(\mathfrak{K},\mathfrak{G})$ of all step functions, generated by all -1_K and 1_G . Fundamental examples of compatible lattices are the systems $\mathfrak{K}(X)$ and $\mathfrak{G}(X)$ of compact, respectively open subsets of a Hausdorff space X, and, for Stonian cones $\mathcal{S}$ of functions, the systems $\mathfrak{K}(\mathcal{S})$ and $\mathfrak{G}(\mathcal{S})$ of all sets

$$\{s \le -1\} \text{ , respectively } \{s > 1\}$$

with $s \in \mathcal{S}$. Due to the asymmetry of function cones, the *Stonian condition* takes the form

$$\max(s,-1) \in \mathcal{S} \quad \text{and} \quad \min(s,1) \in \mathcal{S}$$

for all $s \in \mathcal{S}$.

§ 11. We define contents on a lattice $\mathfrak{K}$ of sets as increasing set functions m with $m(\emptyset) = 0$ which are *strongly additive*, i.e. satisfy

$$m(K \cup L) + m(K \cap L) = m(K) + m(L)$$

for all $K,L \in \mathfrak{K}$. In Corollary 11.3, we prove the existence of a bijection between finite contents on $\mathfrak{K}$ and increasing linear functionals on the lattice cone $\mathcal{E}_{-}(\mathfrak{K})$. This yields an *integration theory for contents* in which measurability in the sense of Stone is equivalent to measurability with respect to the lattice of integrable sets, as well as to a generalization of the measurability concept of Loomis (Theorem 11.8), and also to measurability in the sense of Carathéodory (Proposition 11.9).

§ 12. We introduce the concept of a *$\mathfrak{G}$-regular content* on $\mathfrak{K}$, so that the corresponding linear functional on $\mathcal{E}(\mathfrak{K},\mathfrak{G})$ is regular. As in the functional analytic setting, $\mathfrak{G}$-regularity of m is equivalent to $\mathfrak{G}$-boundedness and *semiregularity*, which means that

$$m(L) = m(K) + m_*(L \setminus K)$$

holds for all $K,L \in \mathfrak{K}$ with $K \subset L$, a property often misleadingly termed tightness.

The most important examples of $\mathfrak{G}$-regular contents are *Radon measures* on a Hausdorff space X. They are introduced as $\mathfrak{G}(X)$-regular, i.e. locally bounded semiregular contents on $\mathfrak{K}(X)$, which turns out to be equivalent to Choquet's definition. The functional analytic counterparts of Radon measures are Radon integrals (Theorem 12.5).

§ 13. *Measures on lattices* of sets are introduced in such a way that the corresponding linear functional on the lattice cone of step functions is an integral. For semiregular contents m, this reduces to requiring the *Daniell property at the empty set*, i.e. $\inf m(K_n) = 0$ for every decreasing sequence (K_n) in $\mathfrak{K}$ with empty intersection. We prove in Theorem 13.8 that essential integration theory in the sense of Daniell for a semiregular measure coincides with that in the abstract

Riemann sense for its canonical extension to a measure on the generated δ-lattice.

§ 14. We apply the abstract Riesz representation theorems to obtain *representation by contents*. Under the usual measurability assumption and a natural boundedness condition, the main result (Theorem 14.11) characterizes the semiregular representable functionals by a *separation condition* and *tightness*. The classical representation theorems of *Riesz, Markoff, Alexandroff,* and *Bauer* are immediate consequences.

§ 15. Representations by measures are studied, the theorem of *Daniell-Stone* being the central part of this section.

An appendix rounds off our exposition.

§ 16. We make some historical comments and discuss the relations to previous work of *Bauer, Bhaskara Rao, König, Loomis, Pollard, Topsoe* and others.

§ 17. Finally, we include a general Hahn–Banach theorem for cones, studying minimal extensions of linear functionals in the sense of *Andenaes*, which in particular is used for the Representation Theorem 7.7.

Most of the material can be covered in a one–year (about 100 hours) course. A one–semester lecture has been based on § 1 to § 6 and § 8, together with those parts of § 12 concerning the relationship between Radon measures and Radon integrals. A short version devoted only to the topological setting (cf. the references *Anger and Portenier* [1991]) is intended to give a quick approach to a functional analytic treatment of Radon integrals.

CHAPTER I.

GENERAL INTEGRATION THEORY

§ 0 NOTATIONS AND TERMINOLOGY

0.1 **The theory of integration is permeated by the interaction of two different structures, the canonical order and multiplication in the extended real line**

$$\bar{\mathbb{R}} := \mathbb{R} \cup \{\pm \infty\}$$

and the canonical conoid structure, i.e. addition and multiplication by positive scalars, in

$$\tilde{\mathbb{R}} := \mathbb{R} \cup \{\infty\} \quad \text{and} \quad -\tilde{\mathbb{R}} .$$

As usual, we define

$$-\infty \le \alpha \le \infty$$

and

$$\alpha \cdot (\pm \infty) = (\pm \infty) \cdot \alpha = (\pm \operatorname{sgn} \alpha) \cdot \infty$$

for $\alpha \in \bar{\mathbb{R}}$**, with**

$$\operatorname{sgn} 0 = 0 \quad \text{and} \quad 0 \cdot \infty = 0 .$$

Furthermore,

$$\alpha + \infty = \infty + \alpha = \infty ,$$

$$\alpha - \beta = \alpha + (-\beta) \quad \text{and} \quad \beta + (-\alpha) = -(\alpha - \beta)$$

for $\alpha \in \tilde{\mathbb{R}}$ **and** $-\beta \in \tilde{\mathbb{R}}$**. We also use the conventions**

$$\frac{1}{0} = \infty \quad \text{and} \quad \frac{1}{\infty} = 0 .$$

Note that multiplication in $\bar{\mathbb{R}}$ **and addition in** $\pm \tilde{\mathbb{R}}$**, i.e. in** $\tilde{\mathbb{R}}$ **or** $-\tilde{\mathbb{R}}$**, is compatible with the order structure. Typical examples of the interplay between the conoid structure in** $\pm \tilde{\mathbb{R}}$ **and the order structure in** $\bar{\mathbb{R}}$ **are the following :**

For $\alpha, \beta, \gamma \in \pm \tilde{\mathbb{R}}$**, we have the equality**

$$\min(\alpha + \beta, \alpha + \gamma) = \alpha + \min(\beta, \gamma) ,$$

and the equivalence of the inequalities

$$-\alpha \le \beta \quad \text{and} \quad \alpha + \beta \ge 0 .$$

Concerning the additive structure, a certain asymmetry in the basic set of numbers is inevitable. For symmetry and technical reasons, it is convenient to consider both $\tilde{\mathbb{R}}$ and $-\tilde{\mathbb{R}}$. For function values however, $\tilde{\mathbb{R}}$ is more important, since approximation from above and upper envelopes occur quite naturally.

Recall that

$$\mathbb{R}^* = \mathbb{R} \setminus \{0\} \ , \ \mathbb{R}_+ = \{\alpha \in \mathbb{R} : \alpha \geq 0\} \ \text{ and } \ \mathbb{R}_- = \{\alpha \in \mathbb{R} : \alpha \leq 0\} \ .$$

So $\mathbb{R}_+^*$ is the set of all *strictly positive* real numbers.

0.2 In integration theory, the behaviour of certain functionals on functions is studied. By a *function* we always mean a mapping from a *basic set* X into the set $\bar{\mathbb{R}}$ of extended real numbers, i.e. an element of $\bar{\mathbb{R}}^X$. We denote by $f_{|A}$ and 1_A respectively the *restriction* of the function f to the subset A of X and the *indicator function* of A, equal to 1 on A and to 0 on the *complement* $\complement A$ of A in X.

A *functional* is a function on a set of functions. If the indicator function of A is in the domain of a functional ρ, then we often use the notation $\rho(A)$ instead of $\rho(1_A)$.

0.3 All operations and relations between functions are defined pointwise. For instance, if $f,g \in \bar{\mathbb{R}}^X$, then

$$f \leq g \text{ means } f(x) \leq g(x) \text{ for all } x \in X \ ,$$

and the functions $\max(f,g)$ and $f \cdot g$ have the values $\max(f(x),g(x))$ and $f(x) \cdot g(x)$ at $x \in X$ respectively. As usual,

$$\min(f,g) = -\max(-f,-g) \ , \ |f| = \max(f,-f) \ ,$$

$$f^+ = \max(f,0) \quad \text{and} \quad f^- = \max(-f,0) \ .$$

We find it convenient to introduce

$$f_- := \min(f,0) = -f^- .$$

Truncation of a function f by functions $g \geq h$ yields the function

$$\operatorname{med}(f,g,h) := \max(\min(f,g),h) = \min(\max(f,h),g) \ .$$

This truncation process will prove to be of importance when the concepts of measurability and essential integration are discussed.

For $f,g \in \pm \tilde{\mathbb{R}}^X$, the function $f + g \in \pm \tilde{\mathbb{R}}^X$ has by definition the value $f(x) + g(x)$ at $x \in X$. We have

$$f = f^+ + f_- \quad \text{and} \quad f + g = \min(f,g) + \max(f,g) .$$

0.4 For every set of functions $\mathcal{S} \subset \bar{\mathbb{R}}^X$, we use the abbreviations

$$\tilde{\mathcal{S}} := \mathcal{S} \cap \tilde{\mathbb{R}}^X , \ \mathcal{S}_{\mathbb{R}} := \mathcal{S} \cap \mathbb{R}^X \quad \text{and} \quad \mathcal{S}^b$$

respectively for the sets of $\tilde{\mathbb{R}}$-valued, $\mathbb{R}$-valued, and bounded functions in $\mathcal{S}$. As usual, we denote by

$$\mathcal{S}_+ := \mathcal{S} \cap \tilde{\mathbb{R}}^X_+$$

the set of *positive* functions in $\mathcal{S}$ (by which we mean functions $s \in \mathcal{S}$ with $s \geq 0$), whereas we use the notation

$$\mathcal{S}_- := \mathcal{S} \cap \mathbb{R}^X_-$$

to denote the set of all *negative real-valued* functions in $\mathcal{S}$ (by which we mean functions $s \in \mathcal{S}_{\mathbb{R}}$ with $s \leq 0$). The sets

$$\mathcal{S}_{max} \quad \text{and} \quad \mathcal{S}^{\uparrow}$$

consist respectively of all functions which are upper envelopes of finitely many functions in $\mathcal{S}$, and of all functions admitting a minorant in $\mathcal{S}$. Finally, let

$$\mathcal{S}_{\sigma} \quad \text{and} \quad \mathcal{S}_{\phi}$$

respectively denote the sets of *upper envelopes* $\sup s_i$ of increasing sequences and of upward directed families of functions $s_i \in \mathcal{S}$. As usual, we say that a sequence $(s_i)_{i \in \mathbb{N}}$ is *increasing* if $s_i \leq s_{i+1}$ for all $i \in \mathbb{N}$, and a family $(s_i)_{i \in I}$ is *upward directed* if for $i,j \in I$ there exists an index $k \in I$ such that $\max(s_i,s_j) \leq s_k$.

0.5 For a property P of the elements in X, we denote by $\{P\}$ the set of those elements in X which have that property. For instance, if f is a function and $\alpha \in \bar{\mathbb{R}}$, then the set of all $x \in X$ with $f(x) = \alpha$ is denoted by $\{f = \alpha\}$, or for a functional ν on $\bar{\mathbb{R}}^X$, the set of all $f \in \tilde{\mathbb{R}}^X$ with $\nu(f) > -\infty$ is denoted by $\{\nu > -\infty\}^{\sim}$.

If X is a topological space, then the *support* supp(f) of a function f is the closure of the set $\{f \neq 0\}$.

For all other terminology, we refer to Bourbaki.

§ 1 FUNCTION CONES AND SUBLINEAR FUNCTIONALS

In this section, the basic tools for an abstract theory of integration are presented. The classical upper Darboux functional is a good prototype and will serve to orient us.

1.1 We begin, in the spirit of Darboux, with a rapid review of the classical Riemann integral in $\mathbb{R}$.

Let $\mathcal{E}$ denote the set of all *elementary functions* in $\mathbb{R}$, i.e. of all functions e of the type

$$e = \sum_{j=1}^{n} \alpha_j 1_{I_j}$$

with real α_j and bounded intervals I_j in $\mathbb{R}$. If we define

$$\iota(e) := \sum_{j=1}^{n} \alpha_j (b_j - a_j) ,$$

where $]a_j,b_j[\ \subset I_j \subset [a_j,b_j]$, then the number $\iota(e)$ is independent of the representation chosen for the elementary function e, and called the *Riemann integral* of e.

Let f be a bounded real-valued function on $\mathbb{R}$ with compact support. To every finite subdivision $U = (x_j)$ of the support of f, by which is meant finitely many points

$$x_0 < x_1 < \ldots < x_n$$

with $\mathrm{supp}(f) \subset [x_0,x_n]$, there correspond the upper and lower Darboux sums of f, given by

$$\overline{S}_U(f) := \sum_{j=1}^{n} \sup f([x_{j-1},x_j])(x_j - x_{j-1})$$

and

$$\underline{S}_U(f) := \sum_{j=1}^{n} \inf f([x_{j-1},x_j])(x_j - x_{j-1}) \,.$$

The *upper* and *lower* Darboux functionals are defined respectively by

$$\overline{S}(f) := \inf_U \overline{S}_U(f)$$

and

$$\underline{S}(f) := \sup_U \underline{S}_U(f) \,,$$

where U runs through all subdivisions of $\operatorname{supp}(f)$. We definitely do not use the terms upper and lower integral, as these are reserved for other concepts (cf. Definition 5.1).

It is easy to see that

$$\overline{S}(f) = \inf_{e\in\mathcal{E},\, e\geq f} \iota(e)$$

and

$$\underline{S}(f) = \sup_{e\in\mathcal{E},\, e\leq f} \iota(e) \,.$$

If the upper and the lower Darboux functionals coincide at f, this function is called *Riemann integrable*, and the (finite) value

$$\underline{S}(f) = \overline{S}(f)$$

is called the *Riemann integral* of f.

Since $\mathcal{E}$ contains $-e$ whenever it contains e, the lower functional may be expressed in terms of the upper functional via the identity

$$\underline{S}(f) = -\overline{S}(-f) \,.$$

In the very general situations to be encountered later it will prove to be more useful to adopt this identity as the definition of the lower functional.

IN ALL THAT FOLLOWS, X IS A FIXED NON-EMPTY SET.

1.2 Let $\mathcal{S} \subset \overline{\mathbb{R}}^X$ be a set of functions. For any functional $\mu : \mathcal{S} \longrightarrow \overline{\mathbb{R}}$ and any function f, we define

$$\mu^*(f) := \inf_{s\in\mathcal{S},\, s\geq f} \mu(s)$$

and

$$\mu_*(f) := -\mu^*(-f) \,,$$

hence

$$\mu_*(f) = \sup_{t\in\mathcal{S},\ -t\le f} -\mu(t)\ .$$

REMARK. Note that in contrast to the situation considered in 1.1, we do *not* assume that along with t also $-t$ belongs to $\mathcal{S}$. This is the reason for the minus signs in the second expression given for $\mu_*(f)$.

It is exactly this asymmetric point of view which allows us to introduce the central concept of an upper functional in § 2 in such a way that it provides a suitable base for a universal Riemannian theory of integration in which concepts such as essential integrability (cf. § 3), measurability (§ 4), and regularity (§ 6) can be treated. Furthermore, the theories of Bourbaki and Lebesgue occur as special cases in which additional convergence properties are available (cf. § 5, § 8 and § 9).

1.3 In general, for any functional μ on $\mathcal{S}$ we have

$$\{\mu_* > -\infty\} = (-\{\mu < \infty\})^{\uparrow} = -\{\mu^* < \infty\}\ .$$

Note that

$$\mu^*(f) = \inf_{s\in\{\mu<\infty\},\ s\ge f} \mu(s)$$

and

$$\mu_*(f) = \sup_{t\in\{\mu<\infty\},\ -t\le f} -\mu(t)$$

for any function f.

The set $\mathcal{S}$ and the functional μ are called *homogeneous* (resp. *positively homogeneous*) if $\alpha\cdot s \in \mathcal{S}$ and $\mu(\alpha\cdot s) = \alpha\cdot\mu(s)$ for every $s \in \mathcal{S}$ and $\alpha \in \mathbb{R}$ (resp. $\alpha \in \mathbb{R}_+$). We then have $0 \in \mathcal{S}$ and $\mu(0) = 0$.

A functional μ on $\mathcal{S}$ is *positive* if $\mu(s) \ge 0$ for $s \in \mathcal{S}_+$, and *increasing* if $\mu(s) \le \mu(t)$ for $s,t \in \mathcal{S}$ with $s \le t$.

Obviously, an increasing positively homogeneous functional is positive.

LEMMA. *The functionals μ^* and μ_* are increasing on $\overline{\mathbb{R}}^X$, and μ^* extends μ iff μ is increasing.*

Let μ be positively homogeneous. Then μ^* and μ_* are positively homogeneous iff μ is positive.

For the last part, one has to note that μ^* and μ_* are always strictly positively homogeneous and that $\mu^*(0) \le 0$, $\mu_*(0) \ge 0$. □

Functionals μ which are also extended by μ_* will play a vital role in our subsequent studies (cf. 1.10, Examples 4 and 5) :

DEFINITION. A functional μ defined on a set $\mathcal{S}$ of functions is called *regular* if $\mu = \mu_*$ on $\mathcal{S}$, i.e. if

$$\mu(s) = \sup_{t \in \mathcal{S},\ -t \leq s} - \mu(t)$$

holds for every $s \in \mathcal{S}$.

Obviously, every regular functional is increasing.

REMARK. Note that on a set $\mathcal{S}$ of positive functions with $0 \in \mathcal{S}$, the only regular functional μ with $\mu(0) = 0$ is the trivial functional 0 . This shows that regularity requires a sufficiently rich domain, containing with positive also negative functions.

1.4 A set of functions $\mathcal{S} \subset \pm \tilde{\mathbb{R}}^X$ is called a *function cone* or a *cone of functions* if $\mathcal{S}$ is positively homogeneous and stable under addition.

With the exception of § 1 and § 2, all function cones will be in $\tilde{\mathbb{R}}^X$, cones in $-\tilde{\mathbb{R}}^X$ being studied only for symmetry and technical reasons.

Obviously, $\tilde{\mathbb{R}}^X$ and $-\tilde{\mathbb{R}}^X$ are function cones. If $\mathcal{S}$ is a function cone in $\tilde{\mathbb{R}}^X$, then the sets $\mathcal{S}_{max}$, $\mathcal{S}^{\uparrow}$, $\mathcal{S}_{\sigma}$ and $\mathcal{S}_{\phi}$ are function cones, too.

A vector space $\mathcal{S} \subset \mathbb{R}^X$ of real-valued functions is called a *function space* .

A functional μ defined on a function cone $\mathcal{S}$ is called *sublinear* if it is positively homogeneous and if the set $\{\mu < \infty\}$ is a function cone on which μ is *subadditive* , i.e. satisfies

$$\mu(s + t) \leq \mu(s) + \mu(t) .$$

A functional μ is called *superlinear* if $-\mu$ is sublinear.

The following lemma is trivial and will be used without reference.

LEMMA. *If μ is sublinear, then for all $s,t \in \{\mu > -\infty\}$ we have*

$$\mu(s + t) \leq \mu(s) + \mu(t) .$$

Due to this fact, a $\pm\tilde{\mathbb{R}}$-valued functional μ on $\mathcal{S}$ is sub- and superlinear iff it is positively homogeneous and ***additive***, i.e. satisfies

$$\mu(s+t) = \mu(s) + \mu(t)$$

for all $s,t \in \mathcal{S}$. For our purposes however, it will be convenient to restrict the concept of a linear functional by the following

DEFINITION. If $\mathcal{S}$ is a function cone in $\tilde{\mathbb{R}}^X$, then a positively homogeneous and additive functional $\mu : \mathcal{S} \longrightarrow \tilde{\mathbb{R}}$ is called ***linear***.

Let μ be linear and s a real-valued function such that $\pm s \in \mathcal{S}$. Then

$$-\infty < \mu(s) = -\mu(-s) < \infty,$$

since $0 = \mu(0) = \mu(s) + \mu(-s)$.

Therefore, a linear functional μ defined on a function space $\mathcal{S}$ is always homogeneous and finite, i.e. is always a ***linear form***. Obviously, it is an increasing functional iff it is positive.

PROPOSITION. ***Every positive linear form μ defined on a function space $\mathcal{S}$ is regular.***

For $s \in \mathcal{S}$ we have

$$\mu_*(s) = -\mu(-s) = \mu(s). \quad \square$$

Let μ be a linear functional defined on a function cone $\mathcal{S}$. Then there exists a unique linear form $\tilde{\mu}$ defined on the function space $\mathcal{S}_- - \mathcal{S}_-$ which coincides with μ on $\mathcal{S}_-$. It is given by the formula

$$\tilde{\mu}(s-t) := \mu(s) - \mu(t),$$

and is positive iff μ is increasing on $\mathcal{S}_-$.

For an increasing linear functional μ we define

$$\mu^\times := \tilde{\mu}^* \quad\text{and}\quad \mu_\times := \tilde{\mu}_*,$$

i.e.

$$\mu^\times(f) = \inf_{s,t\in\mathcal{S}_-,\ s-t\geq f} \mu(s) - \mu(t)$$

and

$$\mu_\times(f) = \sup_{s,t\in\mathcal{S}_-,\ s-t\leq f} \mu(s) - \mu(t).$$

We always have

$$\mu_\times \leq \mu \leq \mu^\times \text{ on } \mathcal{S} \text{ and } \tilde{\mu} = \mu_\times = \mu^\times \text{ on } \mathcal{S}_- - \mathcal{S}_- .$$

In fact, for $s \in \mathcal{S}$ and $u,v \in \mathcal{S}_-$ with $u - v \leq s$, we obtain

$$\mu(u) \leq \mu(s + v) = \mu(s) + \mu(v) ,$$

hence $\mu_\times(s) \leq \mu(s)$, and by analogy $\mu(s) \leq \mu^\times(s)$. The second assertion is trivial. □

REMARK. The integration theory for increasing linear functionals μ which we are going to develop will be based on the functional μ^* in the regular case, but in general on the functional $\mu^\times$. Only those functionals which are extended by $\mu_\times$ will be of interest. We shall call them *difference-regular*. Every regular linear functional is of this type (cf. 6.5). The functional μ_*, however, will in any case be of fundamental importance in formulating regularity conditions (cf. § 6).

If $\mathcal{S}$ is a function space with $\mathcal{S} = \mathcal{S}_- - \mathcal{S}_-$, then obviously μ coincides with $\tilde{\mu}$, and hence μ^* coincides with $\mu^\times$.

1.5 LEMMA. *Let μ be a positive sublinear functional defined on a function cone $\mathcal{S}$. Then*

$$\mu_* \leq \mu^* ,$$

μ^* is sublinear and μ_* is superlinear on $\pm \tilde{\mathbb{R}}^X$.

Let f be any function and consider $s,t \in \{\mu < \infty\}$ with $-t \leq f \leq s$. Then we have $s + t \geq 0$, hence

$$0 \leq \mu(s + t) \leq \mu(s) + \mu(t) .$$

This gives $-\mu(t) \leq \mu(s)$, and taking suprema and infima we get $\mu_*(f) \leq \mu^*(f)$.

By Lemma 1.3, μ^* and $\{\mu^* < \infty\}$ are positively homogeneous. For functions $f,g \in \{\mu^* < \infty\} \cap \pm \tilde{\mathbb{R}}^X$, there exist $s,t \in \{\mu < \infty\}$ with $s \geq f$ and $t \geq g$. For all such s,t we have $s + t \in \{\mu < \infty\}$ and $s + t \geq f + g$, hence

$$\mu^*(f + g) \leq \mu(s + t) \leq \mu(s) + \mu(t) .$$

This gives $f + g \in \{\mu^* < \infty\} \cap \pm \tilde{\mathbb{R}}^X$ and

$$\mu^*(f+g) \le \mu^*(f) + \mu^*(g).$$

The part concerning μ_* follows by symmetry since

$$\{\mu_* > -\infty\} \cap \pm\tilde{\mathbb{R}}^X = -(\{\mu^* < \infty\} \cap \mp\tilde{\mathbb{R}}^X). \quad \square$$

1.6 **LEMMA.** *Let* μ *be a sublinear functional defined on a function cone* $\mathcal{S}$. *Then the following inequalities hold:*

(i) $$\mu_*(f+g) \le \mu_*(f) + \mu^*(g) \le \mu^*(f+g)$$

for all $f,g \in \pm\tilde{\mathbb{R}}^X$ *with* $\mu_*(f), \mu^*(g) \in \pm\tilde{\mathbb{R}}$.

(ii) *If* f *is any function with* $\mu^*(f^+) < \infty$ *or* $\mu^*(f_-) > -\infty$, *then*

$$\mu^*(f) \le \mu^*(f^+) + \mu^*(f_-),$$

and if $\mu_*(f^+) < \infty$ *or* $\mu_*(f_-) > -\infty$, *then*

$$\mu_*(f^+) + \mu_*(f_-) \le \mu_*(f).$$

For the second inequality in (i), we may assume that $\mu^*(f+g) < \infty$ and $\mu_*(f), \mu^*(g) > -\infty$, and consider $s,t \in \{\mu < \infty\}$ with $s \ge f+g$ and $-t \le f$. If $\mathcal{S} \subset \pm\tilde{\mathbb{R}}^X$, then $t + f \ge 0$, hence

$$g \le g + t + f \le s + t,$$

and if $\mathcal{S} \subset \mp\tilde{\mathbb{R}}^X$, then

$$0 \le s - f - g \le s + t - g.$$

In any case, we have $g \le s + t$ and therefore

$$\mu^*(g) \le \mu(s+t) \le \mu(s) + \mu(t).$$

This yields

$$-\mu(t) + \mu^*(g) \le \mu(s),$$

from which the result follows taking suprema and infima.

For the first inequality in (ii), we may assume that $\mu^*(f^+) < \infty$. Then there exist $s,t \in \{\mu < \infty\}$ with $s \ge f^+$ and $t \ge f_-$. For all such s,t we have $f \le s + t$, hence

$$\mu^*(f) \le \mu(s+t) \le \mu(s) + \mu(t),$$

from which the result follows taking infima.

The other inequalities in (i) and (ii) follow by symmetry. □

1.7 A set $\mathcal{S}$ of functions is said to be *min-stable* (resp. *max-stable*, $|\cdot|$*-stable*) if for all $s,t \in \mathcal{S}$ the function $\min(s,t)$ (resp. $\max(s,t)$, $|s|$) belongs to $\mathcal{S}$. A set of functions, a function cone or a function space which is both min- and max-stable will respectively be called a *lattice*, a *lattice cone* or a *vector lattice* of functions.

Obviously, $\bar{\mathbb{R}}^X$ is a lattice of functions. The sets $\mathcal{S}_{max}$, $\mathcal{S}^\uparrow$, $\mathcal{S}_\sigma$ and $\mathcal{S}_\phi$ are min-stable whenever $\mathcal{S}$ is. The sets $\mathcal{S}_{max}$ and $\mathcal{S}^\uparrow$ are always max-stable; $\mathcal{S}_\sigma$ and $\mathcal{S}_\phi$ are max-stable whenever $\mathcal{S}$ is.

PROPOSITION.

(i) *A function space* $\mathcal{S}$ *is a vector lattice iff* $\mathcal{S}$ *is either min-stable, max-stable or* $|\cdot|$*-stable.*

(ii) *For every min-stable function cone* $\mathcal{S}$, *the set* $\mathcal{S}_- - \mathcal{S}_-$ *is a vector lattice of real-valued functions.*

(iii) *A min-stable function cone* $\mathcal{S}$ *is a vector lattice of real-valued functions iff*

$$\mathcal{S} = \mathcal{S}_- - \mathcal{S}_- .$$

All this follows immediately from the formulas

$$\max(s,t) = -\min(-s,-t)\ ,\ |s| = \max(s,-s)\ ,$$

$$\min(s,t) = \tfrac{1}{2}\cdot(s + t - |s - t|)\ ,\ \max(s,t) = \tfrac{1}{2}\cdot(s + t + |s - t|)\ ,$$

$$\min(s - t, u - v) = \min(s + v, u + t) - t - v$$

and

$$s = s_- + s^+ = s_- - (-s)_- . \quad \square$$

A functional μ on a lattice is called *submodular* if $\{\mu < \infty\}$ is a lattice on which μ satisfies

$$\mu(\min(s,t)) + \mu(\max(s,t)) \le \mu(s) + \mu(t) .$$

A functional μ is called *supermodular* if $-\mu$ is submodular, and *modular* if μ is submodular and supermodular.

The following lemma is trivial and will be used without reference.

Lemma. *If μ is increasing and submodular on a lattice $\mathcal{S}$, then*

$$\mu(\min(s,t)) + \mu(\max(s,t)) \leq \mu(s) + \mu(t)$$

holds for all $s,t \in \mathcal{S}$ with $\mu((\min(s,t)) > -\infty$.

A functional μ defined on a lattice cone $\mathcal{S}$ of functions is sublinear and submodular (resp. superlinear and supermodular) iff μ is positively homogeneous and satisfies the inequalities

$$\mu(s+t) \leq \mu(\min(s,t)) + \mu(\max(s,t)) \leq \mu(s) + \mu(t) \quad (\text{resp. } \geq)$$

for all $s,t \in \{\mu < \infty\}$ (resp. $\{\mu > -\infty\}$). In this case, μ will be called *strongly sublinear* (resp. *strongly superlinear*).

Obviously, every linear functional on a lattice cone is modular, hence strongly submodular and strongly supermodular.

1.8. Lemma. *Let μ be a submodular functional defined on a lattice $\mathcal{S}$ of functions. Then μ^* is submodular and μ_* is supermodular.*

For $f,g \in \{\mu^* < \infty\}$ there exist $s,t \in \{\mu < \infty\}$ such that $s \geq f$ and $t \geq g$. For all such s,t we have $\min(s,t)$, $\max(s,t) \in \{\mu < \infty\}$ and

$$\min(s,t) \geq \min(f,g) \ , \ \max(s,t) \geq \max(f,g) .$$

Therefore,

$$\mu^*(\min(f,g)) + \mu^*(\max(f,g)) \leq \mu(\min(s,t)) + \mu(\max(s,t)) \leq \mu(s) + \mu(t) .$$

This proves the first statement. The second one concerning μ_* follows by symmetry. □

1.9 Lemma. *Let $\mathcal{S}$ be a min-stable function cone in $\bar{\mathbb{R}}^X$. Two increasing $\bar{\mathbb{R}}$-valued functionals μ and τ on $\mathcal{S}_{max}$ coincide if they coincide on $\mathcal{S}$ and if either of the following conditions is fulfilled :*

(i) *μ is submodular, τ is supermodular, and $\tau \leq \mu$.*

(ii) *μ and τ are linear.*

In the first case, let $\mathcal{T}$ denote the set of all functions $t \in \mathcal{S}_{max}$ with $\mu(t) = \tau(t)$. As $\mathcal{S} \subset \mathcal{T}$, it suffices to prove that $\mathcal{T}$ is max-stable. For $s,t \in \mathcal{T}$ we have

$$- \infty < \mu(s) + \mu(t) = \tau(s) + \tau(t) \leq \tau(\min(s,t)) + \tau(\max(s,t)) \leq$$

$$\leq \mu(\min(s,t)) + \mu(\max(s,t)) \leq \mu(s) + \mu(t) .$$

If we had $\tau(\max(s,t)) < \mu(\max(s,t))$, then we would have $\tau(\min(s,t)) < \infty$, in contradiction to the equality just proved.

In the second case, let $\mathcal{S}_n$ denote the set of upper envelopes of families involving at most n functions in $\mathcal{S}$. By hypothesis, μ and τ coincide on $\mathcal{S}_1$. If they also coincide on $\mathcal{S}_n$ and if $\max(t,s)$ is an arbitrary element in $\mathcal{S}_{n+1}$, where $t \in \mathcal{S}_n$ and $s \in \mathcal{S}$, then $\min(t,s) \in \mathcal{S}_n$ and consequently, as μ and τ are linear, we have $\mu(\max(t,s)) = \tau(\max(t,s))$ if $\mu(s)$ and $\mu(t)$ are finite. In the other case, this follows because μ and τ are increasing. Therefore, (ii) is proved by induction. □

COROLLARY. *Let $\mu : \mathcal{S} \longrightarrow \tilde{\mathbb{R}}$ be a regular sublinear functional defined on a min-stable function cone $\mathcal{S}$ in $\tilde{\mathbb{R}}^X$. Then $\mu_* = \mu^*$ is the only increasing supermodular extension τ of μ to $\mathcal{S}_{max}$. It is linear and regular with $\tau^* = \mu^*$.*

Let $f,g \in \mathcal{S}_{max}$ and $s,t \in \mathcal{S}$ be such that $s \geq f$, $t \geq g$. From Lemma 1.6.(i) we obtain

$$\mu^*(\min(f,g)) + \mu^*(\max(f,g)) \leq \mu(\min(s,t)) + \mu^*(\max(s,t)) =$$

$$= \mu_*(\min(s,t)) + \mu^*(\max(s,t)) \leq \mu^*(\min(s,t) + \max(s,t)) =$$

$$= \mu(s + t) \leq \mu(s) + \mu(t) ,$$

since $\min(s,t) \in \mathcal{S}$ and μ is regular. This proves that μ^* is submodular on $\mathcal{S}_{max}$.

Let τ be the restriction of μ^* to $\mathcal{S}_{max}$. For every function f, we have

$$\mu^*(f) \geq \tau^*(f) = \inf_{s\in\mathcal{S}_{max},\ s\geq f} \mu^*(s) \geq \mu^*(f) .$$

Hence $\mu_* = \tau_*$ is supermodular on $\mathcal{S}_{max} \subset \{\tau_* > -\infty\}$ by Lemma 1.8. Thus

$$\tau_* = \mu_* = \mu^* = \tau \quad \text{on } \mathcal{S}_{max}$$

by (i), i.e. τ is regular and, by Lemma 1.5, linear.

If ρ is any increasing supermodular extension of μ, then $\rho = \rho^* \leq \mu^*$, hence $\rho = \mu^*$ on $\mathcal{S}_{max}$. □

1.10 EXAMPLES. We close this section with some basic examples of function cones and linear functionals.

(1) As is well-known, the set $\mathcal{E}$ of elementary functions on $\mathbb{R}$, defined in 1.1, is a vector lattice of real-valued functions, and the Riemann integral ι is a positive linear form on $\mathcal{E}$. By definition,

$$\overline{S}(f) = \iota^*(f) \quad \text{and} \quad \underline{S}(f) = \iota_*(f)$$

for every bounded real-valued function f with compact support.

It will be shown in Example 13.4.1 that ι is a regular integral in the sense of Definition 9.3.

(2) In order to define the Lebesgue integral on $\mathbb{R}^n$ in Example 8.2, we introduce the following definitions and notations.

A *grid* on $\mathbb{R}^n$ is a family (x_j^k) in $\mathbb{R}$ with

$$x_j^0 < x_j^1 < \ldots < x_j^{q(j)}$$

for all $j = 1,\ldots,n$ and some $q(j) \in \mathbb{N}$. We associate with this grid the compact rectangle

$$Q := \prod_{j=1}^{n} \left[x_j^0 , x_j^{q(j)}\right],$$

and with every multi-index $l = (l(j))$ such that $0 \leq l(j) < q(j)$ the open rectangle

$$Q_l := \prod_{j=1}^{n} \left]x_j^{l(j)} , x_j^{l(j)+1}\right[$$

with its *volume*

$$vol(Q_l) := \prod_{j=1}^{n} \left[x_j^{l(j)+1} - x_j^{l(j)}\right].$$

A lower semicontinuous function t on $\mathbb{R}^n$ which takes only finitely many real values is an *elementary function* if for some grid it vanishes outside the associated rectangle Q and takes constant values α_l on the rectangles Q_l.

The set $\mathcal{T}(\mathbb{R}^n)$ of all elementary functions is a lattice cone, containing -1_K and 1_G for every compact rectangle K and every bounded open rectangle G. The volume functional

$$v : t \longmapsto \sum \alpha_l \, vol(Q_l)$$

is a regular linear functional on $\mathcal{T}(\mathbb{R}^n)$.

The first part is obvious since any two grids have a common refinement. For the second part, note that adding to the grid one, and by induction finitely many points does not alter the above value. This proves that v is well-defined and linear.

To prove that v is regular, let $t \in \mathcal{T}(\mathbb{R}^n)$ with $t \neq 0$ and some defining grid be given, and let $K := Q \setminus \bigcup Q_l$. Splitting every point x_j^k into a slightly smaller and a slightly larger one gives a new grid with open rectangles (R_p). For $\varepsilon > 0$, we can assume that

$$\sum_{R_p \cap K \neq \emptyset} vol(R_p) \leq \varepsilon/2\|t\|_\infty \quad \text{and} \quad \sum_{\alpha_l > 0} \alpha_l \cdot [vol(Q_l) - vol(R_{p_l})] \leq \varepsilon/2 ,$$

where p_l is the multi-index with $\overline{R}_{p_l} \subset Q_l$. If we define $g \in \mathcal{T}(\mathbb{R}^n)$ to take the value $\|t\|_\infty$ on the open set $R^\circ \setminus \bigcup_l \overline{R}_{p_l}$ and 0 elsewhere, then

$$s := \sum_{\alpha_l < 0} (-\alpha_l) \cdot 1_{Q_l} + g + \sum_{\alpha_l > 0} \alpha_l \cdot (-1_{\overline{R}_{p_l}}) \in \mathcal{T}(\mathbb{R}^n) ,$$

and we have $-s \leq t$ as well as

$$v(s) + v(t) = v(g) + \sum_{\alpha_l > 0} \alpha_l \cdot [vol(Q_l) - vol(R_{p_l})] \leq \varepsilon . \qquad \square$$

Note that every elementary function in $\mathcal{T}(\mathbb{R})$ is an elementary function in the sense of 1.1, and that the Riemann integral ι extends the volume functional v on $\mathcal{T}(\mathbb{R})$ to $\mathcal{E}$.

One might prefer to consider the lattice cone generated by the above functions -1_K and 1_G. It is unfortunately more complicated to prove that it coincides with the lattice cone $\mathcal{E}(\mathfrak{L},\mathfrak{H})$ (cf. Example 5), where $\mathfrak{L}$ and $\mathfrak{H}$ are the compatible lattices of all finite unions of compact, respectively bounded open, rectangles in $\mathbb{R}^n$ (cf. Example 14.13.1).

(3) Let X be a Hausdorff space. As usual, $\mathscr{C}(X)$ and $\mathscr{C}^b(X)$ denote respectively the vector lattices of continuous real-valued and of continuous bounded functions on X, whereas $\mathscr{C}^0(X)$ and $\mathscr{K}(X)$ denote respectively the vector sublattices of those continuous real-valued functions on X vanishing at infinity (i.e. converging to 0 along the filter generated by the complements of compact subsets of X), and of those having compact support.

If X is locally compact, then positive linear forms on $\mathscr{K}(X)$ are often called Radon measures. We will use this term only in the set-theoretic sense (cf. Example 5). The usual extension of μ to the lattice cone $\mathscr{K}(X)_\phi$ is a regular linear functional by Theorem 8.1, thus a Radon integral in the sense of the subsequent general definition.

(4) For an arbitrary Hausdorff space X, we introduce the following

TOPOLOGICAL STANDARD EXAMPLE. We denote by

$$\mathscr{S}(X)$$

the lattice cone of all lower semicontinuous functions on X which do not attain the value $-\infty$ and which are positive outside some compact subset of X.

If X is locally compact, then $\mathscr{S}(X) = \mathscr{K}(X)_\phi$.

DEFINITION. A regular linear functional on $\mathscr{S}(X)$ is called a *Radon integral on X.*

In Corollary 8.2 it will be proved that a Radon integral is a Bourbaki integral in the sense of Definition 8.2.

(5) The following example will show how classical abstract measure theory gets incorporated into our functional analytic framework :

MEASURE–THEORETIC STANDARD EXAMPLE. We denote by

$$\mathscr{E}(\mathfrak{K},\mathfrak{G})$$

the lattice cone of all *step functions* w.r.t. compatible lattices of sets $\mathfrak{K}$ and $\mathfrak{G}$, which will be discussed in detail in § 10.

It consists of all functions of the type

$$\sum_{i=1}^{n} \alpha_i \cdot 1_{G_i} - \sum_{j=1}^{m} \beta_j \cdot 1_{K_j}$$

with finitely many $G_i \in \mathfrak{G}$, $K_j \in \mathfrak{K}$ and $\alpha_i, \beta_j \in \mathbb{R}_+$. We prove in Proposition 10.5 that $\mathscr{E}(\mathfrak{K},\mathfrak{G})$ is a lattice cone of functions. Obviously, it consists of the usual step functions in the case of a ring $\mathfrak{K} = \mathfrak{G}$ (cf. Remark 10.5.3). In Definition 12.2, we introduce the concept of a $\mathfrak{G}$*–regular content*, and prove in Corollary 12.3 that there exists a bijection between $\mathfrak{G}$–regular contents on $\mathfrak{K}$ and regular linear functionals on $\mathscr{E}(\mathfrak{K},\mathfrak{G})$.

If X is a Hausdorff space and if $\mathfrak{K}(X)$ and $\mathfrak{G}(X)$ denote respectively the compatible lattices of compact and open subsets of X, then the $\mathfrak{G}(X)$–regular contents on $\mathfrak{K}(X)$ are by definition the Radon measures on X. We prove in Theorem 12.5 that there exists a bijection between Radon measures and Radon integrals on X.

In the special case $\mathbb{R}^n$, we will see in Example 14.13.1 that moreover there is a bijection between Radon measures and finite contents defined on the lattice of all compact rectangles and regular w.r.t. the compatible lattice of all bounded open rectangles. Furthermore, in the one-dimensional case (cf. Example 14.13.2) these contents can be described as usual by certain increasing functions on $\mathbb{R}$.

(6) Let $\mathfrak{F}$ be a filter on X and $w : X \longrightarrow \mathbb{R}_+^*$. Denote by $\mathscr{S}$ the set of all functions $s : X \longrightarrow \widetilde{\mathbb{R}}$ such that

$$\mu(s) := \lim_{\mathfrak{F}} \frac{s}{w}$$

exists in $\widetilde{\mathbb{R}}$.

$\mathscr{S}$ is a lattice cone of functions containing w, and μ is an increasing linear functional on $\mathscr{S}$ with $\mu(w) = 1$.

For every function f *one has*

$$\mu^*(f) = \limsup_{\mathfrak{F}} \frac{f}{w} \quad \textit{and} \quad \mu_*(f) = \liminf_{\mathfrak{F}} \frac{f}{w}.$$

In particular, μ *is regular.*

Indeed, the first part and the inequality $\mu^*(f) \geq \limsup_{\mathfrak{F}} \frac{f}{w}$ are trivial. For the reverse inequality we may suppose that $\limsup_{\mathfrak{F}} \frac{f}{w} < \infty$. For all $\alpha \in \mathbb{R}$ with

$$\alpha > \limsup_{\mathfrak{F}} \frac{f}{w}$$

there exists $F \in \mathfrak{F}$ such that

$$\sup_{x \in F} \frac{f(x)}{w(x)} \leq \alpha .$$

Denote by s_F the function equal to w on F and to ∞ elsewhere. Then $s_F \in \mathcal{S}$, $\mu(s_F) = 1$ and $\alpha s_F \geq f$, which shows that $\mu^*(f) \leq \alpha$ and proves the inequality $\mu^*(f) \leq \limsup_{\mathfrak{F}} \frac{f}{w}$. The rest is obvious. □

The special case $\{x\} \in \mathfrak{F}$ for some $x \in X$ leads to the Dirac integral with density $\frac{1}{w(x)}$, denoted by

$$\frac{1}{w(x)} \cdot \varepsilon_x .$$

The upper functional is given by

$$f \longmapsto \frac{f(x)}{w(x)} .$$

Another special case leads to the theory of converging sequences in $\mathbb{R}$, taking the filter of sections on $\mathbb{N}$, and $w = 1$ (cf. Example 2.7).

(7) Let X be a Hausdorff space. A *weight* is a lower semicontinuous function $w : X \longrightarrow \overline{\mathbb{R}}^*_+$, and a lower semicontinuous function $s \in \tilde{\overline{\mathbb{R}}}^X$ is said to be *w-dominated* if for every $\varepsilon > 0$ the set $\{s \leq -\varepsilon w\}$ is compact.

Every w-dominated function s *is w-bounded*, i.e. $s \geq -\alpha w$ for some $\alpha \in \mathbb{R}_+$.

Indeed, the set $K := \{s \leq -w\}$ is compact, thus $w \geq \varepsilon > 0$ and $s \geq -\beta$ on K for some $\varepsilon, \beta \in \mathbb{R}_+$. Then choose $\alpha := \max(\frac{\beta}{\varepsilon}, 1)$. □

If $\mathcal{W}$ is a set of weights, then a function is said to be *$\mathcal{W}$-dominated* if it is w-dominated for every $w \in \mathcal{W}$.

By $\mathcal{S}^{\mathcal{W}}(X)$ we denote the set of all lower semicontinuous $\mathcal{W}$-dominated functions on X, and by $\mathcal{C}^{\mathcal{W}}(X)$ the set of all continuous functions s on X such that $-|s|$ is $\mathcal{W}$-dominated.

It is easy to see that $\mathcal{S}^{\mathcal{W}}(X)$ is a lattice cone and $\mathcal{C}^{\mathcal{W}}(X)$ a vector lattice. Note that

$$\mathcal{W} \subset \mathcal{S}(X) \subset \mathcal{S}^{\mathcal{W}}(X) = \mathcal{S}^{\mathcal{W}}(X)_{\phi} \quad \text{and} \quad \mathcal{C}^{\{1\}}(X) = \mathcal{C}^{0}(X)\,.$$

An increasing linear functional τ on a lattice cone $\mathcal{T}$ of w- or $\mathcal{W}$-dominated functions is called w-, respectively *$\mathcal{W}$-bounded* if $\tau_*(w) < \infty$, respectively if this holds for some $w \in \mathcal{W}$.

In Theorem 8.4 and Lemma 8.6 we prove that if $\mathcal{S}_-(X) \subset \mathcal{T}_{\phi}$ then every $\mathcal{W}$-bounded regular linear functional on $\mathcal{T}$ is represented by a unique Radon integral on X.

§ 2 UPPER FUNCTIONALS AND INTEGRABILITY

Depending on just which properties are satisfied by the increasing linear functional μ, different theories of integration may emerge. Basic to all these theories is a functional ν defined for all functions, e.g. $\nu = \mu^*$ for the regular and $\nu = \mu^\times$ for the general theory of integration (cf. § 6), $\nu = \int^* \cdot d\mu$ or $\nu = \int^\times \cdot d\mu$ for the theories of Bourbaki and Daniell (cf. § 8 and § 9), and $\nu^\bullet$ for the corresponding theories of essential integration (cf. § 3). The desire to achieve both economy of effort and clarity of exposition lead (cf. Definition 2.2) to the concept of an upper functional which captures the essential features of all the above examples.

2.1 Let ν be a functional on $\bar{\mathbb{R}}^X$. Then ν_* denotes the functional defined on $\bar{\mathbb{R}}^X$ by

$$\nu_*(f) := -\nu(-f) .$$

With this notation, we have $\mu_* = (\mu^*)_*$ and $\mu_\times = (\mu^\times)_*$.

DEFINITION. A functional ν on $\bar{\mathbb{R}}^X$ is said to be *determined* by a set of functions $\mathcal{S}$ if

$$\nu(f) = \inf_{s \in \mathcal{S},\, s \geq f} \nu(s)$$

holds for every function $f \in \bar{\mathbb{R}}^X$.

If μ denotes the restriction of ν to $\mathcal{S}$, this requirement is equivalent to the assumption that ν coincides with μ^*, hence ν_* coincides with μ_*. In particular, ν and ν_* are then increasing. In contrast to the setting discussed in § 1, the functional ν defined on $\bar{\mathbb{R}}^X$ will now be the central object of our studies, whereas μ and $\mathcal{S}$ will merely be of technical importance, and μ will not necessarily be linear.

2.2 DEFINITION. A functional ν defined on $\tilde{\mathbb{R}}^X$ is called an *upper functional* if $\{\nu < \infty\}^\sim$ is a lattice cone on which ν is strongly sublinear and which determines ν.

The functional ν_* is called the *lower functional* (associated with ν).

EXAMPLE. *Let μ be an increasing strongly sublinear functional defined on a lattice cone of $\tilde{\mathbb{R}}$-valued functions, e.g. a linear functional. Then μ^* is an upper functional.*

In particular, the functionals ι^* and v^* associated with the Riemann integral ι on $\mathscr{E}$ and the volume functional v on $\mathcal{T}(\mathbb{R}^n)$ (cf. Examples 1 and 2 of 1.10) are upper functionals.

This example shows that ν is an upper functional iff it is determined by a lattice cone in $\tilde{\mathbb{R}}^X$ on which ν is strongly sublinear.

THEOREM. *Every upper functional ν has the following properties:*

(i) ν *and* ν_* *are positively homogeneous, increasing, and* $\nu_* \leq \nu$.

(ii) ν *is submodular and* ν_* *is supermodular.*

(iii) ν *is strongly sublinear and* ν_* *is strongly superlinear on* $\pm\,\tilde{\mathbb{R}}^X$.

(iv) *If f is a function with $\nu(f^+) < \infty$ or $\nu(f_-) > -\infty$, resp. $\nu_*(f^+) < \infty$ or $\nu_*(f_-) > -\infty$, then*

$$\nu(f) = \nu(f^+) + \nu(f_-)\,, \quad \textit{resp.} \quad \nu_*(f^+) + \nu_*(f_-) = \nu_*(f)\ .$$

(v) *For all functions $f, g \in \pm\,\tilde{\mathbb{R}}^X$ with $\nu_*(f), \nu(g) \in \pm\,\tilde{\mathbb{R}}$, we have*

$$\nu_*(f + g) \leq \nu_*(f) + \nu(g) \leq \nu(f + g)\,.$$

This follows immediately from Lemmas 1.3, 1.5, 1.8 and 1.6.(i), except (iv). But by Lemma 1.6.(ii) and the above assertion (ii), we get

$$\nu(f) \leq \nu(f^+) + \nu(f_-) = \nu(\max(f, 0)) + \nu(\min(f, 0)) \leq \nu(f) + \nu(0) = \nu(f)\,.$$

The proof for the assertion concerning ν_* follows by symmetry. □

REMARK. Our convention, that an upper functional be determined by a lattice cone in $\tilde{\mathbb{R}}^X$ and hence by $\bar{\mathbb{R}}^X$, permits properties of $\bar{\mathbb{R}}^X$ (generally linked to the order structure) to be combined with properties of $\tilde{\mathbb{R}}^X$ (generally linked to the cone structure). Instances of this occur in particular in the proof of the second statement of Lemma 1.6 and in that of Proposition 2.9.

2.3 PROPOSITION. *Let $\mu : \mathcal{S} \longrightarrow \tilde{\mathbb{R}}$ be a regular linear functional defined on a min-stable function cone $\mathcal{S}$ in $\tilde{\mathbb{R}}^X$. Then μ^* is an upper functional.*

By Corollary 1.9, μ^* is linear on $\mathcal{S}_{max}$ and is determined by this lattice cone. □

COROLLARY. *Let μ be an increasing linear functional defined on a min-stable function cone $\mathcal{S}$. Then $\mu^\times$ is an upper functional.*

With the notations from 1.4 we have $\mu^\times = \tilde{\mu}^*$, and $\tilde{\mu}$ is a (regular) linear functional on the vector lattice $\mathcal{S}_- - \mathcal{S}_-$ (cf. Proposition 1.7). □

REMARK. Let μ be an increasing linear functional defined on a function cone $\mathcal{S}$. Then one can prove that μ^* is an upper functional iff μ^* is linear on $\mathcal{S}_{max}$, which is equivalent to the formally weaker requirement that μ^* be submodular on $\mathcal{S}_{max}$.

In general, for an increasing linear functional μ defined on a min-stable function cone $\mathcal{S}$, the functional μ^* will not be an upper functional. This is shown in the following

EXAMPLE. Let $\mathcal{S}$ be the min-stable cone of all concave real-valued functions on $[-1,1]$ and consider the Dirac integral ε_0 on $\mathcal{S}$, i.e. the evaluation at 0. We have

$$\varepsilon_0(id) = \varepsilon_0(-id) = \varepsilon_0(\min(id,-id)) = 0 \quad \text{and} \quad \varepsilon_0^*(\max(id,-id)) = 1\,,$$

which shows that ε_0^* is not submodular on $\mathcal{S}_{max}$.

IN ALL THAT FOLLOWS, ν IS AN UPPER FUNCTIONAL.

2.4 **PROPOSITION.** *A function f satisfies $\nu(|f|) < \infty$ iff*

$$-\infty < \nu_*(f) \leq \nu(f) < \infty .$$

Since $\pm f \leq |f|$, the necessity follows since ν is increasing, whereas the sufficiency is a consequence of the max–stability of $\{\nu < \infty\}$. □

2.5 We now introduce the basic concept of integrability with respect to upper functionals. Again, we are guided by 1.1.

DEFINITION. A function f is said to be *integrable* (w.r.t. ν) if

$$-\infty < \nu_*(f) = \nu(f) < \infty .$$

This number is called the *integral* of f. The set of integrable functions will be denoted by

$$\mathcal{I}(\nu)$$

or $\mathcal{I}$ for short.

REMARK. *If ν is determined by a set $\mathcal{S}$ of functions, then f is integrable iff for every $\varepsilon > 0$ there exist $s,t \in \mathcal{S}$ with*

$$-t \leq f \leq s \ , \quad \nu(s),\nu(t) \in \mathbb{R} \quad \textit{and} \quad \nu(s) + \nu(t) \leq \varepsilon .$$

PROPOSITION. *Let μ be a regular linear functional defined on a min–stable function cone $\mathcal{S}$. A function f is integrable w.r.t. μ^* iff for every $\varepsilon > 0$ there exist $s,t \in \mathcal{S}$ with*

$$-t \leq f \leq s \quad \textit{and} \quad \mu(s + t) \leq \varepsilon .$$

Every function $s \in \mathcal{S}$ with $\mu(s) < \infty$, e.g. $s \in \mathcal{S}_-$, is integrable. In particular, if μ is a positive linear form on a vector lattice of functions $\mathcal{S}$, then every function in $\mathcal{S}$ is integrable.

This follows directly from the above remark and the definition of regularity. □

Concerning the integration theory w.r.t. $\mu^{\times}$ for arbitrary increasing linear functionals μ, we refer to § 6.

EXAMPLES.

(1) For the Riemann integral ι on $\mathbb{R}$, every continuous function with compact support is integrable w.r.t. ι^*.

In fact, approximation is possible since these functions are uniformly continuous.

(2) Consider the volume functional v on $\mathcal{T}(\mathbb{R}^n)$. The indicator function 1_A of every set A, with $Q \subset A \subset \overline{Q}$ for some bounded open rectangle Q, is integrable w.r.t. v^*.

This follows by enlarging $\overline{Q}$ to an open and shrinking Q to a compact rectangle with approximately the same volume.

Every elementary function e on $\mathbb{R}$ (cf. 1.1) is hence integrable w.r.t. v^*. Since $v^*(e) = \iota(e)$, we have $v^* \le \iota^*$. The converse inequality is trivial, ι being an extension of v. So

$$v^* = \iota^* .$$

2.6 THEOREM. *Integrable functions enjoy the following stability properties:*

(i) $\mathfrak{I}(\nu)$ *is a homogeneous lattice of functions on which* ν *is finite, increasing, homogeneous and modular.*

(ii) $\tilde{\mathfrak{I}}(\nu)$ *is a lattice cone of functions on which* ν *is finite, increasing and linear.*

(iii) $\mathfrak{I}_{\mathbb{R}}(\nu)$ *is a vector lattice of real-valued functions on which* ν *is a positive linear form.*

By Theorem 2.2, for $f,g \in \mathfrak{I}$ and $\alpha \in \mathbb{R}_+$ we have

$$\begin{aligned} -\infty < \nu_*(f) + \nu_*(g) &\le \nu_*(\min(f,g)) + \nu_*(\max(f,g)) \le \\ &\le \nu(\min(f,g)) + \nu(\max(f,g)) \le \\ &\le \nu(f) + \nu(g) = \nu_*(f) + \nu_*(g) < \infty , \end{aligned}$$

$$-\infty < \alpha\cdot\nu_*(f) = \nu_*(\alpha\cdot f) \le \nu(\alpha\cdot f) = \alpha\cdot\nu(f) = \alpha\cdot\nu_*(f) < \infty ,$$

and

$$-\infty < -\nu(f) = \nu_*(-f) \le \nu(-f) = -\nu_*(f) = -\nu(f) < \infty .$$

Finally, for $f,g \in \tilde{\mathfrak{I}}$ we obtain

$$-\infty < \nu_*(f) + \nu_*(g) \le \nu_*(f+g) \le \nu(f+g)$$
$$\le \nu(f) + \nu(g) = \nu_*(f) + \nu_*(g) < \infty . \quad \square$$

REMARK. *The set of all functions f in $\bar{\mathbb{R}}^X$ (resp. in $\tilde{\mathbb{R}}^X$) such that*

$$-\infty < \nu_*(f) = \nu(f)$$

is max-stable (resp. a max-stable function cone).

This follows from the monotonicity of ν if $\nu(f) = \infty$ or $\nu(g) = \infty$, and otherwise from (i). The proof of stability with respect to addition is similar. $\square$

COROLLARY. *For any function f, we have:*

(i) $|f|$ *is integrable whenever f is, and then*

$$|\nu(f)| \le \nu(|f|) .$$

(ii) f *is integrable iff f^+ and f_- (equivalently f^+ and f^-) are each integrable.*

Indeed, if f is integrable, then f^+, f_-, f^- and $|f|$ are also integrable, and the formula in (i) follows from the inequality $\pm f \le |f|$. Assertion (ii) is a consequence of Theorem 2.2.(iv). $\square$

2.7 PROPOSITION. *If at least one of the functions $f,g \in \pm\tilde{\mathbb{R}}^X$ is integrable, then*

$$\nu(f+g) = \nu(f) + \nu(g) \quad \textit{and} \quad \nu_*(f+g) = \nu_*(f) + \nu_*(g) .$$

For instance, if $g \in \tilde{\mathfrak{I}}$, then applying Theorem 2.2, (iii) and (v), one gets

$$\nu(f+g) \le \nu(f) + \nu(g) = \nu(f) + \nu_*(g) \le \nu(f+g) .$$

The second equality follows by symmetry. $\square$

Example. Consider the special case of Example 1.10 with the filter of sections on $\mathbb{N}$ and $w = 1$. A sequence in $\bar{\mathbb{R}}$ is integrable w.r.t. the upper functional

$$\nu : f \longmapsto \limsup_{n\to\infty} f(n)$$

on $\bar{\mathbb{R}}^{\mathbb{N}}$ iff it is convergent in $\mathbb{R}$. Theorem 2.6 yields the calculus of convergent sequences, and Theorem 2.2.(v), respectively the proposition, yields the useful results

$$\liminf\,[f(n) + g(n)] \le \liminf f(n) + \limsup g(n) \le \limsup\,[f(n) + g(n)] \;,$$

respectively

$$\liminf\,[f(n) + g(n)] = \lim f(n) + \liminf g(n)$$

for all sequences f,g in $\pm\,\tilde{\mathbb{R}}$ with $\liminf f(n), \limsup g(n) \in \pm\,\tilde{\mathbb{R}}$, respectively f convergent in $\mathbb{R}$ and $\liminf g(n) \in \pm\,\tilde{\mathbb{R}}$.

2.8 We now introduce an equivalence relation denoted by " mod ν ". If f and g are real-valued functions on X, then by definition $f = g \mod \nu$ if $\nu(|f - g|) = 0$. In order to be able to extend this relation to $\bar{\mathbb{R}}^X$, we find it convenient to introduce the following

Notation. For functions f and g, let $(f - g)^+$ denote the function defined on X by

$$(f - g)^+(x) := f(x) - g(x)\,, \;\text{ if } f(x) > g(x)\,, \;\text{ and } 0 \text{ otherwise.}$$

Furthermore, let

$$|f - g| := \max[(f - g)^+, (g - f)^+]\,.$$

These notations are consistent with the usual meanings of these symbols when f and g are real-valued.

Definition. The relations defined on $\bar{\mathbb{R}}^X$ by the equalities

$$\nu([f - g]^+) = 0 \quad\text{and}\quad \nu(|f - g|) = 0$$

will respectively be denoted by

$$" f \le g \mod \nu " \quad\text{and}\quad " f = g \mod \nu "\,.$$

It is easy to see that the first relation defines a preorder on $\bar{\mathbb{R}}^X$, since

$$(f - h)^+ \leq (f - g)^+ + (g - h)^+$$

for all functions f, g and h. The second one is the equivalence relation associated with this preorder, since $\nu(|f - g|) = 0$ is equivalent to

$$\nu([f - g]^+) = \nu([g - f]^+) = 0 .$$

2.9 **Proposition.** *If f and g are functions with $f \leq g \bmod \nu$, then*

$$\nu(f) \leq \nu(g) \quad \textit{and} \quad \nu_*(f) \leq \nu_*(g) .$$

In particular, if $f = g \bmod \nu$, then

$$\nu(f) = \nu(g) \quad \textit{and} \quad \nu_*(f) = \nu_*(g) ,$$

and g is integrable iff f is.

For every $s \in \{\nu < \infty\}^\sim$ with $s \geq g$, we have $s + (f - g)^+ \geq f$, hence

$$\nu(f) \leq \nu(s + [f - g]^+) \leq \nu(s) + \nu([f - g]^+) = \nu(s)$$

by Theorem 2.2, (i) and (iii), and therefore $\nu(f) \leq \nu(g)$. As $-g \leq -f \bmod \nu$, we also get $\nu_*(f) \leq \nu_*(g)$. This immediately implies the remaining conclusions. □

2.10 The equivalence relation introduced in Definition 2.8 is closely related to the notion of null set, which we are going to define now. For a subset A of X, we denote by ∞_A the function defined on X as ∞ on A and 0 outside A.

Definition. A set $A \subset X$ is called a *null set* (w.r.t. ν) if $\nu(\infty_A) = 0$. A property which holds except for the points of a null set is said to hold *almost everywhere*, abbreviated *a.e.*

Obviously, every subset of a null set, as well as the union of finitely many null sets, is a null set itself. Furthermore, for any null set A, the function ∞_A is integrable.

The relation " $f \leq g$ almost everywhere " defines a preorder on $\overline{\mathbb{R}}^X$. The corresponding equivalence relation is given by " $f = g$ almost everywhere ". We are now going to compare this preorder with the one introduced in Definition 2.8.

2.11 PROPOSITION. *For any functions f and g, we have :*

(i) *If $f \leq g$ a.e., then $f \leq g \bmod \nu$. In particular,*

$$f = g \text{ a.e. implies } f = g \bmod \nu .$$

(ii) *If $\nu(f) < \infty$ or $\nu_*(f) > -\infty$, then respectively*

$$f < \infty \text{ a.e. or } f > -\infty \text{ a.e.}$$

In particular, an integrable function is finite almost everywhere.

Let A be a null set and suppose that $f \leq g$ outside A. Then

$$(f - g)^+ \leq \infty_A$$

and therefore $\nu([f - g]^+) = 0$. This proves (i).

If $\nu(f) < \infty$, then $\nu(f^+) < \infty$ since $\{\nu < \infty\}$ is max-stable. As

$$\infty_{\{f=\infty\}} \leq \alpha f^+ \text{ for every } \alpha > 0 ,$$

we have

$$0 \leq \nu(\infty_{\{f=\infty\}}) \leq \inf_{\alpha>0} \nu(\alpha f^+) = 0 ,$$

hence $f < \infty$ a.e. The assertion concerning ν_* follows replacing f by $-f$. □

COROLLARY. *Modifying a function on a null set changes neither its value of the upper functional nor that of the lower functional.*

In particular, if $\nu(f) < \infty$, $\nu_*(f) > -\infty$ or $-\infty < \nu_*(f) \leq \nu(f) < \infty$, then f can be modified respectively to

$$1_{\{f<\infty\}} \cdot f \ , \ 1_{\{f>-\infty\}} \cdot f \quad or \quad 1_{\{f\in\mathbb{R}\}} \cdot f \ ,$$

without altering the values of the upper and lower functionals. If f is integrable, then the modified functions are integrable, too.

This results immediately from Proposition 2.9. □

REMARKS.

(1) The corollary shows that the quotient of $\mathcal{J}(\nu)$ modulo ν can be identified with that of $\mathcal{J}_{\mathbb{R}}(\nu)$ which is a vector space.

(2) The reasonable conjecture that the equivalence relation " mod ν " on $\bar{\mathbb{R}}^X$ coincides with " equality a.e. " is correct if ν has the so-called Daniell property (cf. Definition 5.1. and Proposition 5.3). In general, however, the equivalence relation defined by null sets will be strictly finer :

In the Example 1.10.1 of the Riemann integral ι on $\mathbb{R}$, let f denote the function on $\mathbb{R}$ which takes the value $\frac{1}{q}$ at $x \in]0,1]$ whenever $x = \frac{p}{q}$ is a rational number with p and q prime to each other, and which vanishes elsewhere. Then f is integrable (w.r.t. ι^*) and its integral is 0. Consequently, we have $f = 0$ mod ι^*, whereas $\iota^*(1_{\mathbb{Q}\cap]0,1]}) = 1$, hence $\iota^*(\infty_{\mathbb{Q}\cap]0,1]}) = \infty$. Therefore, we do not have $f = 0$ almost everywhere w.r.t. ι^*.

This example shows as well that in the following proposition " mod ν " may not, without additional hypotheses, be replaced by " a.e. w.r.t. ν ".

2.12 If $g \geq 0$ is a positive function, then by definition, $g = 0$ mod ν iff $\nu(g) = 0$. More generally, we have the following

PROPOSITION. *Let f and g be integrable functions. If*

$$\nu(f) = \nu(g) \quad \textit{and} \quad f \leq g \bmod \nu ,$$

then $f = g \bmod \nu$.

By Proposition 2.11.(ii), the set $A := \{f \notin \mathbb{R}\} \cup \{g \notin \mathbb{R}\}$ is a null set. The real-valued functions $f_0 := 1_{CA} \cdot f$ and $g_0 := 1_{CA} \cdot g$ are integrable by Corollary 2.11, and

$$\nu(g_0 - f_0) = \nu(g_0) - \nu(f_0) = \nu(g) - \nu(f) = 0 .$$

As

$$(g_0 - f_0)^+ = g_0 - f_0 + (f_0 - g_0)^+ \leq g_0 - f_0 + (f - g)^+ ,$$

it follows that

$$\nu([g_0 - f_0]^+) \leq \nu(g_0 - f_0) + \nu([f - g]^+) = 0 .$$

From

$$(g - f)^+ \le \infty_A + (g_0 - f_0)^+$$

we finally get

$$\nu([g - f]^+) \le \nu(\infty_A) + \nu([g_0 - f_0]^+) = 0\ ,$$

hence $g \le f \mod \nu$ and therefore $g = f \mod \nu$. □

2.13 We now study the set $\mathcal{F}(\nu)$ of all functions f with $\nu(|f|) < \infty$. Since

$$|f - g| \le |f - h| + |h - g|\ ,$$

we have $\nu(|f - g|) \in \mathbb{R}_+$ for $f,g \in \mathcal{F}(\nu)$, and

$$(f,g) \longmapsto \nu(|f - g|)$$

is a pseudo-metric on $\mathcal{F}(\nu)$, called the pseudo-metric of *convergence in mean*.

Proposition. *For $f,g \in \mathcal{F}(\nu)$, the inequality*

$$|\nu(f) - \nu(g)| \le \nu(|f - g|)$$

holds. In particular, ν is uniformly continuous on $\mathcal{F}(\nu)$.

In view of Corollary 2.11 we may suppose f and g to be finite. Then

$$f \le g + |f - g|\ ,$$

hence

$$\nu(f) \le \nu(g + |f - g|) \le \nu(g) + \nu(|f - g|)\ .$$

This yields $\nu(f) - \nu(g) \le \nu(|f - g|)$. If we interchange f and g, we get the stated inequality. □

Corollary. *$\mathcal{I}(\nu)$ is closed in $\mathcal{F}(\nu)$, i.e. if (f_n) is a sequence of integrable functions converging in mean to $f \in \mathcal{F}(\nu)$, then f is integrable, and moreover*

$$\nu(f) = \lim \nu(f_n)\ .$$

The formula follows from the continuity of ν on $\mathcal{F}$. Furthermore,

$$\nu(f) - \nu(f_n) \le \nu(|f - f_n|)$$

and

$$\nu(f_n) - \nu_*(f) = -\nu(-f_n) + \nu(-f) \le \nu(|-f + f_n|) = \nu(|f - f_n|)\ ,$$

the functions f_n being integrable. This implies that

$$\nu(f_n) - \nu(|f - f_n|) \le \nu_*(f) \le \nu(f) \le \nu(f_n) + \nu(|f - f_n|),$$

and the integrability of f emerges upon taking the limit on n. □

2.14 If for an upper functional ρ we have

$$\mathfrak{I}(\rho) = \mathfrak{I}(\nu) \quad \text{and} \quad \rho = \nu \text{ on } \mathfrak{I}(\nu),$$

then we say that the *theories of integration* for ρ and ν *coincide*.

We shall now introduce a class of functionals for which this coincidence implies in fact equality.

Definition. An upper functional is said to be *auto-determined* if it is determined by a set of integrable functions.

It follows from Corollary 2.11 that these functions may be chosen to be $\tilde{\mathbb{R}}$-valued, and then plainly ν is determined by $\tilde{\mathfrak{I}}(\nu)$.

Example. *Let μ be a regular linear functional defined on a min-stable cone of functions. Then the upper functional μ^* is auto-determined.*

This follows from Proposition 2.5 since μ is determined by $\{\mu < \infty\}$. □

Proposition. *If for every function f we define*

$$\nu^{\#}(f) := \inf_{g \in \tilde{\mathfrak{I}},\, g \ge f} \nu(g),$$

then $\nu^{\#}$ is the only auto-determined upper functional whose integration theory coincides with that of ν.

We have $\nu \le \nu^{\#}$, and moreover $\nu = \nu^{\#}$ iff ν is auto-determined.

Obviously, $\nu^{\#}$ coincides with ν on $\tilde{\mathfrak{I}}(\nu)$, and this lattice cone of functions determines $\nu^{\#}$. This proves that $\nu^{\#}$ is an upper functional. Plainly, $\nu \le \nu^{\#}$ and hence $\mathfrak{I}(\nu^{\#}) \subset \mathfrak{I}(\nu)$.

If A is a null set w.r.t. ν, then $\infty_A \in \tilde{\mathfrak{I}}(\nu)$, hence

$$\nu^{\#}(\infty_A) = \nu(\infty_A) = 0,$$

i.e. A is a null set w.r.t. $\nu^\#$.

Let now $f \in \mathfrak{I}(\nu)$ and modify f to $g := 1_{\{f \in \mathbb{R}\}} \cdot f$. Then $\pm g \in \tilde{\mathfrak{I}}(\nu)$, hence

$$\nu^\#_*(g) = -\nu^\#(-g) = -\nu(-g) = \nu(g) = \nu^\#(g),$$

i.e. $g \in \mathfrak{I}(\nu^\#)$. Consequently $f \in \mathfrak{I}(\nu^\#)$ by Corollary 2.11, since $f = g$ a.e. w.r.t. ν, hence w.r.t. $\nu^\#$. By the same argument,

$$\nu^\#(f) = \nu^\#(g) = \nu(g) = \nu(f).$$

This proves that the integration theories for ν and $\nu^\#$ coincide and that $\nu^\#$ is auto-determined. The uniqueness property and the last assertion are obvious. □

2.15 We conclude this section with a discussion of some operations on upper functionals.

DEFINITION. Let ν and ρ be any functionals on $\bar{\mathbb{R}}^X$. The *sum* $\nu + \rho$ is defined by

$$(\nu + \rho)(f) := \nu(f) + \rho(f) \quad \text{if} \quad \nu(f), \rho(f) < \infty,$$

and by ∞ else.

PROPOSITION. *If ν and ρ are upper functionals, then $\nu + \rho$ is an upper functional, for which a function is integrable iff it is integrable w.r.t. ν and ρ.*

The functional $\nu + \rho$ is determined by

$$\{\nu + \rho < \infty\}^{\sim} = \{\nu < \infty\}^{\sim} \cap \{\rho < \infty\}^{\sim}$$

since for $f \in \{\nu + \rho < \infty\}$ and $s \in \{\nu < \infty\}^{\sim}$, $t \in \{\rho < \infty\}^{\sim}$ with $s,t \geq f$, the function $\min(s,t) \in \tilde{\mathbb{R}}^X$ satisfies $\min(s,t) \geq f$ and

$$(\nu + \rho)(\min(s,t)) \leq \nu(s) + \rho(t) < \infty.$$

This proves that $\nu + \rho$ is an upper functional. Indeed, $\{\nu + \rho < \infty\}^{\sim}$ is obviously a lattice cone on which $\nu + \rho$ is strongly sublinear, since

$$(\nu + \rho)(s + t) = \nu(s + t) + \rho(s + t) \leq$$

$$\leq \nu(s) + \nu(t) + \rho(s) + \rho(t) =$$

$$= (\nu + \rho)(s) + (\nu + \rho)(t)$$

and

$$(\nu + \rho)(\min(s,t)) + (\nu + \rho)(\max(s,t)) =$$

$$= \nu(\min(s,t)) + \rho(\min(s,t)) + \nu(\max(s,t)) + \rho(\max(s,t)) \leq$$

$$\leq \nu(s) + \nu(t) + \rho(s) + \rho(t) =$$

$$= (\nu + \rho)(s) + (\nu + \rho)(t)$$

for all $s,t \in \{\nu + \rho < \infty\}^{\sim}$.

Finally, we have

$$(\nu + \rho)(f) = \nu(f) + \rho(f) \text{ and } (\nu + \rho)_*(f) = \nu_*(f) + \rho_*(f)$$

whenever f belongs to $\mathcal{I}(\nu + \rho)$ or $\mathcal{I}(\nu) \cap \mathcal{I}(\rho)$, which proves the integrability assertion. □

2.16 DEFINITION. Let ν be any functional on $\bar{\mathbb{R}}^X$ and $w \in \bar{\mathbb{R}}_+^X$. We say that the functional $w\cdot\nu$, defined by

$$w\cdot\nu(f) := \nu(wf),$$

has the *density* w w.r.t. ν.

NOTATION. For a function $f \in \bar{\mathbb{R}}^X$, let

$$\frac{f}{w} := \frac{1}{w}\cdot f.$$

We have

$$w\cdot\frac{f}{w} = 1_{\{0<w<\infty\}}\cdot f.$$

PROPOSITION. *If ν is an upper functional, then $w\cdot\nu$ is an upper functional, for which a function f is integrable iff wf is integrable w.r.t. ν.*

Note first that $w\cdot\nu$ is determined by the set $\{w\cdot\nu < \infty\}^{\sim}$. Indeed, for $f \in \{w\cdot\nu < \infty\}$ and $s \in \{\nu < \infty\}^{\sim}$ with $s \geq wf$, the function

$$t := \frac{s}{w} + \infty_{\{w=0\}} + 1_{\{w=\infty\}}\cdot\max(s_-,f) \in \tilde{\mathbb{R}}^X$$

satisfies $t \geq f$ and $wt \leq s$, since $s \geq 0$ on $\{w = 0\}$.

Obviously, $\{w\cdot\nu < \infty\}^{\sim}$ is a lattice, multiplication with elements of $\bar{\mathbb{R}}_+$ respecting the order structure of $\bar{\mathbb{R}}$. Moreover, it is a function cone since

$$w\cdot\nu(s+t) \leq \nu(w(s^+ + t^+)) \leq \nu((ws)^+) + \nu((wt)^+) < \infty$$

for $s,t \in \{w\cdot\nu < \infty\}^{\sim}$. This proves that $w\cdot\nu$ is an upper functional, since it is strongly sublinear on $\{w\cdot\nu < \infty\}^{\sim}$. Indeed, according to Corollary 2.11, ws and wt can respectively be replaced by $1_{\{ws<\infty\}}\cdot ws$ and $1_{\{wt<\infty\}}\cdot wt$, and ν is strongly sublinear on $-\tilde{\mathbb{R}}^X$.

The last part is trivial. □

2.17 DEFINITION. Let $p : X \longrightarrow Y$ be a mapping from X into a set Y and ν any functional on $\bar{\mathbb{R}}^X$. We say that the functional $p(\nu)$, defined on $\bar{\mathbb{R}}^Y$ by

$$p(\nu)(g) := \nu(g\circ p)\,,$$

is the *image* of ν under p.

NOTATION. For any function $f \in \bar{\mathbb{R}}^X$, let f^p and f_p denote the functions defined on Y by

$$f^p(y) := \sup_{x\in p^{-1}(y)} f(x) \quad \text{and} \quad f_p(y) := \inf_{x\in p^{-1}(y)} f(x)\,.$$

We have

$$f_p = -(-f)^p \quad \text{and} \quad f_p\circ p \leq f \leq f^p\circ p\,,$$

as well as

$$(g\circ p)^p \leq g \quad \text{and} \quad (g\circ p)^p = g \text{ on } p(X)$$

for any function $g \in \bar{\mathbb{R}}^Y$.

PROPOSITION. *If ν is an upper functional determined by a set $\mathcal{S} \subset \tilde{\mathbb{R}}^X$ with $s_p \in \tilde{\mathbb{R}}^Y$ for all $s \in \mathcal{S}$, then $p(\nu)$ is an upper functional.*

If $p(\nu)$ is an upper functional, then a function g on Y is integrable w.r.t. $p(\nu)$ iff $g\circ p$ is integrable w.r.t. ν.

Note that the functional $p(\nu)$ is determined by the set $\{p(\nu) < \infty\}^{\sim}$. Indeed, for $g \in \{p(\nu) < \infty\}$ and $s \in \mathcal{S} \cap \{\nu < \infty\}^{\sim}$ with $s \geq g\circ p$, the function

$t := s_p \in \tilde{\overline{\mathbb{R}}}^Y$ satisfies $t \geq g$ and $t \circ p \leq s$.

This proves that $p(\nu)$ is an upper functional, since $\{p(\nu) < \infty\}^{\sim}$ is the set of all $t \in \tilde{\overline{\mathbb{R}}}^Y$ with $\nu(t \circ p) < \infty$ and hence a lattice cone on which $p(\nu)$ is strongly sublinear. The last part is trivial. □

Remark. The following example shows that in general $p(\nu)$ is not an upper functional. In order to avoid this, one might however generalize the definition of upper functionals using the function cone $\{\nu < \infty\} \cap - \tilde{\overline{\mathbb{R}}}^X$ both for sublinearity and for almost everywhere majorization in the determination property. But this is not worth while since then functionals of the following kind would occur :

Example. Consider on $\mathbb{R}_+$ the vector lattice $\mathcal{S} := \mathbb{R} \cdot id$ and the positive linear form μ defined by $\mu(id) = 1$. Then μ^* is an upper functional, but for $p : \mathbb{R}_+ \longrightarrow 1$, the functional $p(\mu^*)$ is not. Indeed, for $g \in \overline{\mathbb{R}}$ the function $g \circ p$ is constant and $p(\mu)^*(g) = \mu^*(g \circ p)$ is equal to ∞ if $g > 0$, to 0 if $0 \geq g > -\infty$ and to $-\infty$ if $g = -\infty$.

2.18 Proposition. *Let ρ be an upper functional on a set Y and p a mapping from X to Y. There exists an upper functional ν on X with $p(\nu) = \rho$ iff $Y \setminus p(X)$ is a null set w.r.t. ρ. In this case, the inverse image of ρ under p, defined by*

$$p^{-1}(\rho)(f) := \rho(f^p),$$

is an upper functional on X with

$$p(p^{-1}(\rho)) = \rho.$$

For any upper functional ν on X we have

$$\nu \leq p^{-1}(\rho) \quad iff \quad p(\nu) \leq \rho.$$

Furthermore, a function f on X is integrable w.r.t. $p^{-1}(\rho)$ iff f^p and f_p are integrable w.r.t. ρ and have the same integral. A function g on Y is integrable w.r.t. ρ iff $g \circ p$ is integrable w.r.t. $p^{-1}(\rho)$.

The condition is necessary since

$$\rho(\infty_{\complement p(X)}) = p(\nu)(\infty_{\complement p(X)}) = \nu(\infty_{\complement p(X)} \circ p) = 0 .$$

Conversely, note first that for all functions g on Y we have

$$p^{-1}(\rho)(g \circ p) = \rho((g \circ p)^p) = \rho(g) ,$$

since $(g \circ p)^p = g$ on $p(X)$, hence a.e. w.r.t. ρ. This proves that $p(p^{-1}(\rho)) = \rho$.

The functional $p^{-1}(\rho)$ is determined by the set $\{\rho < \infty\}^{\sim} \circ p$ of all $t \circ p$ with $t \in \{\rho < \infty\}^{\sim}$. Indeed, for $f \in \{p^{-1}(\rho) < \infty\}$ and $t \in \{\rho < \infty\}^{\sim}$ with $t \geq f^p$, we have $t \circ p \geq f$ and $(t \circ p)^p = t$ a.e. w.r.t. ρ. Thus $p^{-1}(\rho)$ is an upper functional, since $\{\rho < \infty\}^{\sim} \circ p$ is a lattice cone on which $p^{-1}(\rho)$ is strongly sublinear by the above formula. This also yields the characterization of integrability for g, as well as the equivalence assertion of the first part. In fact, by the determination property,

$$\nu \leq p^{-1}(\rho) \quad \text{iff} \quad \nu(t \circ p) \leq p^{-1}(\rho)(t \circ p) \quad \text{for all } t \in \{\rho < \infty\}^{\sim} .$$

Finally, for a function f on X we have

$$p^{-1}(\rho)_*(f) = - p^{-1}(\rho)(-f) = - \rho((-f)^p) = \rho_*(f_p)$$

and $f_p \leq f^p$ a.e. w.r.t. ρ. The characterization of integrability of f now follows immediately. □

§ 3 ESSENTIAL INTEGRATION

3.1 In the example of the classical Riemann integral ι on $\mathbb{R}$, a function which is integrable w.r.t. ι^* has to be bounded and to have compact support. In order to define an integral for a function which is unbounded or whose support is not compact, one may proceed as follows :

A function f on $\mathbb{R}$ is called *essentially Riemann integrable* if for all $m,n \in \mathbb{N}$ the truncated functions

$$f_{mn} := \text{med}(f, m\cdot 1_{[-m,m]}, -n\cdot 1_{[-n,n]})$$

are Riemann integrable and if the sequence $(\iota^*(f_{mn}))$ of their integrals converges in $\mathbb{R}$ as m and n tend to infinity independently. In this case, the limit is equal to

$$\inf_n \sup_m \iota^*(f_{mn}) = \sup_m \inf_n \iota^*(f_{mn}) .$$

It is called the *essential Riemann integral* of f. One says that this integral, usually denoted by

$$\int_{-\infty}^{\infty} f(x)dx ,$$

is *absolutely convergent*, since

$$\sup_n \iota^*(\min(|f|, n\cdot 1_{[-n,n]})) < \infty .$$

It is of great importance that this concept of integration is amenable to description as integration with respect to an upper functional and that such a concept may be introduced not only for ι^* but for any upper functional ν. We shall give the precise definition of the associated essential upper functional $\nu^\bullet$ below and prove in Corollary 3.7 that integrability w.r.t. $\nu^\bullet$ can be described as in the classical case.

3.2 According to our general conventions, $\mathcal{I}_- = \mathcal{I}_-(\nu)$ will denote the set of all negative *real-valued* functions which are integrable w.r.t. ν. For every function f we define

$$\nu^\bullet(f) := \inf_{u\in\mathcal{J}_-} \sup_{v\in\mathcal{J}_-} \nu(\mathrm{med}(f,-v,u))$$

and

$$\nu_\bullet := (\nu^\bullet)_* .$$

Alternatively viewed,

$$\nu_\bullet(f) = \sup_{v\in\mathcal{J}_-} \inf_{u\in\mathcal{J}_-} \nu_*(\mathrm{med}(f,-v,u)) .$$

To prove the last equality, one has to note that

$$-\,\mathrm{med}(-f,-v,u) = \mathrm{med}(f,-u,v) .$$

REMARK. *At any place in the formulas for* $\nu^\bullet$ *and* $\nu_\bullet$ *we could have replaced* $\mathcal{J}_-$ *by the set* $\bar{\mathcal{J}}_-$ *of all* $\bar{\mathbb{R}}$*-valued negative functions which are integrable w.r.t.* ν .

This follows immediately by successive application of Corollary 2.11. □

LEMMA.

(i) *For every function* f *with an integrable minorant, we have*

$$\nu^\bullet(f) = \sup_{v\in\mathcal{J}_-} \nu(\min(f,-v)) \le \nu(f)$$

and

$$\nu_\bullet(f) = \sup_{v\in\mathcal{J}_-} \nu_*(\min(f,-v)) \le \nu_*(f) .$$

In particular, $\nu^\bullet(f) = \nu(f)$ *if* f *is bounded by integrable functions.*

(ii) *For every function* f *, we have*

$$\nu^\bullet(f) = \inf_{u\in\mathcal{J}_-} \nu^\bullet(\max(f,u)) .$$

The first assertion follows from the definition by recourse to the preceding remark. The formula in (ii) is proved by taking the infimum in the equalities

$$\nu^\bullet(\max(f,u)) = \sup_{v\in\mathcal{J}_-} \nu(\min[\max(f,u),-v]) = \sup_{v\in\mathcal{J}_-} \nu(\mathrm{med}(f,-v,u)) . \quad \square$$

3.3 LEMMA. *The inequality*

$$\min(f_1 + f_2, -v) \le \min[f_1, -(v+u_2)] + \min[f_2, -(v+u_1)]$$

holds for all $f_1,f_2 \in \tilde{\mathbb{R}}^X$ *and all* $u_1,u_2,v \in \mathbb{R}_-^X$ *which satisfy* $u_1 \le f_1$ *and* $u_2 \le f_2$.

Since for $i = 1,2$ we have

$$f_i - u_i - v \geq -v \quad \text{and} \quad -(2v + u_1 + u_2) = -v - (v + u_1 + u_2) \geq -v\,,$$

we get indeed

$$\min[f_1, -(v + u_2)] + \min[f_2, -(v + u_1)] =$$
$$= \min[f_1 + f_2, f_1 - (v + u_1), f_2 - (v + u_2), -(2v + u_1 + u_2)] \geq$$
$$\geq \min(f_1 + f_2, -v)\,. \quad \Box$$

PROPOSITION. *$\nu^\bullet$ is an upper functional, determined by the lattice cone $\mathcal{I}_-^\uparrow(\nu)$ of all functions admitting a real-valued integrable minorant.*

It follows from Lemma 3.2.(ii) that $\nu^\bullet$ is determined by $\mathcal{I}_-^\uparrow$. It is therefore sufficient to prove that $\nu^\bullet$ is strongly sublinear on this lattice cone. For the calculation of $\nu^\bullet$, we refer to Lemma 3.2.(i).

Let $f_1, f_2 \in \mathcal{I}_-^\uparrow$ and $u_1, u_2 \in \mathcal{I}_-$ be such that $u_n \leq f_n$ for $n = 1,2$. Since ν is sublinear on $\mathcal{I}_-^\uparrow$, the preceding lemma shows that for $v \in \mathcal{I}_-$ we have

$$\nu(\min[f_1 + f_2, -v]) \leq \nu(\min[f_1, -(v + u_2)]) + \nu(\min[f_2, -(v + u_1)]) \leq$$
$$\leq \nu^\bullet(f_1) + \nu^\bullet(f_2).$$

Because $\nu^\bullet$ is plainly positively homogeneous, this establishes the sublinearity of $\nu^\bullet$ on $\mathcal{I}_-^\uparrow$.

Since ν is submodular on $\mathcal{I}_-^\uparrow$, for $v_1, v_2 \in \mathcal{I}_-$ and $v := v_1 + v_2$ we have

$$\nu(\min[\min(f_1,f_2),-v_1]) + \nu(\min[\max(f_1,f_2),-v_2]) \leq$$
$$\leq \nu(\min[\min(f_1,-v),\min(f_2,-v)]) + \nu(\max[\min(f_1,-v),\min(f_2,-v)]) \leq$$
$$\leq \nu(\min(f_1,-v)) + \nu(\min(f_2,-v)) \leq \nu^\bullet(f_1) + \nu^\bullet(f_2)\,.$$

This proves that $\nu^\bullet$ is submodular on $\mathcal{I}_-^\uparrow$. $\Box$

DEFINITION. $\nu^\bullet$ and $\nu_\bullet$ are respectively called the *essential upper functional* and the *essential lower functional* associated with ν.

Let μ be a regular linear functional defined on a min-stable cone of functions. Then $(\mu^*)^\bullet$ is an upper functional, denoted by

$$\mu^\bullet .$$

The corresponding essential lower functional will be denoted by $\mu_\bullet$.

REMARK. To be able to realize that $\nu^\bullet$ is an upper functional, even in the special case of $\mu^\bullet$ for a positive linear form μ , we are compelled to consider lattice cones instead of vector lattices of functions in the role of determining sets for upper functionals. This is one of the reasons which led us to the asymmetric point of view presented in Remark 1.2.

3.4 DEFINITION. A function f is called *essentially integrable* (w.r.t. ν) if f is integrable w.r.t. $\nu^\bullet$, i.e. if

$$-\infty < \nu_\bullet(f) = \nu^\bullet(f) < \infty .$$

This number is called the *essential integral* of f . We denote by

$$\mathcal{I}^\bullet(\nu)$$

(or $\mathcal{I}^\bullet$ for short) the set $\mathcal{I}(\nu^\bullet)$ of all essentially integrable functions.

Notice that the results from § 2 apply in particular to $\nu^\bullet$. Instead of null sets and properties holding almost everywhere w.r.t. $\nu^\bullet$, we speak of *essentially null sets* and properties holding *essentially almost everywhere* (w.r.t. ν).

The next result is an immediate consequence of the definitions and of Lemma 3.2.(i) :

PROPOSITION.

(i) *Every integrable function is essentially integrable, and its integral coincides with its essential integral.*

(ii) *Every essentially integrable function which is bounded by integrable functions is integrable.*

(iii) *Every null set is an essentially null set, and the relation " $\leq$ mod ν " implies the relation " $\leq$ mod $\nu^\bullet$ ".*

3.5 The special role played by the functions $-n\cdot 1_{[-n,n]}$ in the Example 3.1 of the classical Riemann integral on $\mathbb{R}$ is an instance of the following general concept :

DEFINITION. Let $\mathcal{T}$ be a downward directed and $\mathcal{S}$ an arbitrary set of functions.

If every function in $\mathcal{S}$ is minorized by some function in $\mathcal{T}$, i.e. $\mathcal{S} \subset \mathcal{T}^{\uparrow}$, then $\mathcal{T}$ is said to be *coinitial to* $\mathcal{S}$.

More generally, if the formula

$$\nu(s) = \inf_{t\in\mathcal{T}} \nu(\max(s,t))$$

holds for all $s \in \mathcal{S}$, then $\mathcal{T}$ is said to be *almost coinitial to* $\mathcal{S}$ (w.r.t. ν).

If moreover $\mathcal{T}$ is a subset of $\mathcal{S}$, then $\mathcal{T}$ is said to be coinitial, respectively almost coinitial, *in* $\mathcal{S}$.

When no confusion can result, we omit the reference to the upper functional, e.g. $\mathcal{T}$ is said to be almost coinitial in $\mathcal{I}_-(\nu)$ if this holds w.r.t. ν.

EXAMPLE. *$\mathcal{I}_-(\nu)$ is almost coinitial in $\overline{\mathbb{R}}^X$ w.r.t. $\nu^\bullet$, in particular almost coinitial in $\mathcal{I}_-^\bullet(\nu)$.*

This follows from Lemma 3.2.(ii) and Proposition 3.4.(i). □

LEMMA. *Let $\mathcal{T}$ be almost coinitial to $\mathcal{S}$ w.r.t. ν. If $\nu > -\infty$ on $\mathcal{S}$, then for any family (f_i) of functions we have*

$$\inf_{t\in\mathcal{T}} \sup_i \nu(\max(f_i,t)) \leq \inf_{s\in\mathcal{S}} \sup_i \nu(\max(f_i,s)) .$$

If $\mathcal{T}$ and $\mathcal{S}$ are almost coinitial to each other and $\nu > -\infty$ on both of them, then equality holds.

We may assume $\nu(s)$ to be finite for some $s \in \mathcal{S}$. Since $\nu > -\infty$ on $\mathcal{S}^{\uparrow}$ and hence is submodular there, we get

$$\nu(\max(f_i,t)) + \nu(s) \leq$$

$$\leq \nu(\max[\max(f_i,s),\max(s,t)]) + \nu(\min[\max(f_i,s),\max(s,t)]) \leq$$

$$\leq \nu(\max(f_i,s)) + \nu(\max(s,t)) .$$

Taking the supremum over i, the infimum on $\mathcal{T}$ and using coinitiality, this yields the formula by finally taking the infimum on $\mathcal{S}$. □

PROPOSITION.

(i) *Let* $\mathcal{T}$ *be almost coinitial to* $\mathcal{S}$ *and* $\mathcal{S}$ *be almost coinitial to* $\mathcal{U}$. *If* $\nu > -\infty$ *on* $\mathcal{S}$, *then* $\mathcal{T}$ *is also almost coinitial to* $\mathcal{U}$.

(ii) $\mathcal{T}$ *is almost coinitial in* $\bar{\mathbb{R}}^X$ *w.r.t.* ν *iff* ν *is determined by* $\mathcal{T}^{\uparrow}$.

To prove (i), let $u \in \mathcal{U}$ be given. We then have

$$\nu(u) \leq \inf_{t\in\mathcal{T}} \nu(\max(u,t)) \leq \inf_{s\in\mathcal{S}} \nu(\max(u,s)) = \nu(u)$$

by the lemma, which proves that $\mathcal{T}$ is almost coinitial to $\mathcal{U}$.

If ν is determined by $\mathcal{T}^{\uparrow}$ and if f is any function, then for every $g \in \mathcal{T}^{\uparrow}$ with $g \geq f$ there exists $t \in \mathcal{T}$ with $t \leq g$, and consequently $g \geq \max(f,t)$. This yields

$$\nu(f) \leq \inf_{t\in\mathcal{T}} \nu(\max(f,t)) \leq \inf_{g\in\mathcal{T}^{\uparrow},\, g\geq f} \nu(g) = \nu(f) .$$

The converse statement in (ii) is trivial. □

COROLLARY. *Let* $\mathcal{T}$ *be a subset of* $\mathcal{I}_-(\nu)$. *Then* $\mathcal{T}$ *is almost coinitial in* $\mathcal{I}_-(\nu)$ *w.r.t.* ν *iff* $\mathcal{T}$ *is almost coinitial in* $\bar{\mathbb{R}}^X$ *w.r.t.* $\nu^{\bullet}$.

This follows by transitivity since $\mathcal{I}_-$ is almost coinitial in $\bar{\mathbb{R}}^X$ w.r.t. $\nu^{\bullet}$, and $\nu^{\bullet}$ coincides with ν on $\mathcal{I}_-$. □

3.6 LEMMA. *Let* $\mathcal{S} \subset \mathcal{I}_-(\nu)$, *and* $\mathcal{T} \subset \bar{\mathbb{R}}^X$ *be almost coinitial to* $\mathcal{S}$ *w.r.t.* ν. *For any* $f \in \bar{\mathbb{R}}^X$, *we have*

$$\sup_{s\in\mathcal{S}} \nu(\min(f,-s)) \leq \sup_{t\in\mathcal{T}} \nu(\min(f,-t)) .$$

If $\mathcal{T}$ *and* $\mathcal{S}$ *are almost coinitial to each other and contained in* $\mathcal{I}_-(\nu)$, *then equality holds.*

In fact, we have the inequality

$$\min(f,-s) + s - \max(s,t) \le \min(f,-t) ,$$

since

$$\min[f + s - \max(s,t), \min(-s,-t)] \le \min(f,-t)$$

is meaningful and holds on $\{f < \infty\}$. Theorem 2.2.(v) and Proposition 2.7 now give

$$\nu(\min(f,-s)) + \nu(s) - \nu(\max(s,t)) =$$

$$= \nu(\min(f,-s)) + \nu_*(s - \max(s,t)) \le \nu(\min(f,-t)) .$$

Taking the supremum on $\mathcal{T}$, using coinitiality and then taking the supremum on $\mathcal{S}$, the inequality follows. □

THEOREM. *Let* $\mathcal{T}$ *be almost coinitial in* $\mathcal{I}_-(\nu)$ *.*

(i) *If a function* f *admits an integrable minorant, then*

$$\nu^\bullet(f) = \sup_{t\in\mathcal{T}} \nu(\min(f,-t))$$

and

$$\nu_\bullet(f) = \sup_{t\in\mathcal{T}} \nu_*(\min(f,-t)) .$$

(ii) *For every function* f *, we have*

$$\nu^\bullet(f) = \inf_{s\in\mathcal{T}} \sup_{t\in\mathcal{T}} \nu(\mathrm{med}(f,-t,s)) = \lim_{s\in\mathcal{T}} \lim_{t\in\mathcal{T}} \nu(\mathrm{med}(f,-t,s)) ,$$

where the limits are to be taken in $\bar{\mathbb{R}}$ *along the downward directed set* $\mathcal{T}$ *.*

By Lemma 3.2.(i), the first formula in (i) follows from the preceding lemma, and the second one from Lemma 3.5, applied to $-f$.

By the lemma, applied to $\max(f,s)$ for $s \in \mathcal{T}$, we have

$$\sup_{t\in\mathcal{T}} \nu(\mathrm{med}(f,-t,s)) = \sup_{v\in\mathcal{I}_-} \nu(\mathrm{med}(f,-v,s)) .$$

This yields

$$\nu^\bullet(f) = \inf_{u\in\mathcal{I}_-} \sup_{v\in\mathcal{I}_-} \nu(\mathrm{med}(f,-v,u)) = \inf_{s\in\mathcal{T}} \sup_{v\in\mathcal{I}_-} \nu(\mathrm{med}(f,-v,s)) =$$

$$= \inf_{s\in\mathcal{T}} \sup_{t\in\mathcal{T}} \nu(\mathrm{med}(f,-t,s))$$

by Lemma 3.5, applied to $(\min(f,-v))_{v\in\mathcal{I}_-}$. □

COROLLARY. *Essential integration is idempotent, i.e.*

$$(\nu^\bullet)^\bullet = \nu^\bullet .$$

As $\mathcal{I}_-$ is almost coinitial in $\mathcal{I}_-^\bullet$ (cf. Example 3.5), we can apply (ii) w.r.t. $\nu^\bullet$; this gives

$$(\nu^\bullet)^\bullet(f) = \inf_{u\in\mathcal{I}_-} \sup_{v\in\mathcal{I}_-} \nu^\bullet(\mathrm{med}(f,-v,u)) = \nu^\bullet(f)\,,$$

recalling that $\nu^\bullet(\mathrm{med}(f,-v,u)) = \nu(\mathrm{med}(f,-v,u))$ by Lemma 3.2.(i). □

3.7 **Corollary.** *Let* $\mathcal{T}$ *be almost coinitial in* $\mathcal{I}_-(\nu)$. *A function* f *is essentially integrable iff* $\mathrm{med}(f,-t,s)$ *is integrable for all* $s,t \in \mathcal{T}$ *and*

$$\lim_{(s,t)\in\mathcal{T}\times\mathcal{T}} \nu(\mathrm{med}(f,-t,s))$$

exists in $\mathbb{R}$ *along the downward directed set* $\mathcal{T}\times\mathcal{T}$.

In this case, the essential integral of f *is equal to this limit and also coincides with*

$$\inf_{s\in\mathcal{T}} \sup_{t\in\mathcal{T}} \nu(\mathrm{med}(f,-t,s)) = \sup_{t\in\mathcal{T}} \inf_{s\in\mathcal{T}} \nu(\mathrm{med}(f,-t,s))\,.$$

Let f be essentially integrable. Then $\mathrm{med}(f,-t,s)$ is essentially integrable and is bounded by integrable functions, hence is integrable by Proposition 3.4.(ii). Let $\varepsilon > 0$ be given. Since $\nu^\bullet(f)$ is finite, by Theorem 3.6.(ii) there exists $s_0 \in \mathcal{T}$ such that for all $s \in \mathcal{T}$ with $s \le s_0$ we have

$$\sup_{t\in\mathcal{T}} \nu(\mathrm{med}(f,-t,s)) \le \nu^\bullet(f) + \varepsilon\,.$$

By the same argument, there exists $t_0 \in \mathcal{T}$ such that for all $t \in \mathcal{T}$ with $t \le t_0$ we have

$$\inf_{s\in\mathcal{T}} \nu(\mathrm{med}(f,-t,s)) = \inf_{s\in\mathcal{T}} \nu_*(\mathrm{med}(f,-t,s)) \ge \nu_\bullet(f) - \varepsilon\,.$$

Since $\nu_\bullet(f) = \nu^\bullet(f)$, we get

$$\nu^\bullet(f) - \varepsilon \le \nu(\mathrm{med}(f,-t,s)) \le \nu^\bullet(f) + \varepsilon$$

for all $s,t \in \mathcal{T}$ with $s \le s_0$ and $t \le t_0$. This proves that the condition is necessary and that the limit is equal to the essential integral of f .

Conversely, if the limit exists, then it coincides with both numbers given in the formulas. Moreover, since $\mathrm{med}(f,-t,s)$ is integrable, the numbers on the left and on the right hand sides coincide respectively with $\nu^\bullet(f)$ and $\nu_\bullet(f)$. Therefore, f is essentially integrable. □

Remark. *If ρ is an upper functional whose integration theory coincides with that of ν, then the theories of essential integration for ρ and ν also coincide, i.e.*

$$\mathcal{I}^\bullet(\rho) = \mathcal{I}^\bullet(\nu) \quad \textit{and} \quad \rho^\bullet = \nu^\bullet \quad \textit{on } \mathcal{I}^\bullet(\nu) .$$

This is an immediate consequence of the corollary. □

3.8 Examples.

(1) In the example of the classical Riemann integral ι on $\mathbb{R}$, the set of functions $-n\cdot 1_{[-n,n]}$ is coinitial in $\mathcal{I}_-(\iota^*)$. Therefore, for any function f, we have

$$\iota^\bullet(f) = \inf_n \sup_m \iota^*(f_{mn}) .$$

The preceding corollary shows that f is essentially integrable w.r.t. ι^* iff f is essentially Riemann integrable in the sense of 3.1.

(2) Consider the topological standard example.

Let μ be a Radon integral. The set $\mathcal{T}$ of functions $-n\cdot 1_K$, as n runs through $\mathbb{N}$ and K through the system of compact subsets of X, is almost coinitial in $\mathcal{I}_-(\mu^*)$. We have

$$\mu^\bullet(f) = \sup_{n,K} \mu^*(\min(f, n\cdot 1_K))$$

for every function $f \geq 0$.

Since $\mathcal{K}(X) \subset \mathcal{T}^\uparrow$, the first part follows from Example 2.5 and Proposition 3.5.(ii). Applying Theorem 3.6.(i), we get the formula. □

We will prove later in Corollary 8.2 that μ is a Bourbaki integral, and thus μ^* is an upper integral by Theorem 5.2. This implies that

$$\mu^\bullet(f) = \sup_K \mu^*(1_K \cdot f) ,$$

which is the classical definition of the essential upper integral associated with a Radon integral.

The concept of essential integration is of central importance not only in the (finitely additive) integration theory in the abstract Riemann sense, but also in the non σ-finite (σ-additive) integration theory in the sense of Daniell.

3.9 PROPOSITION. *If ν is determined by $\mathcal{T}_-^{\uparrow}(\nu)$, then*

$$\nu^{\bullet} \leq \nu ,$$

and the equality

$$\nu_*(f) = \nu_{\bullet}(f)$$

holds for every function f which admits an integrable minorant.

Proposition 3.5.(ii) shows that $\mathcal{T}_-$ is almost coinitial in $\bar{\mathbb{R}}^X$ w.r.t. ν. This yields

$$\nu(f) = \inf_{u\in\mathcal{T}_-} \nu(\max(f,u)) \geq \inf_{u\in\mathcal{T}_-} \sup_{v\in\mathcal{T}_-} \nu(\mathrm{med}(f,-v,u)) = \nu^{\bullet}(f)$$

for any function f. The second formula follows from the first one and Lemma 3.2.(i). □

COROLLARY. *If ν is auto-determined, then*

$$\nu^{\bullet} \leq \nu ,$$

and the equality

$$\nu_*(f) = \nu_{\bullet}(f)$$

holds for every function f with $\nu_*(f) > -\infty$.

We only have to remark that by hypothesis $\{\nu_* > -\infty\}$ is the set of all functions admitting an integrable minorant. □

3.10 PROPOSITION. *The functional $\nu^{\bullet\#}$ is the smallest auto-determined upper functional whose theory of essential integration coincides with that of ν.*

For any such functional ρ we even have

$$\rho^{\bullet} = \nu^{\bullet\#\bullet} = \nu^{\#\bullet} \geq \nu^{\bullet} .$$

By Proposition 2.14 and Remark 3.7.1, the theories of essential integration for $\nu^{\bullet\#}$ and $\nu^{\bullet}$, hence for $\nu^{\bullet\#}$ and ν, coincide (cf. Corollary 3.6).

By hypothesis on ρ and since $\mathcal{T}(\rho) \subset \mathcal{T}^{\bullet}(\rho)$, we have

$$\nu^{\bullet\#} = \rho^{\bullet\#} \leq \rho^{\#} = \rho ,$$

which proves the first part.

Since $\mathcal{T}_-(\rho)$ is almost coinitial in $\mathcal{T}_-^{\bullet}(\rho)$ w.r.t. $\rho^{\bullet}$ (cf. Example 3.5),

$\mathfrak{I}\_(\rho)$ is also almost coinitial in $\mathfrak{I}\_(\nu^{\bullet\#})$ w.r.t. $\nu^{\bullet\#}$. In fact, the theories of essential integration for ρ and ν coincide, hence the theories of integration for $\rho^{\bullet}$, $\nu^{\bullet}$ and $\nu^{\bullet\#}$ coincide. Moreover, for every function f with $\rho(f) < \infty$, Corollary 3.9 yields

$$\rho(f) = \inf_{g\in\tilde{\mathfrak{I}}(\rho),\, g\geq f} \rho(g) = \inf_{g\in\tilde{\mathfrak{I}}^{\bullet}(\rho),\, g\geq f} \rho^{\bullet}(g) =$$

$$= \inf_{g\in\tilde{\mathfrak{I}}^{\bullet}(\nu),\, g\geq f} \nu^{\bullet}(g) = \nu^{\bullet\#}(f) .$$

Thus, for arbitrary f, we infer from Theorem 3.6.(ii) that

$$\rho^{\bullet}(f) = \inf_{u\in\mathfrak{I}\_(\rho)} \sup_{v\in\mathfrak{I}\_(\rho)} \rho(\mathrm{med}(f,-v,u)) =$$

$$= \inf_{u\in\mathfrak{I}\_(\rho)} \sup_{v\in\mathfrak{I}\_(\rho)} \nu^{\bullet\#}(\mathrm{med}(f,-v,u)) = \nu^{\bullet\#\bullet}(f) ,$$

since $\mathrm{med}(f,-v,u) \leq -v \in \tilde{\mathfrak{I}}(\rho)$.

Finally, note that $\nu^{\#}$ is an auto-determined upper functional whose theory of essential integration coincides with that of ν by Proposition 2.14 and Remark 3.7.1. By the above, we have $\nu^{\#\bullet} = \nu^{\bullet\#\bullet}$. Since $\nu^{\#} \geq \nu$, we infer from Theorem 3.6.(ii) that $\nu^{\#\bullet} \geq \nu^{\bullet}$, which concludes the proof. □

§ 4 MEASURABILITY

4.1 In the present section, we discuss the concept of measurability in the sense of Stone and give some integrability criteria.

DEFINITION. A function f is said to be *measurable* (w.r.t. ν) if the truncated function $\mathrm{med}(f,-v,u)$ is integrable for all $u,v \in \mathcal{I}_-(\nu)$. We denote by

$$\mathcal{I}^0(\nu)$$

(or $\mathcal{I}^0$ for short) the set of all measurable functions.

REMARK. *If f is measurable, then* $\mathrm{med}(f,-v,u)$ *is integrable for all* $u,v \in \overline{\mathcal{I}}_-(\nu)$. *In particular, a measurable function bounded by integrable functions is integrable.*

This follows immediately from Corollary 2.11. □

PROPOSITION. *Every essentially integrable, and in particular every integrable function, is measurable.*

Let f be essentially integrable and choose $u,v \in \mathcal{I}_-$. Since $\mathcal{I}_- \subset \mathcal{I}_-^\bullet$, the truncated function $\mathrm{med}(f,-v,u)$ is essentially integrable and therefore integrable by Proposition 3.4.(ii). □

4.2 **PROPOSITION.** $\mathcal{I}^0(\nu)$ *is a homogeneous lattice of functions. A function f is measurable iff f^+ and f_- (equivalently f^+ and f^-) are each measurable.*

The first result follows immediately from the formulas

$$\mathrm{med}(-f,-v,u) = -\,\mathrm{med}(f,-u,v)$$

and

$$\mathrm{med}[\max(f,g),-v,u] = \max[\mathrm{med}(f,-v,u),\mathrm{med}(g,-v,u)]\ .$$

Therefore, the condition of the second assertion is obviously necessary. By

Corollary 2.6.(ii), it is also sufficient since

$$\mathrm{med}(f,-v,u)_-^+ = \mathrm{med}(f_-^+,-v,u)\ . \quad \square$$

REMARK. In Theorem 4.10 we shall prove in particular that $\tilde{\mathcal{I}}_+^0(\nu)$ is a function cone. In general, however, this is not true for $\tilde{\mathcal{I}}^0(\nu)$. We shall illustrate the bizarre nature of this set by Example 4.7 which shows that $\tilde{\mathcal{I}}^0(\nu)$ may fail to be stable with respect to addition. However, if ν has the so-called Daniell property, i.e. ν is an upper integral in the sense of Definition 5.1, then such pathology does not arise (cf. Proposition 5.10.(iii)).

4.3 PROPOSITION. *Let $\mathcal{T}$ be almost coinitial in $\mathcal{I}_-(\nu)$. A function f is measurable iff, for all $t \in \mathcal{T}$, the truncated function* $\mathrm{med}(f,-t,t)$ *is integrable.*

The condition is obviously necessary. To prove its sufficiency, we may suppose $f \geq 0$ by Corollary 2.6.(ii) and Proposition 4.2. It then remains to prove that, for all $v \in \mathcal{I}_-$, the truncated function $g := \min(f,-v)$ is integrable, i.e. essentially integrable (bearing in mind that this function is bounded by integrable functions). However, for $t \in \mathcal{T}$ we have

$$\min(g,-t) = \min[\min(f,-t),-v] \in \mathcal{I}$$

and therefore, by Theorem 3.6.(i),

$$0 \leq \nu_\bullet(g) = \sup_{t\in\mathcal{T}} \nu_*(\min(g,-t)) = \sup_{t\in\mathcal{T}} \nu(\min(g,-t)) = \nu^\bullet(g) < \infty\ . \quad \square$$

COROLLARY. *The concepts of measurability for ν and $\nu^\bullet$ coincide.*

One has just to notice that $\mathcal{I}_-$ is almost coinitial in $\mathcal{I}_-^\bullet$ w.r.t. $\nu^\bullet$. $\square$

4.4 The concept of measurability enables us to prove the following *Essential Integrability Criterion* :

THEOREM. *A function f is essentially integrable iff f is measurable and*

$$\nu^\bullet(|f|) < \infty\ ,$$

or equivalently

$$-\infty < \nu_{\bullet}(f) \le \nu^{\bullet}(f) < \infty .$$

The necessity follows from Propositions 4.1 and 2.4. To prove the sufficiency, we may suppose $f \ge 0$ by Corollary 2.6.(ii) and Proposition 4.2. But then

$$0 \le \nu_{\bullet}(f) = \sup_{v \in \mathcal{J}_-} \nu_*(\min(f,-v)) =$$

$$= \sup_{v \in \mathcal{J}_-} \nu(\min(f,-v)) = \nu^{\bullet}(f) < \infty . \quad \square$$

Remark. Measurability can be tested by means of Proposition 4.3. The criterion is of practical interest only when measurable functions have sufficient stability properties (cf. Proposition 5.10) or in the case where measurability can be described as usual set–theoretically (cf. Theorems 11.8 and 13.6.(iv)).

Example. Consider the topological standard example and a Radon integral μ. Then the set of functions $-n \cdot 1_K$, where n runs through $\mathbb{N}$ and K through the compact subsets of X, is almost coinitial in $\mathcal{J}_-(\mu^*)$ (cf. Example 3.8.2).

A function f is measurable, respectively essentially integrable w.r.t. μ^* iff each function $1_K \cdot \mathrm{med}(f,n,-n)$ is integrable, respectively if moreover

$$\sup_{n,K} \mu^*(\inf(|f|, n \cdot 1_K)) < \infty .$$

This is an immediate consequence of Propositions 4.3 and 4.4. $\square$

As in Example 3.8.2, we use now the fact, to be proved later, that μ^* is an upper integral.

A function f is measurable, respectively essentially integrable iff each function $1_K \cdot f$ is measurable, respectively integrable and

$$\sup_K \mu^*(1_K \cdot |f|) < \infty .$$

Just the asserted integrability of $1_K \cdot f$ is not obvious. It will follow from our next proposition. $\square$

4.5 In its more familiar form, the *Integrability Criterion* is true only when additional hypotheses are imposed on the upper functional, in particular if it is auto-determined.

PROPOSITION. *Let* ν *be auto-determined. Then a function* f *is integrable iff* f *is measurable and*

$$\nu(|f|) < \infty ,$$

or equivalently

$$-\infty < \nu_*(f) \le \nu(f) < \infty .$$

Indeed, if $\nu(|f|) < \infty$, then f is bounded by integrable functions. Therefore, the measurability of f implies its integrability. □

REMARK. This integrability criterion shows that the integration theory for an upper functional ρ whose theory of essential integration coincides with that of ν is easily determined from the last one. In fact, the integrability of a function f w.r.t. ρ simply means that f is measurable or essentially integrable and that $\rho(f)$ is finite.

Therefore one could be led to concider only essential integration. But having applications in mind, one should note that the upper functionals linked directly to the considered problem are in general easier to handle and more informative.

4.6 **PROPOSITION.** *Let* f *be a measurable function and let* g *be a function with* $g = f \mod \nu^\bullet$ *. Then* g *is measurable.*

Note that

$$|\mathrm{med}(f,-v,v) - \mathrm{med}(g,-v,v)| \le |f - g|$$

for $v \in \mathcal{T}^\bullet_-$. If we consider Corollary 4.3, the assertion follows from Proposition 2.9 applied to $\nu^\bullet$. □

4.7 The following special example (2) shows that in general $\widetilde{\mathcal{T}}^0(\nu)$ fails to be a function cone.

EXAMPLES.

(1) Consider the regular linear functional μ of Example 1.10.6 defined by

$$\mu(s) := \lim_{\mathfrak{F}} \frac{s}{w}.$$

A function f is integrable iff $\lim_{\mathfrak{F}} \frac{f}{w}$ exists in $\mathbb{R}$, in particular

$$\mathfrak{I}(\mu^*) = \mathcal{S} \cap \{\mu < \infty\}.$$

The set of functions $-nw$ with $n \in \mathbb{N}$ is almost coinitial in $\mathfrak{I}_-(\mu^*) = \mathcal{S}_-$, and a function f is measurable iff $\lim_{\mathfrak{F}} \frac{f}{w}$ exists in $\overline{\mathbb{R}}$.

In fact, since $\mu(s) = \mu(\max(s,-nw))$ for n large enough, we have almost coinitiality. Therefore, by Proposition 4.3, a function f is measurable iff

$$\lim_{\mathfrak{F}} \frac{1}{w}\cdot\mathrm{med}(f,nw,-nw) = \lim_{\mathfrak{F}} \mathrm{med}(\frac{f}{w},n,-n)$$

exists in $\mathbb{R}$ for all $n \in \mathbb{N}$, which is equivalent to $\lim_{\mathfrak{F}} \frac{f}{w}$ exists in $\overline{\mathbb{R}}$.

(2) Consider the special case of the above example with basic set $[0,1]$ and μ defined by

$$\mu(s) := \lim_{x\to 0+} s(x),$$

where $x \to 0+$ means convergence along the filter $\mathfrak{F}$ on $[0,1]$ generated by the sets $]0,\varepsilon[$ with $\varepsilon > 0$. Note that $\lim_{\mathfrak{F}} f$ exists iff f admits a right hand limit at 0. The functions f and g, defined by 0 at 0 and elsewhere by

$$f(x) := \frac{1}{x} + \sin\frac{1}{x} \quad \text{and} \quad g(x) := -\frac{1}{x},$$

are measurable. However, $f + g = \sin\frac{1}{\cdot}$ is not measurable.

The functions h and k, defined by 0 at 0 and elsewhere by

$$h(x) = \frac{1}{x} + \sin\frac{1}{x} + 1 \quad \text{and} \quad k(x) = \frac{1}{x},$$

illustrate that even if h and k are real-valued measurable functions with $0 \le k \le h$, the function $h - k$ will not be measurable in general.

In contrast, if we consider the Dirac integral ε_0, then every function is measurable, and hence $\tilde{\mathfrak{I}}^0(\varepsilon_0^*) = \tilde{\mathbb{R}}^X$ is a function cone.

4.8 Next we give a characterization of measurable functions which allows us to show that a substantial subclass of these functions will indeed form a function cone. The following concept, though somewhat technical, will prove to be valuable also in connection with representation theorems (cf. § 7) and in the description of measurability in the set-theoretic sense (cf. § 11).

DEFINITION. Let $\mathcal{S}$ be a set of negative real-valued functions. We denote by

$$\mathcal{A}(\mathcal{S})$$

the set of all functions which are *one-sided relatively uniform limits* of functions in $\mathcal{S}$, i.e. all functions f for which there exists a sequence (s_n) in $\mathcal{S}$ and a null sequence (ε_n) in $\mathbb{R}_+$ such that

$$s_n + \varepsilon_n s_0 \leq f \leq s_n$$

for $n \geq 1$. Furthermore,

$$\hat{\mathcal{S}}$$

denotes the *one-sided relatively uniform closure* of $\mathcal{S}_{max}$, i.e. the smallest subset $\mathcal{T} \subset \mathbb{R}_-^X$ with the properties $\mathcal{S}_{max} \subset \mathcal{T}$ and $\mathcal{A}(\mathcal{T}) = \mathcal{T}$.

We always have the following inclusions :

$$\mathcal{S} \subset \mathcal{A}(\mathcal{S}) \subset \hat{\mathcal{S}} = (\mathcal{S}_{max})\hat{} = \hat{\mathcal{S}}\hat{} \ .$$

PROPOSITION. *If $\mathcal{S} \subset \mathbb{R}_-^X$ is min-stable, a lattice or a function cone, so is $\mathcal{A}(\mathcal{S})$. If $\mathcal{S}$ is a min-stable cone, then $\hat{\mathcal{S}}$ is a lattice cone of functions with*

$$\hat{\mathcal{S}} \subset (\mathcal{S}_{max})_\sigma = \hat{\mathcal{S}}_\sigma \subset \mathcal{S}^\uparrow .$$

For $f,g \in \mathcal{A}(\mathcal{S})$, there exist sequences $(s_n), (t_n)$ in $\mathcal{S}$ and, without loss of generality, a common null sequence (ε_n) in $\mathbb{R}_+$ such that

$$s_n + \varepsilon_n s_0 \leq f \leq s_n \quad \text{and} \quad t_n + \varepsilon_n t_0 \leq g \leq t_n .$$

Then

$$\min(s_n,t_n) + \varepsilon_n \min(s_0,t_0) \leq \min(s_n + \varepsilon_n s_0, t_n + \varepsilon_n t_0) \leq \min(f,g) \leq \min(s_n,t_n) .$$

Replacing minima by maxima with the exception of $\varepsilon_n \min(s_0,t_0)$, respectively by sums, this proves the first assertion.

If $\mathcal{S}$ is a min-stable cone, then by Zorn's Lemma, for every $\mathcal{T} \subset \mathbb{R}_-^X$ with

$\mathcal{S}_{max} \subset \mathcal{T} = \mathcal{A}(\mathcal{T})$ there exists a maximal lattice cone $\mathcal{U}$ with

$$\mathcal{S}_{max} \subset \mathcal{U} \subset \mathcal{T} ,$$

hence

$$\mathcal{S}_{max} \subset \mathcal{U} \subset \mathcal{A}(\mathcal{U}) \subset \mathcal{A}(\mathcal{T}) = \mathcal{T} .$$

Since $\mathcal{A}(\mathcal{U})$ is a lattice cone by the first part, we have $\mathcal{U} = \mathcal{A}(\mathcal{U})$ by the maximality of $\mathcal{U}$. This shows that $\hat{\mathcal{S}}$ is the intersection of all lattice cones $\mathcal{T} \subset \overline{\mathbb{R}}^X_-$ with $\mathcal{S}_{max} \subset \mathcal{T} = \mathcal{A}(\mathcal{T})$ and therefore a lattice cone itself. The rest is obvious. □

EXAMPLES.

(1) We always have

$$\hat{\mathcal{I}}_-(\nu) = \mathcal{I}_-(\nu)$$

and, for arbitrary topological Hausdorff spaces X,

$$\hat{\mathcal{S}}_-(X) = \mathcal{S}_-(X) .$$

(2) *Let X be a locally compact Hausdorff space and $\mathcal{S}$ be a min-stable cone uniformly dense in $\mathscr{C}^0_-(X)$. Then*

$$\mathcal{K}_-(X) \subset \hat{\mathcal{S}} .$$

Indeed, let (t_n) be a sequence in $\mathcal{S}$ converging uniformly to $f \in \mathcal{K}_-(X)$. Furthermore, choose $s_0 \in \mathcal{S}$ with $s_0 \le -1_{\mathrm{supp}(f)}$ and define

$$s_n := t_n - \|f - t_n\|_\infty \cdot s_0 \in \mathcal{S} .$$

Then $(\|f - t_n\|_\infty)$ is a null sequence and

$$s_n + 2 \cdot \|f - t_n\|_\infty \cdot s_0 \le f \le s_n . \quad \square$$

4.9 **REMARK.** *A function f is measurable w.r.t. ν iff*

$$[\max(f,u) + v]_- \in \mathcal{I}_-(\nu)$$

for all $u,v \in \mathcal{I}_-(\nu)$.

One just has to note the formula

$$\mathrm{med}(f,-v,u) = \min[\max(f,u),-v\,] = [\max(f,u) + v\,]_- - v . \quad \square$$

This leads to the following

DEFINITION. Let $\mathcal{S}$ be a min-stable function cone in $\tilde{\mathbb{R}}^X$. Then the set of all functions $f \in \tilde{\mathbb{R}}^X$ with $(f+s)_- \in \hat{\mathcal{S}}_-$ for all $s \in \mathcal{S}_-$ will be denoted by

$$\mathcal{G}(\mathcal{S}) .$$

A function f with $\max(f,s) \in \mathcal{G}(\mathcal{S})$ for all $s \in \mathcal{S}_-$ is said to be *$\mathcal{S}$-measurable*; we denote by

$$\mathcal{H}(\mathcal{S})$$

the set of all such functions.

For simplicity, we talk of $\mathcal{S}$-measurability, though this notion only depends on $\mathcal{S}_-$.

EXAMPLE. *If X is a Hausdorff space, then $\mathcal{S}(X) \subset \mathcal{G}(\mathcal{S}(X))$, and every lower semicontinuous function on X is $\mathcal{S}(X)$-measurable.*

In Example 10.10, we prove that conversely every $\mathcal{S}(X)$-measurable function is lower semicontinuous for the finest topology on X inducing the given one on every compact subset.

PROPOSITION. *$\mathcal{G}(\mathcal{S})$ is a lattice cone of functions with*

$$\mathcal{G}_-(\mathcal{S}) = \hat{\mathcal{S}}_- \quad \text{and} \quad \mathcal{S}_{max} \subset \mathcal{G}(\mathcal{S}) = \mathcal{G}(\hat{\mathcal{S}}_-) .$$

$\mathcal{H}(\mathcal{S})$ is a positively homogeneous lattice of functions with

$$\mathcal{H}(\mathcal{S}) = \mathcal{H}(\hat{\mathcal{S}}_-) \quad \text{and} \quad \mathcal{G}(\mathcal{S}) = \mathcal{H}(\mathcal{S}) \cap \mathcal{S}^{\uparrow} .$$

A function f is $\mathcal{S}$-measurable iff f^+ and f_- are each $\mathcal{S}$-measurable. Furthermore,

$$\mathcal{G}(\mathcal{S})_\sigma \subset \mathcal{G}((\mathcal{S}_{max})_\sigma) \quad \text{and} \quad \mathcal{H}(\mathcal{S})_\sigma \subset \mathcal{H}((\mathcal{S}_{max})_\sigma) .$$

For $f,g \in \mathcal{G}(\mathcal{S})$, $\alpha \in \mathbb{R}_+^*$ and $s \in \mathcal{S}_-$, we have

$$(\alpha f + s)_- = \alpha(f + \tfrac{s}{\alpha})_- ,$$

$$[\min(f,g) + s]_- = \min[(f+s)_-,(g+s)_-] ,$$

and

$$[\max(f,g) + s]_- = \max[(f+s)_-,(g+s)_-] ,$$

hence αf, $\min(f,g)$ and $\max(f,g)$ each belong to $\mathcal{G}(\mathcal{S})$.

To prove that $\mathcal{G}(\mathcal{S}) \subset \mathcal{G}(\hat{\mathcal{S}}_-)$, let $g \in \mathcal{G}(\mathcal{S})$ and denote by $\mathcal{S}_g$ the set of all functions $f \in \hat{\mathcal{S}}_-$ such that $(g+f)_- \in \hat{\mathcal{S}}_-$. Then $\mathcal{S}_{-max} \subset \mathcal{S}_g$ because $\hat{\mathcal{S}}_-$ is max-stable. Consider $f \in \mathcal{A}(\mathcal{S}_g)$, a sequence (s_n) in $\mathcal{S}_g$ and a null sequence (ε_n) in $\mathbb{R}_+$ with

$$s_n + \varepsilon_n s_0 \leq f \leq s_n .$$

Then

$$(g + s_n)_- + \varepsilon_n s_0 \leq (g + s_n + \varepsilon_n s_0)_- \leq (g + f)_- \leq (g + s_n)_- ,$$

hence $(g + f)_- \in \mathcal{A}(\mathcal{S}_-^\wedge) = \mathcal{S}_-^\wedge$ and therefore $f \in \mathcal{S}_g$. This shows that $\mathcal{A}(\mathcal{S}_g) = \mathcal{S}_g$, hence $\mathcal{S}_-^\wedge \subset \mathcal{S}_g$ and thus $g \in \mathcal{G}(\mathcal{S}^\wedge)$.

For the first assertions, it remains to show that $\mathcal{G}(\mathcal{S})$ is stable under addition. Indeed, for $f,g \in \mathcal{G}(\mathcal{S})$ the identities

$$(f + g + s)_- = \min[f^+ + (g + f_- + s), f^+, 0] = [f^+ + (g + f_- + s)_-]_- \in \mathcal{S}_-^\wedge$$

prove that $f + g \in \mathcal{G}(\mathcal{S})$.

The properties of $\mathcal{H}(\mathcal{S})$ are obvious. If $f^+, f_- \in \mathcal{H}(\mathcal{S})$, then we have

$$\max(f,s) = f^+ + \max(f_-,s) \in \mathcal{G}(\mathcal{S})$$

for every $s \in \mathcal{S}_-$, which shows that $f \in \mathcal{H}(\mathcal{S})$.

It remains to prove the assertions concerning upper envelopes. Let (g_n) and (t_n) be increasing sequences in $\mathcal{G}(\mathcal{S})$ and $\mathcal{S}_{max-}$ respectively. Then

$$(\sup g_n + \sup t_n)_- = \sup(g_n + t_n)_- \in \mathcal{G}_-(\mathcal{S})_\sigma = \mathcal{S}_{-\sigma}^\wedge \subset (\mathcal{S}_{max})_\sigma ,$$

hence $\sup g_n \in \mathcal{G}((\mathcal{S}_{max})_\sigma)$. From this, the last inclusion follows immediately. □

4.10 THEOREM. *The following assertions hold:*

$$\mathcal{I}^0(\nu) = \mathcal{H}(\mathcal{I}(\nu)) \quad and \quad \mathcal{G}(\mathcal{I}(\nu)) \subset \mathcal{G}(\mathcal{I}^\bullet(\nu)) = \mathcal{I}^0(\nu) \cap \{\nu_\bullet > -\infty\}^\sim .$$

If f is a measurable function with $\nu_\bullet(f) > -\infty$, then

$$\nu_\bullet(f) = \nu^\bullet(f) .$$

In particular, $\nu^\bullet$ is linear on the lattice cone $\mathcal{I}^0(\nu) \cap \{\nu_\bullet > -\infty\}^\sim$.

The first formula follows immediately from Remark 4.9, thus $\mathcal{I}^0 = \mathcal{H}(\mathcal{I}^\bullet)$ by Corollary 4.3. The Essential Integrability Criterion 4.4 and Proposition 4.9 then show that

$$\mathcal{I}^0 \cap \{\nu_\bullet > -\infty\}^\sim = \mathcal{H}(\mathcal{I}^\bullet) \cap \mathcal{I}_-^{\bullet\uparrow} = \mathcal{G}(\mathcal{I}^\bullet) .$$

Furthermore,

$$\mathcal{G}(\mathcal{I}) = \mathcal{I}^0 \cap \mathcal{I}_-^\uparrow \subset \mathcal{I}^0 \cap \{\nu_\bullet > -\infty\}^\sim .$$

Applying Lemma 3.2.(i) with $\nu^\bullet$ instead of ν and bearing in mind Corollaries 3.6 and 4.3, it follows that

$$\nu^\bullet(f) = (\nu^\bullet)^\bullet(f) = \sup_{v \in \mathcal{I}_-^\bullet} \nu^\bullet(\min(f,-v)) =$$

$$= \sup_{v \in \mathcal{I}_-^\bullet} \nu_\bullet(\min(f,-v)) = (\nu_\bullet)_\bullet(f) = \nu_\bullet(f) .$$

In view of Theorem 2.2.(iii), this implies the linearity of $\nu^\bullet$ on $\mathcal{G}(\mathcal{I}^\bullet)$. □

Corollary. *If* ν *is auto-determined, then*

$$\mathcal{I}^0(\nu) \cap \{\nu_* > -\infty\}^\sim = \mathcal{G}(\mathcal{I}(\nu))$$

is a lattice cone of functions, on which both ν *and* $\nu^\bullet$ *are linear.*

For every function f *with* $\nu_*(f) > -\infty$ *, the equality*

$$\nu_*(f) = \nu_\bullet(f)$$

holds. If moreover f *is measurable, then*

$$\nu_*(f) = \nu^\bullet(f) .$$

By analogous considerations as above, one proves the first formula using the Integrability Criterion 4.5. Similarly, to prove the linearity of ν , in view of Proposition 2.7 it is sufficient to show , for $f,g \in \mathcal{G}(\mathcal{I})$, that

$$\nu(f+g) < \infty \text{ implies } \nu(f) < \infty .$$

But this follows immediately from the inequality $f \le (f+g) + g^-$ and the sublinearity of ν (cf. Theorem 2.2.(iii)), g^- being integrable.

The last part follows from Corollary 3.9 and the theorem. □

Remarks.

(1) The coincidence of ν_* and $\nu^\bullet$ on $\mathcal{I}^0(\nu) \cap \{\nu_* > -\infty\}$ can be interpreted as a manifestation of regularity. This will be shown in § 6.

(2) As Example 4.7.2 shows, $\tilde{\mathcal{H}}$ may fail to be a function cone in general.

4.11 For a min-stable function cone $\mathcal{S}$, the cone $\mathcal{G}(\mathcal{S})$ is a reasonably large cone of "universally measurable" functions in the sense of the following

PROPOSITION. *If* $\mathcal{S}_-$ *is almost coinitial in* $\mathcal{I}^\bullet_-(\nu)$, *then*

$$\mathcal{H}(\mathcal{S}) \subset \mathcal{I}^0(\nu) \quad \textit{and} \quad \mathcal{G}(\mathcal{S}) \subset \mathcal{I}^0(\nu) \cap \mathcal{I}^{\bullet\uparrow}_-(\nu) .$$

Since $\mathcal{S}^\wedge_- \subset \mathcal{I}^\bullet_-(\nu)$ by Example 4.8.1, for $f \in \mathcal{H}(\mathcal{S})$ and $s \in \mathcal{S}_-$ we get

$$\text{med}(f,-s,s) = [\max(f,s) + s]_- - s \in \mathcal{I}^\bullet(\nu) .$$

By Proposition and Corollary 4.3, f is measurable. The remaining inclusion is now obvious. □

4.12 We conclude this section with a discussion of operations on upper functionals.

PROPOSITION. *Let* ν *and* ρ *be upper functionals. If* $\mathcal{I}_-(\nu) \cap \mathcal{I}_-(\rho)$ *is almost coinitial in* $\mathcal{I}_-(\nu)$ *as well as in* $\mathcal{I}_-(\rho)$, *then*

$$(\nu + \rho)^\bullet = \nu^\bullet + \rho^\bullet ,$$

and a function is measurable, respectively essentially integrable w.r.t. $\nu + \rho$ *iff it is measurable, respectively essentially integrable w.r.t.* ν *and* ρ .

For every function f, we infer from Theorem 3.6.(ii), using Proposition 2.15 and the hypothesis, that $\nu^\bullet(f), \rho^\bullet(f) < \infty$ iff there exists some $u \in \mathcal{I}_-(\nu+\rho)$ with

$$\sup_{v \in \mathcal{I}_-(\nu+\rho)} \nu(\text{med}(f,-v,u)), \ \sup_{v \in \mathcal{I}_-(\nu+\rho)} \rho(\text{med}(f,-v,u)) < \infty ,$$

i.e.

$$\sup_{v \in \mathcal{I}_-(\nu+\rho)} (\nu + \rho)(\text{med}(f,-v,u)) < \infty ,$$

since the family $(\text{med}(f,-v,u))_{v \in \mathcal{I}_-(\nu+\rho)}$ is upward directed. This is obviously equivalent to $(\nu + \rho)^\bullet(f) < \infty$, and in this case we infer from Theorem 3.6.(ii) that

$$(\nu + \rho)^\bullet(f) = \inf_{u \in \mathcal{I}_-(\nu+\rho)} \sup_{v \in \mathcal{I}_-(\nu+\rho)} (\nu + \rho)(\text{med}(f,-v,u)) =$$

$$= \lim_{u\in\mathcal{J}_-(\nu+\rho)} \lim_{v\in\mathcal{J}_-(\nu+\rho)} \nu(\mathrm{med}(f,-v,u)) + \lim_{u\in\mathcal{J}_-(\nu+\rho)} \lim_{v\in\mathcal{J}_-(\nu+\rho)} \rho(\mathrm{med}(f,-v,u)) =$$

$$= \nu^\bullet(f) + \rho^\bullet(f) = (\nu^\bullet + \rho^\bullet)(f) .$$

The last assertions follow immediately from Propositions 4.3, 2.15 and the Essential Integrability Criterion 4.4. □

Remark. The condition is necessary, since for every function $f \le 0$ we have

$$(\nu + \rho)^\bullet(f) = \inf_{u\in\mathcal{J}_-(\nu+\rho)} (\nu + \rho)^\bullet(\max(f,u)) =$$

$$= \inf_{u\in\mathcal{J}_-(\nu+\rho)} \nu^\bullet(\max(f,u)) + \inf_{u\in\mathcal{J}_-(\nu+\rho)} \rho^\bullet(\max(f,u)) \ge$$

$$\ge \nu^\bullet(f) + \rho^\bullet(f) = (\nu + \rho)^\bullet(f) ,$$

hence

$$\nu(f) = \inf_{u\in\mathcal{J}_-(\nu+\rho)} \nu(\max(f,u)) \quad \text{for } f \in \mathcal{J}_-(\nu)$$

and

$$\rho(f) = \inf_{u\in\mathcal{J}_-(\nu+\rho)} \rho(\max(f,u)) \quad \text{for } f \in \mathcal{J}_-(\rho) . \quad □$$

4.13 **Proposition.** *If* $w \in \bar{\mathbb{R}}_+^X$ *is finite essentially almost everywhere and* $\infty_{\{w>0\}}$ *is measurable w.r.t.* ν *, then*

$$(w\cdot\nu)^\bullet = w\cdot\nu^\bullet ,$$

and a function f *is measurable, respectively essentially integrable w.r.t.* $w\cdot\nu$ *iff* wf *is measurable, respectively essentially integrable w.r.t.* ν .

For any function f and all $u,v \in \mathcal{J}_-(\nu)$ we have

$$\mathrm{med}(wf,-v,u) = \mathrm{med}(wf,-w\cdot\tfrac{v}{w},w\cdot\tfrac{u}{w}) \quad \text{on } \{w < \infty\} ,$$

hence a.e. w.r.t. $\nu^\bullet$. By Proposition 2.9 this yields

$$\nu(\mathrm{med}(wf,-v,u)) = \nu(\mathrm{med}[wf,-w\cdot\tfrac{v}{w},w\cdot\tfrac{u}{w}]) = w\cdot\nu(\mathrm{med}[f,-\tfrac{v}{w},\tfrac{u}{w}]) ,$$

these functions being bounded by integrable ones, since $w\cdot\frac{u}{w} \ge u$.

In particular, for $q \in \mathcal{J}_-(w\cdot\nu)$ we have

$$w\cdot\nu(q) = \nu^\bullet(wq) = \inf_{u\in\mathcal{J}_-(\nu)} \nu(\max(wq,u)) = \inf_{u\in\mathcal{J}_-(\nu)} w\cdot\nu(\max(q,\tfrac{u}{w})) .$$

Proposition 2.16 gives $\frac{u}{w} \in \mathcal{J}\_(w\cdot\nu)$, since $w\cdot\frac{u}{w} = 1_{\{0<w<\infty\}}\cdot u = \max(-\infty_{\{w>0\}},u)$ a.e. w.r.t. $\nu^\bullet$, and hence belongs to $\mathcal{J}\_(\nu)$. This shows that the set of all functions $\frac{u}{w}$ with $u \in \mathcal{J}\_(\nu)$ is almost coinitial in $\mathcal{J}\_(w\cdot\nu)$. Thus we have

$$(w\cdot\nu)^\bullet(f) = \inf_{u\in\mathcal{J}\_(\nu)} \sup_{v\in\mathcal{J}\_(\nu)} w\cdot\nu(\mathrm{med}[f,-\frac{v}{w},\frac{u}{w}]) =$$

$$= \inf_{u\in\mathcal{J}\_(\nu)} \sup_{v\in\mathcal{J}\_(\nu)} \nu(\mathrm{med}(wf,-v,u)) = \nu^\bullet(wf) = w\cdot\nu^\bullet(f)$$

by Theorem 3.6.(ii).

Finally, by Proposition 4.3, a function f is measurable w.r.t. $w\cdot\nu$ iff

$$\mathrm{med}(f,-\frac{u}{w},\frac{u}{w}) \in \mathcal{J}(w\cdot\nu) \quad \text{for all } u \in \mathcal{J}\_(\nu) ,$$

i.e. in view of Proposition 2.16, iff

$$w\cdot\mathrm{med}(f,-\frac{u}{w},\frac{u}{w}) \in \mathcal{J}(\nu) \quad \text{for all } u \in \mathcal{J}\_(\nu).$$

By Proposition 4.6 and the above this means that wf is measurable w.r.t. ν . Proposition 2.16 yields the last assertion. □

Remark. The conditions are necessary. Indeed,

$$\nu^\bullet(\infty_{\{w=\infty\}}) = \nu^\bullet(w\cdot\infty_{\{w=\infty\}}) = (w\cdot\nu)^\bullet(\infty_{\{w=\infty\}}) =$$

$$=\sup_{q\in\mathcal{J}\_(w\cdot\nu)} w\cdot\nu(\min[\infty_{\{w=\infty\}},-q]) =$$

$$=\sup_{q\in\mathcal{J}\_(w\cdot\nu)} \nu(\infty_{\{w=\infty\}\cap\{q<0\}}) = \sup_{q\in\mathcal{J}\_(w\cdot\nu)} \nu(\infty_{\{wq=-\infty\}}) = 0 ,$$

since wq is finite a.e. w.r.t. ν by Propositions 2.16 and 2.11.(ii). Furthermore,

$$w\cdot\nu(\infty_{\{w=0\}}) = \nu(w\cdot\infty_{\{w=0\}}) = 0 ,$$

hence $\infty_{\{w>0\}} = \infty$ a.e. w.r.t. $w\cdot\nu$ and so $\infty_{\{w>0\}} = w\cdot\infty_{\{w>0\}}$ is measurable w.r.t. ν . □

4.14 Proposition. *Let $p : X \longrightarrow Y$ be a mapping from X into a set Y . If $p(\nu)$ is an upper functional, and if $\mathcal{J}\_(p(\nu))\circ p$ is almost coinitial in $\mathcal{J}\_(\nu)$, then*

$$p(\nu)^\bullet = p(\nu^\bullet) .$$

A function g on Y is measurable, respectively essentially integrable w.r.t. $p(\nu)$ iff $g\circ p$ is measurable, respectively essentially integrable w.r.t. ν .

For any function g on Y we infer from Theorem 3.6.(ii) that

$$p(\nu)^{\bullet}(g) = \inf_{q \in \mathcal{J}_{-}(p(\nu))} \sup_{r \in \mathcal{J}_{-}(p(\nu))} p(\nu)(\operatorname{med}(g,-r,q)) =$$

$$= \inf_{q \in \mathcal{J}_{-}(p(\nu))} \sup_{r \in \mathcal{J}_{-}(p(\nu))} \nu(\operatorname{med}(g \circ p,-r \circ p,q \circ p)) =$$

$$= \nu^{\bullet}(g \circ p) = p(\nu^{\bullet})(g) .$$

The last assertions follow immediately from Propositions 4.3 and 2.17. □

Remarks.

(1) Note that $\nu^{\bullet}$ satisfies the condition of Proposition 2.17 since, by Corollary and Proposition 3.5, it is determined by the set $\mathcal{J}_{-}(p(\nu)) \circ p^{\dagger}$.

(2) The almost coinitiality of $\mathcal{J}_{-}(p(\nu)) \circ p$ in $\bar{\mathbb{R}}^{Y} \circ p$ w.r.t. $\nu^{\bullet}$ is a necessary condition.

Indeed, we have

$$p(\nu)^{\bullet}(g) = \inf_{q \in \mathcal{J}_{-}(p(\nu))} p(\nu)^{\bullet}(\max(g,q)) =$$

$$= \inf_{q \in \mathcal{J}_{-}(p(\nu))} \nu^{\bullet}(\max[g \circ p,q \circ p]) \geq$$

$$\geq \inf_{u \in \mathcal{J}_{-}(\nu)} \nu^{\bullet}(\max[g \circ p,u]) = \nu^{\bullet}(g \circ p) = p(\nu)^{\bullet}(g) . \quad \square$$

It is sufficient if $\nu(u_p \circ p) > -\infty$ for all $u \in \mathcal{J}_{-}(\nu)$ by the transitivity property in Proposition 3.5.(i).

§ 5 UPPER INTEGRALS AND CONVERGENCE THEOREMS

5.1 As is well-known, the classical Riemann-integral on $\mathbb{R}$ has the following additional property (cf. also Example 13.4.1):

DEFINITION. An increasing functional μ on a set $\mathcal{S}$ of functions is said to have the *Daniell property* if for every increasing sequence (s_n) in $\mathcal{S}$ with $\sup s_n \in \mathcal{S}$ and $\mu(s_n) > -\infty$ the equality

$$\mu(\sup s_n) = \sup \mu(s_n)$$

holds.

An upper functional ν having the Daniell property is called an *upper integral*. The functional ν_* is then called a *lower integral*.

For an upper integral, a richer theory of integration is available. The theory presented in § 2 – § 4 can be augmented with convergence theorems.

The following technical lemma will be very useful in our subsequent studies (cf. 5.2, 9.2, 9.5, 9.9, and 9.10).

LEMMA. *Let μ be an increasing functional on a set $\mathcal{S}$ of functions, and let $\mathcal{T}$ be a lattice of functions with $\mathcal{S} \cap \mathcal{T}^{\uparrow} \subset \mathcal{T}$ such that μ^* is submodular on $\mathcal{T}$. Then for every increasing sequence (t_n) in $\mathcal{T}$ with $\mu^*(t_n) < \infty$ and every $\varepsilon > 0$, there exists an increasing sequence (s_n) in $\mathcal{S}$ with $t_n \le s_n$ and*

$$\mu(s_n) \le \mu^*(t_n) + \varepsilon .$$

We define (s_n) inductively such that

$$\mu(s_n) \le \mu^*(t_n) + \varepsilon \cdot \sum_{i=1}^{n} \frac{1}{2^i} .$$

For t_1 choose $s_1 \in \mathcal{S}$ such that $s_1 \ge t_1$ and

$$\mu(s_1) \le \mu^*(t_1) + \frac{\varepsilon}{2} .$$

If $s_n \in \mathcal{S}$ is already defined, then $s_n \in \mathcal{T}$, since $\mathcal{S} \cap \mathcal{T}^{\uparrow} \subset \mathcal{T}$, and

$$\mu^*(\min(s_n,t_{n+1})) + \mu^*(\max(s_n,t_{n+1})) \le \mu(s_n) + \mu^*(t_{n+1}) < \infty ,$$

hence $\mu^*(\max(s_n,t_{n+1})) < \infty$. Therefore, there exists $s_{n+1} \in \mathcal{S}$ with

$$s_{n+1} \ge \max(s_n,t_{n+1})$$

and

$$\mu(s_{n+1}) \le \mu^*(\max(s_n,t_{n+1})) + \frac{\varepsilon}{2^{n+1}} .$$

Then

$$\mu^*(t_n) + \mu(s_{n+1}) \le$$

$$\le \mu^*(\min(s_n,t_{n+1})) + \mu(s_{n+1}) \le \mu^*(\min(s_n,t_{n+1})) + \mu^*(\max(s_n,t_{n+1})) + \frac{\varepsilon}{2^{n+1}} \le$$

$$\le \mu(s_n) + \mu^*(t_{n+1}) + \frac{\varepsilon}{2^{n+1}} \le \mu^*(t_n) + \varepsilon \cdot \sum_{i=1}^{n} \frac{1}{2^i} + \mu^*(t_{n+1}) + \frac{\varepsilon}{2^{n+1}} .$$

The extreme inequalities imply that

$$\mu(s_{n+1}) \le \mu^*(t_{n+1}) + \varepsilon \cdot \sum_{i=1}^{n+1} \frac{1}{2^i} ,$$

since $\mu^*(t_n) \in \mathbb{R}$. □

REMARK. If μ is a strongly sublinear functional on a lattice cone $\mathcal{S}$, then the lemma holds for $\mathcal{T} := \{\mu_* > -\infty\}$. In this case, the proof may be simplified. If s_n is already defined, choose $s \in \mathcal{S}$ such that $s \ge t_{n+1}$ and

$$\mu(s) \le \mu^*(t_{n+1}) + \frac{\varepsilon}{2^{n+1}} ,$$

and set

$$s_{n+1} := \max(s,s_n) .$$

5.2 THEOREM. *Let μ be an increasing functional on a set $\mathcal{S}$ of functions such that $\mathcal{S} = \mathcal{S}_\sigma$ and μ^* is an upper functional. If μ has the Daniell property, then μ^* is an upper integral.*

Let (f_n) be an increasing sequence of functions. We may suppose without loss of generality that $\mu^*(f_n)$ is finite for all $n \in \mathbb{N}$. By Theorem 2.2.(ii), μ^* is submodular on $\{f_0\}^\uparrow$. Therefore, by Lemma 5.1, for given $\varepsilon > 0$ there exists an increasing sequence (s_n) in $\mathcal{S}$ with $f_n \le s_n$ and $\mu(s_n) \le \mu^*(f_n) + \varepsilon$. But then we have $\sup s_n \in \mathcal{S}_\sigma = \mathcal{S}$ and

$$\mu^*(\sup f_n) \le \mu(\sup s_n) = \sup \mu(s_n) \le \sup \mu^*(f_n) + \varepsilon \le \mu^*(\sup f_n) + \varepsilon ,$$

which completes the proof. □

COROLLARY. *For every upper integral* ν , *the essential upper functional* $\nu^\bullet$ *is also an upper integral.*

Let μ denote the restriction of $\nu^\bullet$ to the lattice cone $\mathcal{S} := \mathcal{T}_-^{\uparrow}$. Then $\mu^* = \nu^\bullet$ by Proposition 3.3, hence μ^* is an upper functional. According to the theorem it suffices to prove that μ has the Daniell property, since $\mathcal{S} = \mathcal{S}_\sigma$.

To this end, let (s_n) be an increasing sequence in $\mathcal{S}$ with upper envelope s . Then

$$\min(s_n,-v) \in \mathcal{S} \subset \{\nu > -\infty\}$$

for $n \in \mathbb{N}$ and $v \in \mathcal{T}_-(\nu)$, hence

$$\nu(\min(s,-v)) = \sup_n \nu(\min(s_n,-v))$$

by the Daniell property of ν , and therefore, by Lemma 3.2.(i),

$$\mu(s) = \nu^\bullet(s) = \sup_{v\in\mathcal{T}_-} \nu(\min(s,-v)) = \sup_{v\in\mathcal{T}_-} \sup_n \nu(\min(s_n,-v)) =$$
$$= \sup_n \sup_{v\in\mathcal{T}_-} \nu(\min(s_n,-v)) = \sup \nu^\bullet(s_n) = \sup \mu(s_n) . \quad \square$$

THROUGHOUT THE REST OF THIS SECTION, ν IS AN UPPER INTEGRAL.

5.3 We first discuss once more the concept of null sets and show that for upper integrals the relations " mod ν " and " almost everywhere w.r.t. ν " (cf. 2.8 and 2.10) coincide.

PROPOSITION. *For functions* f *and* g , *we have* $f \le g$ mod ν *iff* $f \le g$ *almost everywhere w.r.t.* ν , *and* $f = g$ mod ν *iff* $f = g$ *almost everywhere w.r.t.* ν .

Since $\infty_{\{f>g\}} = \sup n\cdot(f-g)^+$, we have $\nu(\infty_{\{f>g\}}) = \sup n\cdot\nu(f-g)^+$

by the Daniell property, and so $\nu(\infty_{\{f>g\}}) = 0$ iff $\nu((f - g)^{+}) = 0$. The second assertion follows from the first one, interchanging f and g . □

REMARK. Thanks to the proposition, in the statements of Propositions 2.9 and 2.12 for upper integrals we may always replace " mod ν " by " a.e. w.r.t. ν " . In the sequel, we will make such exchanges without commenting each time.

LEMMA. *For every sequence* (f_n) *of positive functions we have*

$$\nu\big(\sum f_n\big) \leq \sum \nu(f_n) \,.$$

Since the sequence of the partial sums is increasing, the assertion follows immediately from the subadditivity and the Daniell property of ν . □

COROLLARY. *The union of countably many null sets is a null set.*

REMARK. Since $\infty_A = \sup n \cdot 1_A$, a set A is a null set iff $\nu(A) = 0$. This characterization is interesting only if 1 is measurable, i.e. if the integration theory for ν can be recovered from the integrable sets (cf. Theorem 11.8 and Corollary 14.6).

5.4 The following *Monotone Convergence Theorem* is an immediate consequence of the Daniell property. It is often also called the

THEOREM OF BEPPO LEVI. *Let* (f_n) *be an increasing sequence of integrable functions. The upper envelope* $\sup f_n$ *is integrable iff* $\sup \nu(f_n)$ *is finite. In this case*

$$\nu(\sup f_n) = \sup \nu(f_n) \,.$$

The assertion follows from the chain of inequalities

$$-\infty < \nu(f_1) \leq \sup \nu(f_n) = \sup \nu_*(f_n) \leq \nu_*(\sup f_n) \leq \nu(\sup f_n) = \sup \nu(f_n) \,. \quad \square$$

ADDENDUM. *An analogous assertion holds for decreasing sequences.*

It follows from the above, replacing (f_n) by $(-f_n)$. □

5.5 As preparation for the proof of Lebesgue's dominated convergence theorem we need the following lemma, which also is of some intrinsic interest :

Lemma of Fatou. *Let* (f_n) *be a sequence of functions having a common integrable minorant. Then*

$$\nu(\liminf f_n) \le \liminf \nu(f_n) .$$

We define an increasing sequence of functions by $g_n := \inf_{m \ge n} f_m$ whose upper envelope is $\liminf f_n$. If g is a common integrable minorant of the f_n, then $g \le g_n$, hence $-\infty < \nu(g) \le \nu(g_n)$ and therefore

$$\nu(\liminf f_n) = \nu(\sup g_n) = \sup \nu(g_n) \le \sup_n \inf_{m \ge n} \nu(f_m) = \liminf \nu(f_n) . \qquad \square$$

Addendum. *If the functions* f_n *have a common integrable majorant, then*

$$\limsup \nu_*(f_n) \le \nu_*(\limsup f_n) .$$

The Lemma of Fatou is the key to proving the *Dominated Convergence*

Theorem of Lebesgue. *Let* (f_n) *be a sequence of integrable functions and suppose that the* $|f_n|$ *have a common almost everywhere majorant* g *which is integrable. If the sequence* (f_n) *converges almost everywhere to a function* f, *then* f *is integrable and*

$$\nu(f) = \lim \nu(f_n) .$$

More precisely,

$$\lim \nu(|f - f_n|) = 0 .$$

In view of Corollary 5.3 and Proposition 2.9 we may suppose, after modification of all functions on a common null set, that majorization and convergence hold everywhere.

For all $n \in \mathbb{N}$ we have $-g \le f_n \le g$. Since $-g$ is integrable too, the Lemma of Fatou and its addendum give

$$-\infty < \nu(-g) \le \limsup \nu(f_n) \le \nu_*(f) \le \nu(f) \le$$

$$\le \liminf \nu(f_n) \le \limsup \nu(f_n) \le \nu(g) < \infty .$$

Therefore, f is integrable and

$$\nu(f) = \lim\inf \nu(f_n) = \lim\sup \nu(f_n) = \lim \nu(f_n) .$$

To prove the last assertion, we may by Corollary 2.11 suppose that f and all f_n are finite. If we define $g_n := |f - f_n|$, then all g_n are integrable, $\lim g_n = 0$ and

$$0 \le g_n \le |f| + |f_n| \le 2g .$$

Applying to (g_n) what has been proved so far, we get the last formula. Note that the last formula implies the first one, since

$$|\nu(f) - \nu(f_n)| \le \nu(|f - f_n|)$$

by Proposition 2.13. □

5.6 As a further consequence of the Daniell property for upper integrals, with the notation $\mathcal{F}(\nu)$ of 2.13 for the set of all functions f with $\nu(|f|) < \infty$, we can prove the following

PROPOSITION. *$\mathcal{F}(\nu)$ is complete w.r.t. convergence in mean, and every sequence (f_n) in $\mathcal{F}(\nu)$ which converges in mean to $f \in \mathcal{F}(\nu)$ has a subsequence which converges almost everywhere to f.*

By Corollary 2.11, without loss of generality we may suppose that (f_n) is a Cauchy sequence in $\mathcal{F}_{\mathbb{R}}$. Since it is sufficient to prove that a subsequence of (f_n) converges in mean and almost everywhere, we may in addition assume that

$$\nu(|f_{n+1} - f_n|) \le \frac{1}{2^n}$$

for all $n \in \mathbb{N}$. By Lemma 5.3, we have

$$g := \sum_{n=1}^{\infty} |f_{n+1} - f_n| \in \mathcal{F} .$$

Consequently, g is finite a.e. and the series $\sum (f_{n+1} - f_n)$ converges absolutely a.e. But since

$$f_{n+1} = f_1 + \sum_{k=1}^{n} (f_{k+1} - f_k) ,$$

the sequence (f_n) converges a.e. to a real-valued function f. From

$$|f_n| \leq g + |f_1|,$$

we get $|f| \leq g + |f_1|$ a.e., hence $f \in \mathcal{F}$. Since moreover

$$|f - f_n| \leq |\sum_{k=n}^{\infty} (f_{k+1} - f_k)| \leq \sum_{k=n}^{\infty} |f_{k+1} - f_k| \text{ a.e.},$$

again from Lemma 5.3 we obtain

$$\nu(|f - f_n|) \leq \sum_{k=n}^{\infty} \nu(|f_{k+1} - f_k|) \leq \frac{1}{2^{n-1}} . \quad \square$$

In view of Corollary 2.13 this immediately implies the following *Riesz–Fischer Theorem*:

COROLLARY. *$\mathcal{I}(\nu)$ is complete w.r.t. convergence in mean.*

5.7 For upper integrals, the relationship between the theories of integration and of essential integration can now be studied in more detail:

PROPOSITION. *A function f is essentially integrable iff f coincides essentially almost everywhere with an integrable function.*

If f coincides essentially a.e. with $g \in \mathcal{I} \subset \mathcal{I}^\bullet$, then $f \in \mathcal{I}^\bullet$ by Proposition 2.9. To prove the converse, by Corollary 2.6.(ii) it is sufficient to consider the case $f \geq 0$. In view of Proposition 3.2.(i) there exists a decreasing sequence (v_n) in $\mathcal{I}_-$ with

$$\nu^\bullet(f) = \sup_n \nu(\min(f,-v_n)) < \infty .$$

The functions $\min(f,-v_n)$ are integrable by Proposition 3.4.(ii); hence

$$g := \sup_n(\min(f,-v_n))$$

is integrable by the Monotone Convergence Theorem 5.4, and

$$\nu^\bullet(g) = \nu(g) = \nu^\bullet(f) .$$

Since $g \leq f$, we finally get $g = f$ essentially a.e. by Proposition 2.12. $\square$

5.8 **PROPOSITION.** *If $\mathcal{T}$ is almost coinitial in $\mathfrak{I}_{\_}(\nu)$, then*

$$\nu^{\bullet}(f) = \sup_{t\in\mathcal{T}} \nu(1_{\{t<0\}}\cdot f)$$

for every positive function f.

Since $\mathbb{N}\cdot\mathcal{T}$ is also almost coinitial in $\mathfrak{I}_{\_}$, we infer from Proposition 3.6.(i) that

$$\sup_{t\in\mathcal{T}} \nu(1_{\{t<0\}}\cdot f) = \sup_{t\in\mathcal{T}} \nu(\sup_n \min(f,-nt)) =$$

$$= \sup_{t\in\mathcal{T},\, n\in\mathbb{N}} \nu(\min(f,-nt)) = \nu^{\bullet}(f) . \quad \square$$

COROLLARY. *A set A is an essentially null set iff $A \cap \{t < 0\}$ is a null set for every $t \in \mathcal{T}$.*

This follows immediately from the proposition, since

$$1_{\{t<0\}}\cdot\infty_A = \infty_{A\cap\{t<0\}} . \quad \square$$

DEFINITION. A *subset* of X is said to be $(\nu-)$*moderated* if it is contained in a set of the type $\{f > 0\}$ for some integrable function f.

A *function* is called *moderated* if it vanishes outside some moderated set, and an *upper integral* is said to be *moderated* if X is moderated.

REMARKS.

(1) *Every integrable function is moderated.*

(2) *A set A is an essentially null set iff $A \cap B$ is a null set for every moderated set B.*

(3) *A set is a null set iff it is moderated and an essentially null set.*

5.9 **PROPOSITION.** *For every moderated function f which has an integrable minorant, we have $\nu^{\bullet}(f) = \nu(f)$.*

Suppose f vanishes outside $\{g > 0\}$ for some integrable function $g \geq 0$. By Lemma 3.2.(i) and the Daniell property we get

$$\nu^\bullet(f) = \nu^\bullet(\sup_n \min(f,ng)) = \sup_n \nu^\bullet(\min(f,ng)) =$$
$$= \sup_n \nu(\min(f,ng)) = \nu(f)\ . \quad \square$$

REMARK. *If* ν *is auto-determined, then* $\nu_*(f) = \nu(f)$ *for every moderated measurable function* f *which admits an integrable minorant.*

This follows from the above proposition and Corollary 4.10. $\square$

COROLLARY. *A function is integrable iff it is moderated and essentially integrable. In particular, if* ν *is moderated, then every essentially integrable function is integrable.*

Obviously, the conditions are necessary. Conversely, for every essentially integrable and moderated function f, by Proposition 5.7 there exists an integrable and hence moderated function g which coincides essentially a.e. with f. Since $|f-g|$ is moderated, we have

$$\nu(|f-g|) = \nu^\bullet(|f-g|) = 0\ ,$$

hence f is integrable by Proposition 2.9. $\square$

5.10 For upper integrals, we can now prove the following additional stability properties of measurable functions (cf. Proposition 4.2) :

PROPOSITION.

(i) *For every sequence* (f_n) *of measurable functions, the functions* $\inf f_n$, $\sup f_n$, $\liminf f_n$, $\limsup f_n$ *are measurable, as is every function* f *to which* (f_n) *converges in* $\bar{\mathbb{R}}$ *essentially almost everywhere.*

(ii) *If* f *and* g *are measurable functions, then*

$$(f-g)^+\ ,\ |f-g|\ ,\ \text{and}\ \infty_{\{f>g\}}$$

are measurable.

(iii) $\tilde{\mathfrak{I}}^0(\nu)$ *is a lattice cone of functions.*

In (i), it is obviously sufficient to prove the measurability of $\sup f_n$ for increasing sequences (f_n). For the last assertion in (i), one also has to consider Proposition 4.6.

For $t \in \mathcal{T}_-$, the sequence of integrable functions $\mathrm{med}(f_n,-t,t)$ is increasing, has $-t$ as common integrable minorant and has $\mathrm{med}(\sup f_n,-t,t)$ as upper envelope. The Monotone Convergence Theorem implies that this function is integrable, hence $\sup f_n$ is measurable by Proposition 4.3.

To prove (ii), let $t \in \mathcal{T}_-$ be given. For every $n \in \mathbb{N}$, the functions

$$f_n := \mathrm{med}(f,-nt,nt) \;,\; g_n := \mathrm{med}(g,-nt,nt)$$

and therefore also $h_n := \min([f_n - g_n]^+,-t)$ are integrable with

$$\min([f - g]^+,-t) = \lim h_n \,.$$

By Lebesgue's Dominated Convergence Theorem, this function is integrable since $|h_n| \le -t$ for all $n \in \mathbb{N}$. Therefore, $(f - g)^+$ is measurable. Assertion (ii) now follows from Proposition 4.2 and the formula

$$\infty_{\{f>g\}} = \sup n\cdot(f - g)^+ \,.$$

The third part is proved in the same way, using $h_n := \mathrm{med}(f_n + g_n,-t,t)$, since

$$\mathrm{med}(f + g,-t,t) = \lim h_n \,. \quad \square$$

CHAPTER II.

FUNCTIONAL ANALYTIC ASPECTS AND RADON INTEGRALS

§ 6 REGULARITY

We next discuss abstract Riemann integration theory for linear functionals. One of our main goals is a detailed treatment of regularity properties. Recall that a linear functional is $\tilde{\mathbb{R}}$–valued and defined on a function cone in $\tilde{\mathbb{R}}^X$, and that only such function cones will be considered in the the sequel.

THROUGHOUT THIS SECTION,
μ IS AN INCREASING LINEAR FUNCTIONAL
ON A MIN–STABLE FUNCTION CONE $\mathcal{S}$.

6.1 **To begin with the most important case, let us first assume that μ is** ***regular*** **. By Definition 1.3 this means that**

$$\mu(s) = \sup_{t \in \mathcal{S},\ -t \leq s} -\mu(t)$$

for every $s \in \mathcal{S}$.

Every function $s \in \mathcal{S}$ with $\mu(s) < \infty$, e.g. $s \in \mathcal{S}_-$, is integrable w.r.t. μ^*. Moreover, for $s \in \mathcal{S}$ we have

$$\mu(s) = \sup_{t \in \{\mu < \infty\},\ t \leq s} \mu(t) .$$

In particular, μ is the smallest increasing extension of its restriction to $\{\mu < \infty\}$.

Indeed, the integrability was already mentioned in Example 2.5. Furthermore, for $t \in \{\mu < \infty\}$ with $-t \leq s$, by the integrability of $-t$ there exists $r \in \{\mu < \infty\}$ with $-t \leq r$, hence $-t \leq \min(s,r) \in \{\mu < \infty\}$ and therefore $-\mu(t) \leq \mu(\min(s,r))$. From this we infer that

$$\mu(s) = \mu_*(s) \leq \sup_{t \in \{\mu < \infty\},\ t \leq s} \mu(t) \leq \mu(s) . \quad \square$$

Conversely, it is easy to see that integrability (w.r.t. μ^*) of all functions $s \in \mathcal{S}$ with $\mu(s) < \infty$, together with the above formula, implies that μ is regular.

These considerations show that only for regular functionals μ, and these are the kind most often encountered, we can expect to achieve an effective theory of integration w.r.t. μ^*. What we get in this case resembles the well-known theory for vector lattices of real-valued functions.

DEFINITION. If μ is a regular linear functional, then we speak of *integrable*, *essentially integrable* and *measurable* functions, if necessary *w.r.t.* μ instead of w.r.t. μ^*, and we denote the sets of these functions by

$$\mathcal{R}^*(\mu)\,,\ \mathcal{R}^\bullet(\mu) \text{ and } \mathcal{R}^0(\mu)$$

respectively.

If f is a function which is integrable, respectively essentially integrable w.r.t. μ, then $\mu^*(f)$ and $\mu^\bullet(f)$ are called the (*abstract Riemann*) *integral*, respectively the *essential* (*abstract Riemann*) *integral* of f w.r.t. μ.

PROPOSITION. *If μ is regular, then μ^* is an auto-determined upper functional, and $\mathcal{S}_- \subset \mathcal{R}^*_-(\mu)$ is almost coinitial in $\bar{\mathbb{R}}^X$ w.r.t. μ^* and $\mu^\bullet$. Moreover, every $\mathcal{S}$-measurable function is measurable w.r.t. μ, and we have*

$$\mu_* = \mu_\bullet \text{ on } \{\mu_* > -\infty\} \quad \text{and} \quad \mu_\bullet = \mu^\bullet \text{ on } \mathcal{R}^0(\mu) \cap \{\mu_\bullet > -\infty\}\,.$$

The first assertion follows from Example 2.14. Since μ^* is determined by $\mathcal{S}_-^\uparrow$, the set $\mathcal{S}_-$ is almost coinitial in $\bar{\mathbb{R}}^X$ w.r.t. μ^* by Proposition 3.5.(ii) and w.r.t. $\mu^\bullet$ by Corollary 3.5. The last assertions follow from Proposition 4.11 and from 4.10. □

REMARKS.

(1) The preceding proposition shows that for functions $f \geq 0$ which are measurable w.r.t. μ or which admit an integrable majorant, we have

$$\mu^\bullet(f) = \sup_{s\in\mathcal{S}_-,\, -s\leq f} -\mu(s) \quad \text{and} \quad \mu^\bullet(f) = \inf_{s\in\mathcal{S}_+,\, s\geq f} \mu(s)$$

respectively. Since $\mu = \mu_\bullet = \mu^\bullet$ on $\mathcal{S}$, this can be interpreted as *inner* and *outer $\mathcal{S}$-regularity* of $\mu^\bullet$.

(2) As we have seen in Corollary 1.9, in the regular case it is always possible to restrict our considerations to lattice cones as long as we deal only with integration problems. However, this may lead to a loss of information concerning regularity properties.

(3) If $\mathcal{T}$ is a min-stable cone of functions with $\mathcal{T}_-$ almost coinitial in $\mathcal{R}_-^\bullet(\mu)$, then every $\mathcal{T}$-measurable function is measurable.

Example 4.9 immediately gives the following

EXAMPLE. *Let μ be a Radon integral on a Hausdorff space X, i.e. a regular linear functional on $\mathcal{S}(X)$. Then every lower semicontinuous function h is measurable w.r.t. μ.*

In this case, the formulas in Remark 1 have a set-theoretic interpretation as inner and outer regularity (cf. Corollary 8.10).

6.2 **LEMMA.** *If μ is regular, then the formula*

$$\mu(s) = \sup_{u\in\mathcal{S}_-} \mu([s+u]_-) - \mu(u)$$

holds for every $s \in \mathcal{S}$. In particular, μ is given by its values on $\mathcal{S}_-$.

In fact, by Theorem 3.6.(i) we have

$$\mu(s) = \mu^\bullet(s) = \sup_{u\in\mathcal{S}_-} \mu^*(\min(s,-u)) = \sup_{u\in\mathcal{S}_-} \mu^*([s+u]_- - u) =$$
$$= \sup_{u\in\mathcal{S}_-} \mu([s+u]_-) - \mu(u) . \quad \square$$

PROPOSITION. *Let ν be an upper functional determined by $\mathcal{S}$. If $\mathcal{S}_- \subset \mathcal{I}_-(\nu)$, then all functions in $\mathcal{S}$ are measurable w.r.t. ν, and*

$$\mu := \nu_* = \nu^\bullet \ \text{on}\ \mathcal{S}$$

is the only regular linear functional which coincides with $\nu^\bullet$ on $\mathcal{S}_-$. Furthermore, we have

$$\nu^\bullet = \nu^{\bullet\#\bullet} = \mu^\bullet \le \nu^{\bullet\#} \le \mu^* \le \nu .$$

Since ν is determined by $\mathcal{S}_-^{\uparrow}$ and $\mathcal{S}_- \subset \mathcal{I}_-$, the set $\mathcal{S}_-$ is almost coinitial in $\mathcal{I}_-^{\bullet}$ by Proposition 3.5.(ii) and Corollary 3.5. Therefore, the functions in $\mathcal{S}$ are measurable (cf. Proposition 4.11). Let $\mu := \nu^{\bullet}$ on $\mathcal{S}$. The Essential Integrability Criterion 4.4 shows that

$$\nu^{\bullet} \leq \nu^{\bullet\#} \leq \mu^{*} .$$

As ν is also determined by $\mathcal{I}_-^{\uparrow}$, we infer from Theorem 3.9 that $\nu^{\bullet} \leq \nu$, and hence $\mu^{*} \leq \nu$ since ν is determined by $\mathcal{S}$. From Theorem 4.10 and Proposition 3.9 we therefore get

$$\mu(s) = \nu^{\bullet}(s) = \nu_{\bullet}(s) = \nu_{*}(s) \leq \mu_{*}(s)$$

for $s \in \mathcal{S}$, which proves that μ is regular and that $\nu_{*} = \nu^{\bullet}$ on $\mathcal{S}$. The uniqueness follows immediately from the above lemma.

Finally, an application of Theorem 3.6.(ii) to the functionals $\nu^{\bullet}, \nu^{\bullet\#}, \mu^{*}$ and ν, and to the set $\mathcal{S}_-$, which is almost coinitial in the classes of integrable functions w.r.t. each of these functionals, gives

$$\nu^{\bullet} = \nu^{\bullet\bullet} \leq \nu^{\bullet\#\bullet} \leq \mu^{\bullet} \leq \nu^{\bullet} ,$$

hence $\nu^{\bullet} = \nu^{\bullet\#\bullet} = \mu^{\bullet}$. □

COROLLARY. *Let $\mathcal{T}$ be a min-stable cone of functions contained in $\mathcal{S}$, and let τ be an increasing sublinear functional on $\mathcal{T}$.*

(i) *If κ is an increasing sublinear extension of τ to $\mathcal{S}$, then we have*

$$\kappa^{*} \leq \tau^{*} , \quad \text{and} \quad \tau_{*} \leq \kappa \leq \tau^{*} \quad \text{on } \mathcal{S} .$$

(ii) *If τ is a regular linear functional and if $\mathcal{S}_- \subset \mathcal{R}_-^{*}(\tau)$, then $\mu := \tau_{*} = \tau^{\bullet}$ is the smallest and the only regular one, whereas τ^{*} is the largest increasing linear extension of τ to $\mathcal{S}$. We have*

$$\mu^{\bullet} = \tau^{\bullet} .$$

If τ_{*} and τ^{*} coincide on $\mathcal{S}$, then μ is the only increasing linear extension of τ to $\mathcal{S}$, and we have

$$\mu^{*} = \tau^{*} .$$

The first inequality in (i) is trivial. By Lemma 1.5, it implies

$$\tau_{*} \leq \kappa_{*} \leq \kappa = \kappa^{*} \leq \tau^{*} \quad \text{on } \mathcal{S} .$$

The first part of (ii) and the equality $\mu^\bullet = \tau^\bullet$ follow applying the proposition to the upper functional τ^* determined by $\mathcal{S}$, and considering (i) as well as Corollary 4.10. Note that τ is linear by Corollary 1.9. To prove the first uniqueness property, let κ be any regular linear extension of τ to $\mathcal{S}$. By (i), κ coincides with μ on $\mathcal{S}_-$, since $\mathcal{S}_- \subset \mathcal{R}^*_-(\tau)$, and by the proposition, κ also coincides with μ on $\mathcal{S}$.

The second uniqueness property follows immediately from (i), and since τ^* is determined by $\mathcal{S}$ and μ coincides with τ^* on $\mathcal{S}$, we get $\mu^* = \tau^*$. □

6.3 As already mentioned in Remark 1.4, integration theory in the non-regular case is based on the upper functional $\mu^\times$.

DEFINITION. For an arbitrary increasing linear functional μ, we define integration with respect to μ as integration with respect to the regular linear functional $\tilde{\mu}$. To avoid tedious repetitions, apart from the set of all *integrable functions w.r.t.* μ (i.e. w.r.t. $\mu^\times$), which will be denoted by

$$\mathcal{R}^\times(\mu),$$

we use the notations of 6.1 with μ instead of $\tilde{\mu}$ when confusion is impossible.

This is justified by the fact that in the regular case there is no significant difference between the abstract Riemann integration theories associated with μ and $\tilde{\mu}$. We will prove in Theorem 6.4 that the essential upper functionals $\mu^\bullet = \mu^{*\bullet}$ and $\mu^{\times\bullet} = \tilde{\mu}^{*\bullet}$ coincide. We therefore use the notation

$$\mu^\bullet$$

for both of them, and $\mu_\bullet$ for the corresponding essential lower functional.

PROPOSITION. *The functional* $\mu^\times$ *is an auto-determined upper functional, and* $\mathcal{S}_- \subset \mathcal{R}^\times_-(\mu)$ *is almost coinitial in* $\overline{\mathbb{R}}^X$ *w.r.t.* $\mu^\times$ *and* $\mu^\bullet$. *Moreover, every* $\mathcal{S}$*-measurable function is measurable w.r.t.* μ, *and we have*

$$\mu^\bullet \le \mu^*,$$

$$\mu_\times = \mu_\bullet \quad on \ \{\mu_\times > -\infty\} \quad and \quad \mu_\bullet = \mu^\bullet \quad on \ \mathcal{R}^0(\mu) \cap \{\mu_\bullet > -\infty\}.$$

By definition, we have $\mu^{\times} = \tilde{\mu}^{*}$ and $\tilde{\mu}$ is regular. Furthermore, $\mathcal{S}_{-}$ is coinitial in $(\mathcal{S}_{-} - \mathcal{S}_{-})_{-}$. Therefore, in view of Proposition 6.1, only the inequality $\mu^{\bullet} \leq \mu^{*}$ remains to be proved. For any function f and any $s \in \mathcal{S}$ with $s \geq f$, we have $\mu(s) \geq \mu_{\times}(s) = \mu^{\bullet}(s) \geq \mu^{\bullet}(f)$ by 1.4, hence $\mu^{*}(f) \geq \mu^{\bullet}(f)$. □

REMARKS.

(1) *Let f be a function such that $\max(f,s) \in \mathcal{S}$ for all $s \in \mathcal{S}_{-}$. If f admits a majorant in $-\mathcal{S}_{-}$, i.e. if $\mu^{\times}(f) < \infty$, then*

$$\mu^{\bullet}(f) = \mu^{\times}(f) = \mu^{*}(f).$$

It only remains to prove that $\mu^{\times}(f) \geq \mu^{*}(f)$. But for $s,t \in \mathcal{S}_{-}$ with $s - t \geq f$ we have $s - t \geq \max(f,s) \in \mathcal{S}$, hence

$$\mu(s) - \mu(t) \geq \mu(\max(f,s)) \geq \mu^{*}(f). \quad \square$$

(2) One might hope that it would be sufficient to treat integration theory only for positive linear forms on function spaces. But in such a restricted theory no interesting regularity properties would be available. These have to be formulated w.r.t. the function cone $\mathcal{S}$ using the lower functional μ_{*} and not w.r.t. the function space $\mathcal{S}_{-} - \mathcal{S}_{-}$ using the lower functional $\tilde{\mu}_{*} = \mu_{\times}$ (cf. Remarks 6.1.1 and 6.8).

(3) If $\mathcal{T}$ is a min-stable cone of functions with $\mathcal{T}_{-}$ almost coinitial in $\mathcal{R}_{-}^{\bullet}(\mu)$, then every $\mathcal{T}$-measurable function is measurable.

6.4 **THEOREM.** *If μ is regular, then the essential upper functionals $\mu^{*\bullet}$ and $\mu^{\times\bullet}$ coincide. With the common notation $\mu^{\bullet}$ for both of them, the inequalities*

$$\mu^{\bullet} \leq \mu^{*} \leq \mu^{\times}$$

hold. In particular, the theories of essential integration and the concepts of measurability w.r.t. μ^{*} and $\mu^{\times}$ coincide, and $\mathcal{R}^{\times}(\mu) \subset \mathcal{R}^{*}(\mu)$ with coincidence of the integrals. Moreover, the equality

$$\mu_{\times} = \mu_{*} \quad \textit{holds on} \quad \{\mu_{\times} > -\infty\}.$$

For all functions f and all $s,t \in \mathcal{S}_-$ with $s - t \leq f$, we conclude from Proposition 2.7 that

$$\mu(s) = \mu_*(s) \leq \mu_*(t + f) = \mu(t) + \mu_*(f),$$

hence $\mu_\times(f) \leq \mu_*(f)$ and therefore $\mu^* \leq \mu^\times$. By Proposition 6.3, this implies that $\mu_\times$ and μ_* coincide on $\{\mu_\times > -\infty\}$.

It remains to prove that $\mu_{*\bullet}(f) = \mu_{\times\bullet}(f)$. For $s,t \in \mathcal{S}_-$ we have

$$\mu_\times(\mathrm{med}(f,-t,s)) \geq \mu_\times(s) = \mu(s) > -\infty,$$

hence

$$\mu_\times(\mathrm{med}(f,-t,s)) = \mu_*(\mathrm{med}(f,-t,s)).$$

Applying Theorem 3.6.(ii) with the set $\mathcal{S}_-$, which is almost coinitial in both $\mathcal{R}^*_-(\mu)$ and $\mathcal{R}^\times_-(\mu)$, this yields

$$\mu_{*\bullet}(f) = \sup_{t\in\mathcal{S}_-} \inf_{s\in\mathcal{S}_-} \mu_*(\mathrm{med}(f,-t,s)) =$$

$$= \sup_{t\in\mathcal{S}_-} \inf_{s\in\mathcal{S}_-} \mu_\times(\mathrm{med}(f,-t,s)) = \mu_{\times\bullet}(f). \quad \square$$

REMARK. Since the two measurability concepts coincide, according to the criterion 4.5, integrability w.r.t. the two integration theories differs only by the fact that the finiteness condition imposed on $\mu^\times$ is (possibly) more restrictive than that on μ^*.

6.5 Since the functions in $\mathcal{S}$ are always measurable, the functionals $\mu_\bullet$ and $\mu^\bullet$ coincide with $\mu_\times$ on $\mathcal{S}$ (cf. Proposition 6.3). However, in general only the inequalities $\mu_\times \leq \mu \leq \mu^\times$ hold. The reasonable requirement that $\mu_\times$ should be an extension of μ turns out to define the weakest possible regularity condition that can be imposed on μ:

DEFINITION. If μ and $\mu_\times$ coincide on $\mathcal{S}$, i.e. if

$$\mu(s) = \sup_{u,v\in\mathcal{S}_-,\ u-v\leq s} \mu(u) - \mu(v)$$

holds for all $s \in \mathcal{S}$, then μ is called *difference-regular*.

Such a functional is increasing and given by its values on $\mathcal{S}_-$. An increasing linear functional is always difference–regular on $\mathcal{S}_-$. If μ is regular, then μ is difference–regular (cf. Theorem 6.4).

Proposition 6.3 together with Remark 6.3.3 is the key to the following *Extension Theorem*:

PROPOSITION. *Let $\mathcal{T}$ be a min–stable cone of functions contained in $\mathcal{S}$, and let τ be an increasing linear functional on $\mathcal{T}$.*

(i) *For every increasing linear functional κ on $\mathcal{S}$ which coincides with τ on $\mathcal{T}_-$, we have*

$$\kappa^\times \leq \tau^\times , \text{ and } \tau_\times \leq \kappa \leq \tau^\times \text{ on } \mathcal{S} .$$

(ii) *If $\mathcal{S}_- \subset \mathcal{R}_-^\times(\tau)$, then $\mu := \tau_\times = \tau^\bullet$ is the smallest and the only difference–regular increasing linear functional on $\mathcal{S}$ which coincides with τ on $\mathcal{T}_-$, whereas $\tau^\times$ is the largest one. We have*

$$\mu^\times = \tau^\times ,$$

and μ extends τ iff τ is difference–regular.

The inequality $\kappa^\times \leq \tau^\times$ results from the inclusion $\mathcal{T}_- \subset \mathcal{S}_-$, and this gives (i) by 1.4.

If $\mathcal{S}_- \subset \mathcal{R}_-^\times(\tau)$, then $\mathcal{S}_- \supset \mathcal{T}_-$ is almost coinitial in $\mathcal{R}_-^\bullet(\tau)$ by Proposition 6.3, hence $\mathcal{S} \subset \mathcal{R}^0(\tau) \cap \{\tau_\times > -\infty\}$ by Remark 6.3.3, and therefore $\tau_\times$ and $\tau^\bullet$ coincide on $\mathcal{S}$. By Corollary 4.10, the functionals μ and $\tau^\times$ are linear on $\mathcal{S}$. Since all functions in $\mathcal{S}_-$ are integrable w.r.t. τ, we have

$$\kappa = \tau_\times = \tau^\times \text{ on } \mathcal{S}_- .$$

This implies that $\kappa^\times \geq \tau^\times$, hence $\kappa^\times = \tau^\times$. Replacing κ by μ, the first part of (ii) is proved. Furthermore, we infer that κ is difference–regular iff $\kappa = \tau_\times$ on $\mathcal{S}$, i.e. iff $\kappa = \mu$.

The last assertion follows since by definition μ coincides with $\tau_\times$ on $\mathcal{S}$ and hence on $\mathcal{T}$. □

As we have seen in the proof of (ii), not only do the theories of integration coincide for μ and τ, but also for any κ of the above type. However, this is of no particular interest since the non-regular theories of integration for μ and κ only depend on $\mathcal{S}_-$ and both functionals coincide on this function cone.

The above proposition will be generalized in Corollary 7.3.

By Proposition 6.3, the function cones $\mathcal{S}_-$ and $\mathcal{S}$ satisfy the requirements in (ii). For difference-regular extensions, the only one of interest in integration theory, we have the following

COROLLARY. *The restriction to $\mathcal{S}_-$ defines a bijection between the difference-regular linear functionals on $\mathcal{S}$ and the increasing linear functionals on $\mathcal{S}_-$.*

In view of Theorem 6.4, the following formula was proved in Lemma 6.2 for the special case of a regular functional. However, it is true for arbitrary increasing linear functionals :

ADDENDUM. *The formula*

$$\mu_\times(s) = \sup_{u\in\mathcal{S}_-} \mu([s+u]_-) - \mu(u) = \lim_{u\in\mathcal{S}_-} \mu([s+u]_-) - \mu(u)$$

holds for all $s \in \mathcal{S}$. Furthermore, in this formula $\mathcal{S}_-$ may be replaced by any set which is coinitial in $\mathcal{S}_-$.

Indeed : Consider $s \in \mathcal{S}$ and $u,v \in \mathcal{S}_-$. If $u \le v$, then

$$(s+v)_- + u = \min(s+v+u,\, u) \le \min(s+u+v,\, v) = (s+u)_- + v\,,$$

hence

$$\mu([s+v]_-) - \mu(v) \le \mu([s+u]_-) - \mu(u)\,;$$

this proves the second equality and the last statement. Now, if $u - v \le s$, then $u \le (s+v)_-$, hence

$$\mu_\times(s) = \sup_{u,v\in\mathcal{S}_-,\ u-v\le s} \mu(u) - \mu(v) \le \sup_{v\in\mathcal{S}_-} \mu([s+v]_-) - \mu(v) \le \mu_\times(s)\,. \quad \square$$

REMARK. From this formula, the linearity of $\mu_\times$ on $\mathcal{S}$ follows directly without recourse to integration theory :

To prove that $\mu_\times$ is superadditive and subadditive one can use respectively the inequalities

$$(s+t+2u)_- \geq (s+u)_- + (t+u)_-$$

and

$$(s+t+u)_- + s_- + t_- + u \leq (s+t_- + u)_- + (s_- + t + u)_-$$

for $s,t \in \mathcal{S}$ and $u \in \mathcal{S}_-$.

Example. *Every difference-regular linear functional on* $\mathcal{S}$ *has a unique difference-regular linear extension to* $\mathcal{G}(\mathcal{S})$.

This follows from the second part of the proposition, since $\mathcal{G}_-(\mathcal{S}) = \mathcal{S}_-^{\wedge}$ consists of integrable functions. □

6.6 Every regular linear functional has the following weak regularity property (cf. the subsequent Remark 1) which leads to a useful characterization of regularity itself (cf. Theorem 6.9):

Definition. An increasing functional $\mu : \mathcal{S} \longrightarrow \tilde{\mathbb{R}}$ for which

$$\mu(t) \leq \mu(s) + \mu_*(t-s)$$

holds for all $s \in \mathcal{S}_-$ and $t \in \mathcal{S}$ with $s \leq t$ is said to be *semiregular*.

Remarks.

(1) *Every regular linear functional is semiregular.*

In fact, by Lemma 1.6.(i),

$$\mu(t) = \mu_*(s + [t-s]) \leq \mu(s) + \mu_*(t-s)$$

holds for $s \in \mathcal{S}_-$ and $t \in \mathcal{S}$ with $s \leq t$. □

(2) *For a semiregular sublinear functional, equality obtains in the above inequality. In particular,* $\mu(t) = \mu_*(t)$ *for* $t \in \mathcal{S}_+$.

To get the reverse inequality use Lemma 1.6.(i):

$$\mu(t) = \mu^*(s + [t-s]) \geq \mu(s) + \mu_*(t-s) . \quad \square$$

(3) ***Every semiregular sublinear functional is linear.***

For $s,t \in \mathcal{S}$, the inequalities

$$s_- + t_- \le s + t\,,\ s_- \le s\,,\ t_- \le t \text{ and } s_- + t_- \le s$$

imply by Remark 2 that

$$\mu(s+t) = \mu(s_- + t_-) + \mu_*(s^+ + t^+) \ge \mu(s_- + t_-) + \mu_*(s^+) + \mu_*(t^+)\,,$$

$$\mu(s) = \mu(s_-) + \mu_*(s^+)\,,\ \mu(t) = \mu(t_-) + \mu_*(t^+)\,,$$

and

$$\mu(s_-) = \mu(s_- + t_-) + \mu_*(-t_-) = \mu(s_- + t_-) - \mu(t_-)\,.$$

Thus

$$\mu(s+t) \ge \mu(s_-) + \mu(t_-) + \mu_*(s^+) + \mu_*(t^+) = \mu(s) + \mu(t) \ge \mu(s+t)\,. \quad \square$$

6.7 PROPOSITION. *The functional* μ *is semiregular iff* μ *is difference-regular and* $\mu_*(f) = \mu_\times(f)$ *for every positive function* f.

Suppose first that μ is semiregular. Then, for all $s,t \in \mathcal{S}_-$ with $s - t \le f$ we have

$$s - \min(s,t) = (s-t)^+ \le f\,,$$

hence

$$\mu(s) - \mu(t) \le \mu(\min(s,t)) + \mu_*(s - \min(s,t)) - \mu(t) \le$$

$$\le \mu(\min(s,t)) - \mu(t) + \mu_*(f) \le \mu_*(f)\,.$$

Consequently, we have $\mu_\times(f) \le \mu_*(f)$, hence $\mu_\times(f) = \mu_*(f)$ by Proposition 6.3. For $s \in \mathcal{S}$, this implies that

$$\mu(s) \le \mu(s_-) + \mu_*(s - s_-) = \mu_\times(s_-) + \mu_\times(s^+) = \mu_\times(s) \le \mu(s)\,,$$

and proves that μ is difference-regular.

Conversely, consider $s \in \mathcal{S}_-$ and $t \in \mathcal{S}$ with $s \le t$. In view of Proposition 2.7, we have

$$\mu(t) = \mu_\times(t) = \mu_\times(s) + \mu_\times(t-s) = \mu(s) + \mu_*(t-s)\,.$$

This proves that μ is semiregular. $\square$

REMARK. μ *is semiregular iff* μ *is difference-regular and the restriction* κ *of* μ *to* $\mathcal{S}_-$ *is semiregular.*

This follows from the proposition since $\mu_\times = \kappa_\times$ and $\mu_*(f) = \kappa_*(f)$ for every positive function f. □

Corollary. *If $(s-t)_- \in \mathcal{S}_-$ holds for all $s,t \in \mathcal{S}_-$, or equivalently $s - t \in \mathcal{S}_-$ for all $s,t \in \mathcal{S}_-$ with $s \le t$, i.e.*

$$(\mathcal{S}_- - \mathcal{S}_-)_- = \mathcal{S}_- ,$$

then every difference-regular linear functional μ on $\mathcal{S}$ is semiregular.

Indeed, let f be a positive function and $s,t \in \mathcal{S}_-$ such that $s - t \le f$. Then

$$s - t \le (s-t)^+ = -(t-s)_- \le f$$

and $(t-s)_- \in \mathcal{S}_-$. This gives

$$\mu(s) - \mu(t) \le -\mu([t-s]_-) \le \mu_*(f)$$

and therefore $\mu_\times(f) \le \mu_*(f)$. By Proposition 6.3, equality follows. □

6.8 Proposition. *If μ is semiregular, then the equality*

$$\mu_* = \mu_\bullet \quad \textit{holds on} \quad \{\mu_* > -\infty\}^\sim .$$

In particular, $\mu(s) = \mu_*(s)$ holds for all $s \in \mathcal{S}$ with $\mu_*(s) > -\infty$.

Let $f \in \tilde{\mathbb{R}}^X$ be a function with $\mu_*(f) > -\infty$. Then $\mu_*(f) \le \mu_\bullet(f)$ by Proposition 6.3. To prove the reverse inequality, choose $s \in \mathcal{S}$ with $-s \le f$ and $\mu(s) < \infty$. In view of Propositions 6.3, 6.7 and Lemma 1.6.(i), we get

$$\mu_\bullet(f) + \mu(s) = \mu_\bullet(f) + \mu_\bullet(s) \le \mu_\bullet(f+s) = \mu_\times(f+s) = \mu_*(f+s) \le$$

$$\le \mu_*(f) + \mu(s) \le \mu_\bullet(f) + \mu(s) ,$$

since $\mu(s) = \mu_\times(s) = \mu_\bullet(s)$ and $f + s \ge 0$. We conclude that $\mu_\bullet(f) = \mu_*(f)$. □

Remarks.

(1) The functional $\mu^\bullet$ has the inner $\mathcal{S}$-regularity property of Remark 6.1.1 : For every positive measurable function f, we have

$$\mu^{\bullet}(f) = \sup_{s \in \mathcal{S}_-,\, -s \leq f} - \mu(s) .$$

But in general it lacks the outer $\mathcal{S}$–regularity property $\mu^{\bullet} = \mu^{*}$. Indeed, if this equality even just holds on $-\mathcal{S}_-$, then we would have

$$\mu_*(s) = - \mu^*(-s) = - \mu^{\bullet}(-s) = \mu(s)$$

for every $s \in \mathcal{S}_-$. From this, regularity of μ would follow, as we will show in the next Theorem 6.9.(ii).

(2) If μ is semiregular and if μ^* is an upper functional (cf. Remark 2.3), then we even have

$$\mu_* = \mu_{\bullet} \quad \text{on } \{\mu_* > -\infty\} ,$$

since in view of Corollary 2.11 we may consider $\tilde{\mathbb{R}}$–valued functions without loss of generality.

6.9 To state the useful characterization of regularity mentioned in 6.6, we shall need the following

DEFINITION. The functional μ is said to be *$\mathcal{S}$–bounded below* if for every $s \in \mathcal{S}_-$ there exists a function $t \in \mathcal{S}$ with $-t \leq s$ and $\mu_*(t) < \infty$.

THEOREM. *The following assertions are equivalent:*

(i) *μ is regular.*

(ii) *μ is difference–regular and coincides with μ_* on $\mathcal{S}_-$.*

(iii) *μ is semiregular and $\mathcal{S}$–bounded below.*

(iv) *μ coincides with μ_* on $\mathcal{S}_+$ and is $\mathcal{S}$–bounded below.*

Obviously, (i) implies (ii). Let $f \geq 0$ and $s,t \in \mathcal{S}_-$ with $s - t \leq f$. From (ii) and Lemma 1.6.(i) we infer that

$$\mu(s) = \mu_*(s) \leq \mu_*(f + t) \leq \mu_*(f) + \mu(t) ,$$

hence $\mu_{\times}(f) = \mu_*(f)$ by Proposition 6.3. This proves that μ is semiregular using Proposition 6.7. Moreover, for every function $s \in \mathcal{S}_-$, there exists $t \in \mathcal{S}$ such that $-t \leq s$ and $\mu(t) < \infty$, hence $\mu_*(t) \leq \mu(t) < \infty$. This completes the proof that (ii) implies (iii).

Finally, (iv) follows from (iii) by Proposition 6.7, and assuming (iv), for every $s \in \mathcal{S}$ there exists a function $t \in \mathcal{S}_+$ with

$$-t \le s_- \le s \text{ and } \mu(t) = \mu_*(t) < \infty .$$

Since $s + t \in \mathcal{S}_+$, we therefore get

$$\mu(s) = \mu(s+t) - \mu(t) = \sup_{u \in \mathcal{S},\, -u \le s+t} -\mu(u) - \mu(t) =$$

$$= \sup_{u \in \mathcal{S},\, -(u+t) \le s} -\mu(u+t) \le \mu_*(s) \le \mu(s) ,$$

which proves (i). □

REMARKS.

(1) Condition (iii) is formally weaker than

$$\mu \text{ is semiregular and } \mu_*(s) > -\infty \text{ for all } s \in \mathcal{S}_- ,$$

and is very useful.

Indeed, $\mu_*(s) > -\infty$ implies the existence of $t \in \mathcal{S}$ with $-t \le s$ and

$$\mu_*(t) \le \mu(t) < \infty . \quad \square$$

(2) Semiregularity is strictly weaker than regularity. In fact, we will show in Example 12.5.1 that a semiregular functional on $\mathcal{S}$ need not even be $\mathcal{G}(\mathcal{S})$-bounded below.

COROLLARY. *Restriction to* $\mathcal{S}_-$ *defines a bijection between the semiregular, respectively regular, linear functionals on* $\mathcal{S}$ *and the linear functionals on* $\mathcal{S}_-$ *which are semiregular, respectively semiregular and* $\mathcal{S}$*-bounded below.*

This follows immediately from Corollary 6.5, Remark 6.7 and the above theorem. □

6.10 DEFINITION. Let μ and τ be increasing linear functionals on $\mathcal{S}$. We say that the increasing linear functional $\mu + \tau$, defined on $\mathcal{S}$ by

$$(\mu + \tau)(s) := \mu(s) + \tau(s) ,$$

is the *sum* of μ and τ.

THEOREM. *We have*

$$(\mu+\tau)^* = \mu^* + \tau^* \ , \ (\mu+\tau)^\times = \mu^\times + \tau^\times \quad \textit{and} \quad (\mu+\tau)^\bullet = \mu^\bullet + \tau^\bullet .$$

A function is integrable, essentially integrable, respectively measurable w.r.t. $\mu+\tau$ *iff it is so w.r.t.* μ *and* τ .

If μ *and* τ *are respectively difference-regular, semiregular or regular, so is* $\mu+\tau$.

Since $\mathcal{S}$ is min-stable, for any function f we have $\mu^*(f),\tau^*(f) < \infty$ iff $(\mu+\tau)^*(f) < \infty$, and the first formula holds because

$$(\mu+\tau)^*(f) = \lim_s [\mu(s)+\tau(s)] = \lim_s \mu(s) + \lim_s \tau(s) = \mu^*(f) + \tau^*(f) ,$$

the limits being taken along the downward directed set of all $s \in \mathcal{S}$ with $s \geq f$. Applied to $\tilde{\mu}$ and $\tilde{\tau}$, we get the second formula. The third one, as well as the second assertion, is then a consequence of Proposition 4.12, since $\mathcal{S}_-$ is contained in $\mathcal{R}_-^\times(\mu) \cap \mathcal{R}_-^\times(\tau)$ and almost coinitial in $\mathcal{R}_-^\times(\mu)$ and $\mathcal{R}_-^\times(\tau)$.

Finally, if μ and τ are difference-regular, we have

$$(\mu+\tau)(s) = \mu_\times(s) + \tau_\times(s) = -[\mu^\times(-s) + \tau^\times(-s)] = -(\mu+\tau)^\times(-s) = (\mu+\tau)_\times(s)$$

for $s \in \mathcal{S}$, which proves that $\mu+\tau$ is difference-regular.

In the case of semiregularity, let $s \in \mathcal{S}_-$ and $t \in \mathcal{S}$ be such that $s \leq t$. Then we get

$$(\mu+\tau)(t) \leq \mu(s) + \mu_*(t-s) + \tau(s) + \tau_*(t-s) = (\mu+\tau)(s) + (\mu+\tau)_*(t-s) .$$

Finally, if μ and τ are regular, so is $\mu+\tau$ since

$$(\mu+\tau)(t) = \mu_*(t) + \tau_*(t) = (\mu+\tau)_*(t) . \quad \square$$

6.11 DEFINITION. For increasing linear functionals μ and τ on $\mathcal{S}$, we define the preorder relation

$$\tau \preceq \mu \quad \text{by} \quad \tau(t) - \tau(s) \leq \mu(t) - \mu(s) \quad \text{for all } s,t \in \mathcal{S}_- \text{ with } s \leq t .$$

From the definitions one easily gets the following

REMARKS.

(1) The relation $\tau \preceq \mu$ is equivalent to $\tilde{\mu} - \tilde{\tau}$ being a positive linear form, or

to one of the following inequalities on $\bar{\mathbb{R}}_+^X$:

$$\tau^\times \le \mu^\times \ , \ \tau_\times \le \mu_\times \ , \ \tau^\bullet \le \mu^\bullet \text{ or } \tau_\bullet \le \mu_\bullet \, .$$

In this case, we also have $\mu \le \tau$ on $\mathcal{S}_-$, hence $\tau_* \le \mu_*$ on $\bar{\mathbb{R}}_+^X$.

(2) If μ and τ are difference-regular with $\tau \preceq \mu$, then

$$\tau \le \mu \text{ on } \mathcal{S}_+ \text{ and } \tau^* \le \mu^* \text{ on } \bar{\mathbb{R}}_+^X \, .$$

Restricted to the set of all difference-regular linear functionals on $\mathcal{S}$, $\preceq$ is an order relation.

PROPOSITION. *Let μ and τ be increasing linear functionals on $\mathcal{S}$.*

(i) *If $\tau \preceq \mu$, then*

$$\mathcal{R}^\times(\mu) \subset \mathcal{R}^\times(\tau) \ , \ \mathcal{R}^\bullet(\mu) \subset \mathcal{R}^\bullet(\tau) \text{ and } \mathcal{R}^0(\mu) \subset \mathcal{R}^0(\tau) \, .$$

(ii) *If τ is semiregular, then $\tau \preceq \mu$ iff $\mu \le \tau$ on $\mathcal{S}_-$.*

(iii) *If τ is difference-regular and μ regular, then $\tau \preceq \mu$ iff $\tau \le \mu$ on $\mathcal{S}_+$.*

(iv) *If τ is semiregular, μ regular and $\tau \preceq \mu$, then τ is regular and*

$$\mathcal{R}^*(\mu) \subset \mathcal{R}^*(\tau) \, .$$

If f is integrable w.r.t. μ , then for any $\varepsilon > 0$ there exist functions $s_i, t_i \in \mathcal{S}_-$ with

$$s_1 - t_1 \le f \le s_2 - t_2 \text{ and } \mu(s_2) - \mu(t_2) - [\mu(s_1) - \mu(t_1)] \le \varepsilon \, .$$

Since $s_1 + t_2 \le s_2 + t_1$, we get

$$\tau(s_2 + t_1) - \tau(s_1 + t_2) \le \mu(s_2 + t_1) - \mu(s_1 + t_2) \le \varepsilon \, .$$

This proves that f is integrable w.r.t. τ . The measurability assertion now follows from the definition, and $\mathcal{R}^\bullet(\mu) \subset \mathcal{R}^\bullet(\tau)$ is a consequence of the Essential Integrability Criterion 4.4 and the inequality $\tau^\bullet \le \mu^\bullet$ on $\bar{\mathbb{R}}_+^X$. This proves (i).

The necessity part of (ii) is trivial. Conversely, if $\mu \le \tau$ on $\mathcal{S}_-$, then $\tau_* \le \mu_*$ on $\bar{\mathbb{R}}_+^X$, hence for all $s,t \in \mathcal{S}_-$ with $s \le t$ we have

$$\tau(t) - \tau(s) = \tau_\times(t - s) = \tau_*(t - s) \le \mu_*(t - s) \le \mu(t) - \mu(s)$$

by Lemma 1.6.(i).

The necessity in (iii) is part of Remark 2. Conversely, if $\tau \le \mu$ on $\mathcal{S}_+$, then $\tau^* \le \mu^*$ on $\overline{\mathbb{R}}_+^X$, hence for all $s,t \in \mathcal{S}_-$ with $s \le t$ we have

$$\tau(t) - \tau(s) = \tau_\times(t-s) \le \tau^*(t-s) \le \mu^*(t-s) = \mu(t) - \mu(s) .$$

Finally, in (iv) the regularity of τ follows from Theorem 6.9, since μ is $\mathcal{S}$–bounded below and $\tau_* \le \mu_*$ on $\mathcal{S}_+$. If f is integrable w.r.t. μ, then for any $\varepsilon > 0$ there exist functions $s,t \in \mathcal{S}$ with

$$-t \le f \le s \quad \text{and} \quad \tau(s+t) \le \mu(s+t) \le \varepsilon ,$$

which proves that f is integrable w.r.t. τ. □

COROLLARY. *Let μ and τ be difference-regular with $\tau \preceq \mu$, and denote by ρ the difference-regular linear extension of the increasing linear functional defined on $\mathcal{S}_-$ by*

$$s \longmapsto \mu(s) - \tau(s) .$$

Then

$$\tau + \rho = \mu ,$$

and if μ is semiregular, respectively regular, so is ρ.

The formula follows from Theorem 6.10, since for $s \in \mathcal{S}$ we get

$$\tau(s) + \rho(s) = \tau_\times(s) + \rho_\times(s) = (\tau+\rho)_\times(s) = \mu_\times(s) = \mu(s) .$$

If μ is semiregular, then for $s,t \in \mathcal{S}_-$ with $s \le t$ we have by Lemma 1.6.(i)

$$\rho(t) = \mu(t) - \tau(t) \le \mu(s) + \mu_*(t-s) - \tau(s) - \tau_*(t-s) = \rho(s) + \rho_*(t-s) ,$$

since by Theorem 6.10

$$\tau_*(t-s) + \rho_*(t-s) = (\tau+\rho)_*(t-s) = \mu_*(t-s) .$$

Finally, if μ is regular, then ρ is regular by (iv). □

6.12 We are now able to define the *upper envelope* of linear functionals.

PROPOSITION. *Let $(\mu_i)_{i \in I}$ be a family of difference-regular linear functionals on $\mathcal{S}$, upward directed w.r.t. $\preceq$, such that the functional*

$$s \longmapsto \inf_{i \in I} \mu_i(s)$$

on $\mathcal{S}_-$ is finite. Then its difference-regular linear extension $\sup \mu_i$ *is, w.r.t $\preceq$,* *the smallest difference–regular linear functional* μ *on* $\mathcal{S}$ *with* $\mu_i \preceq \mu$ *for all* $i \in I$.

For any function $f \geq 0$ *we have*

$$(\sup \mu_i)_\times(f) = \sup \mu_{i\times}(f) ,$$

$$(\sup \mu_i)_\bullet(f) = \sup \mu_{i\bullet}(f) \quad \textit{and} \quad (\sup \mu_i)^\bullet(f) = \sup \mu_i^\bullet(f) .$$

In particular,

$$\mathcal{R}^0(\sup \mu_i) = \bigcap_{i \in I} \mathcal{R}^0(\mu_i) .$$

Since $(\mu_i(s))_{i \in I}$ is downward directed for any $s \in \mathcal{S}_-$, the functional

$$\mu := \sup \mu_i$$

is increasing and linear on $\mathcal{S}_-$, hence on $\mathcal{S}$. For all $s,t \in \mathcal{S}_-$ with $s \leq t$, the family $(\mu_i(t) - \mu_i(s))_{i \in I}$ is upward directed, so we have

$$\mu(t) - \mu(s) = \lim \mu_i(t) - \lim \mu_i(s) = \sup_{i \in I} \; [\mu_i(t) - \mu_i(s)] ,$$

hence $\mu_i \preceq \mu$ for all $i \in I$. On the other hand, for any difference–regular linear functional τ on $\mathcal{S}$ with $\mu_i \preceq \tau$ for all $i \in I$, we also get $\mu \preceq \tau$.

Permuting two suprema, we obtain the first two equalities, and consequently

$$\mu^\times(\min(f,-s)) = \mu^\times([f+s]_-) - \mu(s) = \inf_{i \in I} \; \mu_i^\times([f+s]_-) - \inf_{i \in I} \; \mu_i(s) =$$
$$= \sup_{i \in I} \; \mu_i^\times(\min(f,-s))$$

for $s \in \mathcal{S}_-$, hence

$$\mu^\bullet(f) = \sup_{s \in \mathcal{S}_-} \; \mu^\times(\min(f,-s)) = \sup_{i \in I} \; \mu_i^\bullet(f) .$$

By Proposition 6.11.(i), we have $\mathcal{R}^0(\mu) \subset \mathcal{R}^0(\mu_i)$ for all $i \in I$. Conversely, it is sufficient to prove that for every function $f \geq 0$ which is measurable w.r.t. all μ_i, and any $s \in \mathcal{S}_-$, the function $\min(f,-s)$ is essentially integrable w.r.t. μ. But

$$0 \leq \mu_\bullet(\min(f,-s)) = \sup_{i \in I} \; \mu_{i\bullet}(\min(f,-s)) =$$

$$= \sup_{i \in I} \; \mu_i^\bullet(\min(f,-s)) = \mu^\bullet(\min(f,-s)) \leq \mu^\bullet(-s) < \infty . \quad \square$$

COROLLARY. *If all* μ_i *are semiregular, so is* $\sup \mu_i$, *and*

$$(\sup \mu_i)_*(f) = \sup_{i \in I} \; \mu_{i*}(f)$$

for any function $f \geq 0$ *. Moreover,* $\sup \mu_i$ *exists and is regular iff* $(\mu_i)_{i \in I}$ *is uniformly* $\mathcal{S}$*-bounded below, i.e. for every* $s \in \mathcal{S}_-$ *there exists* $t \in \mathcal{S}_+$ *with* $-t \leq s$ *and*

$$\sup_{i \in I} \mu_{i*}(t) < \infty .$$

For $\mu := \sup \mu_i$ we have

$$\mu_*(f) = \sup_{s \in \mathcal{S}_-, -s \leq f} - \mu(s) = \sup_{s \in \mathcal{S}_-, -s \leq f} \sup_{i \in I} - \mu_i(s) =$$
$$= \sup_{i \in I} \sup_{s \in \mathcal{S}_-, -s \leq f} - \mu_i(s) = \sup_{i \in I} \mu_{i*}(f) .$$

For all $s,t\, \mathcal{S}_-$ with $s \leq t$ this yields

$$\mu(t) = \inf \mu_i(t) \leq \inf_{i \in I} [\mu_i(s) + \mu_{i*}(t - s)] = \mu(s) + \mu_*(t - s) .$$

By Remark 6.7, μ is semiregular.

Finally, the regularity assertion follows from Theorem 6.9. □

6.13 It is now easy to consider the *sum* of a family of linear functionals.

THEOREM. *Let* $(\mu_i)_{i \in I}$ *be a family of difference-regular linear functionals on* $\mathcal{S}$ *with*

$$\sum_{i \in I} \mu_i(s) > -\infty$$

for all $s \in \mathcal{S}_-$ *. Then*

$$\sum_{i \in I} \mu_i := \sup_J \sum_{j \in J} \mu_j ,$$

where J *runs through all finite subsets of* I *, is a difference-regular linear functional on* $\mathcal{S}$ *with*

$$\mathcal{R}^0(\sum_{i \in I} \mu_i) = \bigcap_{i \in I} \mathcal{R}^0(\mu_i) .$$

For any function $f \geq 0$ *we have*

$$(\sum_{i \in I} \mu_i)_\times(f) = \sum_{i \in I} \mu_{i\times}(f) ,$$

$$(\sum_{i \in I} \mu_i)_\bullet(f) = \sum_{i \in I} \mu_{i\bullet}(f) \quad \textit{and} \quad (\sum_{i \in I} \mu_i)^\bullet(f) = \sum_{i \in I} \mu_i^\bullet(f) .$$

For finite subsets J and K of I, we have

$$\sum_{j \in J} \mu_j \preceq \sum_{i \in J \cup K} \mu_i$$

since for $s,t \in \mathcal{S}_-$ with $s \leq t$ obviously

$$\sum_{k \in K} \mu_k(t) - \sum_{k \in K} \mu_k(s) \geq 0 .$$

The assertion now follows from Theorem 6.10 and Proposition 6.12. □

Corollary 6.12 immediately gives the following

COROLLARY. *If all μ_i are semiregular, so is $\sum_{i \in I} \mu_i$, and*

$$\Big(\sum_{i \in I} \mu_i\Big)_*(f) = \sum_{i \in I} \mu_{i*}(f)$$

for any function $f \geq 0$. Moreover, $\sum_{i \in I} \mu_i$ exists and is regular iff for every $s \in \mathcal{S}_-$ there exists $t \in \mathcal{S}_+$ with $-t \leq s$ and

$$\sum_{i \in I} \mu_i(t) < \infty .$$

6.14 **DEFINITION.** Let $w \in \overline{\mathbb{R}}_+^X$ and suppose that

$$ws \in \mathcal{R}^0(\mu) \quad \text{and} \quad \mu_\bullet(ws) > -\infty$$

for all $s \in \mathcal{S}$. We say that the functional $w \cdot \mu$, defined on $\mathcal{S}$ by

$$w \cdot \mu(s) := \mu^\bullet(ws) ,$$

has the *density* w w.r.t. μ.

Recall the notation $\frac{f}{w}$ of 2.16.

THEOREM. *The function w is finite essentially a.e. w.r.t. μ, and the functional $w \cdot \mu$ is increasing and linear.*

If μ is semiregular and

$$\mu^\bullet(1_{\{0<w<\infty\}}\cdot s) = \inf_{t\in\mathcal{S}_-,\ t\geq s/w} \mu^\bullet(wt)$$

holds for all $s \in \mathcal{S}_-$ *, then* $w\cdot\mu$ *is a semiregular linear functional. It is regular iff for every* $s \in \mathcal{S}_-$ *there exists a function* $t \in \mathcal{S}$ *with*

$$-t \leq s \quad \text{and} \quad \mu_*(wt) < \infty\ .$$

If moreover $\infty_{\{w>0\}}$ *is measurable w.r.t.* μ *, then*

$$(w\cdot\mu)^\bullet = w\cdot\mu^\bullet\ ,$$

and a function f *is essentially integrable, resp. measurable w.r.t.* $w\cdot\mu$ *iff* wf *is essentially integrable, resp. measurable w.r.t.* μ .

For all $s \in \mathcal{S}_-$ we have

$$\min[\infty_{\{w=\infty\}},-s] = \min[\infty_{\{w=\infty\}\cap\{s<0\}},-s] \leq \infty_{\{ws=-\infty\}}\ ,$$

hence

$$0 \leq \mu^\times(\min[\infty_{\{w=\infty\}},-s]) = \mu^\bullet(\min[\infty_{\{w=\infty\}},-s]) \leq \mu^\bullet(\infty_{\{ws=-\infty\}}) = 0$$

by Proposition 2.11.(ii). This shows that

$$\mu^\bullet(\infty_{\{w=\infty\}}) = \inf_{s\in\mathcal{S}_-} \mu^\times(\min[\infty_{\{w=\infty\}},-s]) = 0\ .$$

The rest of the first part is obvious by Proposition 6.3.

For the second part, we first show that

$$(w\cdot\mu)^*(f) = \mu^*(wf)$$

for all $f \in \overline{\mathbb{R}}^X_-$. In fact, by Proposition 6.8 we have

$$(w\cdot\mu)^*(f) = \inf_{s\in\mathcal{S}_-,\ s\geq f} \mu^\bullet(ws) \geq \mu^\bullet(wf) =$$

$$= \mu^*(wf) = \inf_{s\in\mathcal{S}_-,\ s\geq wf} \mu(s) = \inf_{s\in\mathcal{S}_-,\ s\geq wf} \mu^\bullet(1_{\{0<w<\infty\}}\cdot s) =$$

$$= \inf_{s\in\mathcal{S}_-,\ s\geq wf}\ \inf_{t\in\mathcal{S}_-,\ t\geq s/w} \mu^\bullet(wt) \geq \inf_{t\in\mathcal{S}_-,\ t\geq f} \mu^\bullet(wt) = (w\cdot\mu)^*(f)\ ,$$

since $s \geq wf$ implies $s = 0$ on $\{w = 0\}$, hence $s = 1_{\{0<w<\infty\}}\cdot s$ essentially a.e. w.r.t. μ , and $t \geq \frac{s}{w}$ implies $t \geq w\cdot\frac{f}{w} = 1_{\{0<w<\infty\}}\cdot f \geq f$.

This enables us to prove that $w\cdot\mu$ is semiregular. Let $s \in \mathcal{S}_-$ and $t \in \mathcal{S}$ with $s \leq t$. Replacing w by $1_{\{0<w<\infty\}}\cdot w$ according to Corollary 2.11, we have

$$w\cdot\mu(t) = \mu^\bullet(wt) = \mu_\bullet(ws + w[t-s]) \le \mu^\bullet(ws) + \mu_\bullet(w[t-s]) =$$
$$= \mu^\bullet(ws) + \mu_*(w[t-s]) = w\cdot\mu(s) + (w\cdot\mu)_*(t-s)$$

by successive application of Proposition 6.3, Theorem 2.2.(v) and Proposition 6.8. The regularity assertion follows immediately from Theorem 6.9.(iii).

We now prove that

$$(w\cdot\mu)^\bullet(f) = \mu^\bullet(wf)$$

holds for all $f \in \bar{\mathbb{R}}^X$.

From the semiregularity of μ and $w\cdot\mu$ we infer that

$$(w\cdot\mu)^\times(f) = (w\cdot\mu)^*(f) = \mu^*(wf) = \mu^\bullet(wf)$$

for $f \in \bar{\mathbb{R}}^X_-$. For arbitrary f and $r,t \in \mathcal{S}_-$, replacing again w by $1_{\{0<w<\infty\}}\cdot w$, we then have

$$(w\cdot\mu)^\times(\mathrm{med}(f,-r,t)) = (w\cdot\mu)^\times([\max(f,t)+r]_-) - (w\cdot\mu)(r) =$$
$$= \mu^\bullet(w[\max(f,t)+r]_-) - \mu^\bullet(wr) = \mu^\bullet(w\cdot\mathrm{med}(f,-r,t))$$

by Proposition 2.7, thus

$$(w\cdot\mu)^\bullet(f) = \inf_{t\in\mathcal{S}_-} \sup_{r\in\mathcal{S}_-} (w\cdot\mu)^\times(\mathrm{med}(f,-r,t)) =$$
$$= \inf_{t\in\mathcal{S}_-} \sup_{r\in\mathcal{S}_-} \mu^\bullet(\mathrm{med}(wf,-wr,wt))\,.$$

In order to calculate the right hand side, we are going to apply Lemmas 3.5 and 3.6. For $s \in \mathcal{S}_-$, note first that $w\cdot\frac{s}{w} = 1_{\{0<w<\infty\}}\cdot s$ is equal essentially a.e. to $\max(s,-\infty_{\{w>0\}})$, and hence essentially integrable w.r.t. μ by the measurability of $-\infty_{\{w>0\}}$. The sets of all functions wt with $t \in \mathcal{S}_-$ and $w\cdot\frac{s}{w}$ with $s \in \mathcal{S}_-$ are almost coinitial to each other w.r.t. $\mu^\bullet$, by assumption for one part, and by the formula

$$\max(wt, w\cdot\tfrac{s}{w}) = \max(wt,s) \text{ essentially a.e. w.r.t. } \mu$$

and the almost coinitiality of $\mathcal{S}_-$ in $\mathcal{R}^\bullet_-(\mu)$ for the other part. Therefore we have

$$(w\cdot\mu)^\bullet(f) = \inf_{t\in\mathcal{S}_-} \sup_{r\in\mathcal{S}_-} \mu^\bullet(\mathrm{med}(wf,-wr,wt)) =$$
$$= \inf_{s\in\mathcal{S}_-} \sup_{r\in\mathcal{S}_-} \mu^\bullet(\mathrm{med}(wf,-wr,w\cdot\tfrac{s}{w})) =$$
$$= \inf_{s\in\mathcal{S}_-} \sup_{t\in\mathcal{S}_-} \mu^\bullet(\mathrm{med}(wf,-w\cdot\tfrac{t}{w},w\cdot\tfrac{s}{w})) =$$

$$= \inf_{s\in\mathcal{S}_-} \sup_{t\in\mathcal{S}_-} \mu^\bullet(\text{med}(wf,-t,s)) = \mu^{\bullet\bullet}(wf) = \mu^\bullet(wf) ,$$

since

$$\text{med}(wf,-w\cdot\tfrac{t}{w},w\cdot\tfrac{s}{w}) = \text{med}(wf,-t,s) \text{ essentially a.e. w.r.t. } \mu .$$

Finally, the last assertions follow immediately from Corollary 4.3 and Proposition 4.13. □

Remark. The conditions are necessary. Indeed, for $s \in \mathcal{S}_-$ we get

$$\mu^\bullet(1_{\{0<w<\infty\}}\cdot s) = \mu^\bullet(w\cdot\tfrac{s}{w}) = (w\cdot\mu)^\bullet(\tfrac{s}{w}) = (w\cdot\mu)^*(\tfrac{s}{w}) = \inf_{t\in\mathcal{S}_-,\ t\geq s/w} \mu^\bullet(wt) .$$

The measurability of $\infty_{\{w>0\}}$ is proved as in Remark 4.13. □

6.15 Definition. Let $p : X \longrightarrow Y$ be a mapping from X into a set Y and suppose that $\mathcal{T} \subset \tilde{\mathbb{R}}^Y$ is an min-stable function cone such that

$$t\circ p \in \mathcal{R}^0(\mu) \quad \text{and} \quad \mu_\bullet(t\circ p) > -\infty$$

for every $t \in \mathcal{T}$. We say that the functional $p(\mu)$, defined on $\mathcal{T}$ by

$$p(\mu)(t) := \mu^\bullet(t\circ p) ,$$

is the *image* of μ under p on $\mathcal{T}$.

Recall the notation f_p of 2.17.

Theorem. *The functional $p(\mu)$ is increasing and linear.*

If μ is semiregular and

$$\mu(s) = \inf_{t\in\mathcal{T}_-,\ t\geq s_{p-}} \mu^\bullet(\max(s,t\circ p))$$

for all $s \in \mathcal{S}_-$, then $p(\mu)$ is a semiregular linear functional. It is regular iff for every $t \in \mathcal{T}_-$ there exists $r \in \mathcal{T}$ with

$$-r \leq t \quad \textit{and} \quad \mu_*(r\circ p) < \infty .$$

Moreover, we have

$$p(\mu)^\bullet = p(\mu^\bullet) ,$$

and a function g on Y is essentially integrable, resp. measurable w.r.t. $p(\mu)$ iff $g\circ p$ is essentially integrable, resp. mesurable w.r.t. μ.

The first part is obvious by Proposition 6.3. For the second part, we first show that

$$p(\mu)^*(g) = \mu^*(g\circ p)$$

for all $g \in \bar{\mathbb{R}}^Y_-$. In fact, by Proposition 6.8 we have

$$p(\mu)^*(g) = \inf_{t\in\mathcal{T}_-,\, t\geq g} \mu^\bullet(t\circ p) \geq \mu^\bullet(g\circ p) =$$

$$=\mu^*(g\circ p) = \inf_{s\in\mathcal{S}_-,\, s\geq g\circ p} \mu(s) = \inf_{s\in\mathcal{S}_-,\, s\geq g\circ p} \inf_{t\in\mathcal{T}_-,\, t\geq s_{p-}} \mu^\bullet(\max(s,t\circ p)) \geq$$

$$\geq \inf_{t\in\mathcal{T}_-,\, t\geq g} \mu^\bullet(t\circ p) = p(\mu)^*(g),$$

since $s \geq g\circ p$ and $t \geq s_{p-}$ imply $t \geq g$, t being 0 on $\complement p(X)$.

This enables us to prove that $p(\mu)$ is semiregular. Let $r \in \mathcal{T}_-$ and $t \in \mathcal{T}$ with $r \leq t$. Then

$$p(\mu)(t) = \mu^\bullet(t\circ p) = \mu_\bullet(r\circ p + [t-r]\circ p) \leq \mu^\bullet(r\circ p) + \mu_\bullet([t-r]\circ p) =$$

$$= \mu^\bullet(r\circ p) + \mu_*([t-r]\circ p) = p(\mu)(r) + p(\mu)_*(t-r)$$

by successive application of Proposition 6.3, Theorem 2.2.(v) and Proposition 6.8. The regularity assertion follows immediately from Theorem 6.9.(iii).

We now show that

$$p(\mu)^\bullet(g) = \mu^\bullet(g\circ p) = p(\mu^\bullet)(g)$$

for all $g \in \bar{\mathbb{R}}^Y$.

From the semiregularity of μ and $p(\mu)$ we infer that

$$p(\mu)^\times(g) = p(\mu)^*(g) = \mu^*(g\circ p) = \mu^\bullet(g\circ p).$$

for all $g \in \bar{\mathbb{R}}^Y_-$. For arbitrary g and $r,t \in \mathcal{T}_-$ we therefore have

$$p(\mu)^\times(\mathrm{med}(g,-t,r)) = p(\mu)^\times([\max(g,r)+t\,]_-) - p(\mu)(t) =$$

$$= \mu^\bullet([\max(g,r)+t\,]_-\circ p) - \mu^\bullet(t\circ p) = \mu^\bullet(\mathrm{med}(g,-t,r)\circ p)$$

by Proposition 2.7. Thus

$$p(\mu)^\bullet(g) = \inf_{r\in\mathcal{T}_-} \sup_{t\in\mathcal{T}_-} p(\mu)^\times(\mathrm{med}(g,-t,r)) =$$

$$= \inf_{r\in\mathcal{T}_-} \sup_{t\in\mathcal{T}_-} \mu^\bullet(\mathrm{med}(g\circ p,-t\circ p,r\circ p)) = \mu^{\bullet\bullet}(g\circ p) = \mu^\bullet(g\circ p)$$

by Theorem 3.6.(ii), since by assumption the set of all $t\circ p$ with $t \in \mathcal{T}_-$ is almost coinitial to $\mathcal{S}_-$ w.r.t. $\mu^\bullet$, hence in $\mathcal{R}^\bullet_-(\mu)$ and $\bar{\mathbb{R}}^X$ by Propositions 6.3 and 3.5.(i).

Finally, this also proves the measurability assertion, using Corollary 4.3

and Proposition 4.14. □

REMARKS.

(1) The condition

$$\mu^{\bullet}(s_p \circ p) = \inf_{t \in \mathcal{T}_-,\ t \geq s_{p-}} \mu^{\bullet}(t \circ p) \quad \text{for } s \in \mathcal{S}_-$$

is necessary.

Indeed, we have

$$\mu^{\bullet}(s_p \circ p) = p(\mu)^{\bullet}(s_{p-}) = p(\mu)^{*}(s_{p-}) = \inf_{t \in \mathcal{T}_-,\ t \geq s_{p-}} p(\mu)(t) =$$

$$= \inf_{t \in \mathcal{T}_-,\ t \geq s_{p-}} \mu^{\bullet}(t \circ p)\ . \quad \square$$

It is sufficient if $\mu^{\bullet}(s_p \circ p) > -\infty$ for $s \in \mathcal{S}_-$, by the transitivity property in Proposition 3.5.(i).

(2) The theorem shows that the so-called image measure catastrophe (cf. *Schwartz* [1973], p. 30) does not occur in the semiregular case.

§ 7 REPRESENTATION THEOREMS

THROUGHOUT THIS SECTION,

$\mathcal{S}$ AND $\mathcal{T}$ ARE MIN-STABLE FUNCTION CONES,

AND τ IS AN INCREASING LINEAR FUNCTIONAL ON $\mathcal{T}$.

7.1 This section deals with the question whether there exists a linear functional μ on $\mathcal{S}$ such that

$$\tau(t) = \mu_{\bullet}(t) = \mu^{\bullet}(t)$$

holds for all $t \in \mathcal{T}$. If this is the case, τ is said to be *represented* by μ.

This implies that all functions in $\{\tau < \infty\}$ are essentially integrable w.r.t. μ, that $\mu^{\bullet}$ is the unique increasing linear extension of τ to $\mathcal{T}_{max}$, and that this extension is represented by μ.

In fact, uniqueness follows from the second and representability from the first part of Lemma 1.9. □

The following concept will prove to be decisive when necessary conditions are being sought :

DEFINITION. An upper functional ν is said to be *$\mathcal{S}$-tight* for a function f if the formula

$$\nu_*(f) = \inf_{s \in \mathcal{S}_-} \nu_*(\max(f,s))$$

holds.

The linear functional τ is said to be *$\mathcal{S}$-tight* if $\tau^{\times}$ (or equivalently $\tau^{\bullet}$) is $\mathcal{S}$-tight on $\mathcal{T}_-$, i.e. if

$$\tau(t) = \inf_{s \in \mathcal{S}_-} \tau_{\times}(\max(t,s))$$

holds for all $t \in \mathcal{T}_-$.

Note that tightness, in contrast to almost coinitiality, uses the lower functional, and is therefore weaker. In case of measurability of $\mathcal{S}_-$ however, both concepts coincide :

Lemma. *If $\mathcal{S}_-$ is almost coinitial to $\mathcal{T}_-$ w.r.t. $\tau^\times$ or $\tau^\bullet$, then τ is $\mathcal{S}$-tight.*

Conversely, if $\mathcal{S}_- \subset \mathcal{R}^0(\tau)$ and τ is $\mathcal{S}$-tight, then $\mathcal{S}_-$ is almost coinitial in $\bar{\mathbb{R}}^X$ w.r.t. $\tau^\bullet$.

Note that in view of Proposition 6.3 we have

$$\tau_\times(\max(t,s)) = \tau_\bullet(\max(t,s)) \le \tau^\bullet(\max(t,s)) = \tau^\times(\max(t,s))$$

for all $t \in \mathcal{T}_-$ and $s \in \mathcal{S}_-$, with equality holding if $\mathcal{S}_-$ consists of measurable functions. This proves the first part, and conversely, that $\mathcal{S}_-$ is almost coinitial to $\mathcal{T}_-$ w.r.t. $\tau^\bullet$. Since $\mathcal{T}_-$ is almost coinitial in $\bar{\mathbb{R}}^X$ w.r.t. $\tau^\bullet$, the remaining assertion follows from the transitivity property in Proposition 3.5.(i). □

Remarks.

(1) *If $f \in \mathcal{I}_-(\nu)$, then ν is $\mathcal{S}$-tight for f, respectively $\mathcal{S}$ is almost coinitial to $\{f\}$, iff*

$$\sup_{s\in\mathcal{S}_-} \nu([f-s]_-) = 0\,, \quad \textit{resp.} \quad \sup_{s\in\mathcal{S}_-} \nu_*([f-s]_-) = 0\,.$$

In particular, τ is $\mathcal{S}$-tight iff

$$\sup_{s\in\mathcal{S}_-} \tau^\times([t-s]_-) = 0$$

holds for every $t \in \mathcal{T}_-$, or for every t in a coinitial subset of $\mathcal{T}$.

This follows from

$$\nu([f-s]_-) = \nu(f - \max(f,s)) = \nu(f) - \nu_*(\max(f,s))\,,$$

which is consequence of Proposition 2.7, and the corresponding formula for the lower functional. □

(2) *If τ is semiregular, then τ is $\mathcal{S}$-tight iff*

$$\sup_{s\in\mathcal{S}_-} \tau^*([t-s]_-) = 0$$

holds for every $t \in \mathcal{T}_-$.

A regular linear functional τ is $\mathcal{S}$-tight iff τ^* is $\mathcal{S}$-tight on $\mathcal{T}_-$, i.e.

$$\tau(t) = \inf_{s \in \mathcal{S}_-} \tau_*(\max(t,s))$$

holds for all $t \in \mathcal{T}_-$.

This follows immediately from (1) and Proposition 6.7, and from Theorem 6.4 for the second part. □

(3) *Let* κ *be an increasing linear functional on a min-stable function cone* $\mathcal{K}$ *such that* $\mathcal{T}_- \subset \mathcal{K}$ *and* κ *coincides with* τ *on* $\mathcal{T}_-$. *If* κ *is* $\mathcal{S}$*-tight, then* τ *is* $\mathcal{S}$*-tight.*

Conversely, if $\mathcal{S}_-$ *is almost coinitial to* $\mathcal{T}_-$ *w.r.t.* $\tau^\times$ *and* $\mathcal{T}_-$ *is almost coinitial to* $\mathcal{K}_-$ *w.r.t.* $\kappa^\times$, *then* $\mathcal{S}_-$ *is almost coinitial to* $\mathcal{K}_-$ *w.r.t.* $\kappa^\times$, *in particular* κ *is* $\mathcal{S}$*-tight.*

The first part follows from Remark 1 since $\kappa^\times \le \tau^\times$. Conversely, by the same argument, $\mathcal{S}_-$ is almost coinitial to $\mathcal{T}_-$ w.r.t. $\kappa^\times$, so the result is a consequence of the transitivity property in Proposition 3.5.(i). □

EXAMPLES.

(1) $\mathcal{T}$ is said to be $\mathcal{S}$*-adapted* if for all $t \in \mathcal{T}_-$ there exists $u \in \mathcal{T}_-$ such that the following condition holds :

For every $\varepsilon > 0$ there is $s \in \mathcal{S}_-$ with $t \ge \varepsilon u + s$.

If $\mathcal{T}$ *is* $\mathcal{S}$*-adapted, then every increasing linear functional on* $\mathcal{T}$ *is* $\mathcal{S}$*-tight, and* $\mathcal{S}_-$ *is even almost coinitial in* $\bar{\mathbb{R}}^X$ *w.r.t.* $\tau^\bullet$.

Indeed, we have

$$\tau_\times([t - s]_-) \ge \varepsilon \cdot \tau(u) . \quad \square$$

If X *is a Hausdorff space, then* $\mathcal{C}^0(X)$ *is* $\mathcal{K}(X)$*–, and therefore* $\mathcal{S}(X)$*–adapted.*

We only have to choose $u := -\sqrt{-t} \in \mathcal{C}^0(X)$ and $s := (t - \varepsilon u)_- \in \mathcal{K}(X)$, since $\{t - \varepsilon u < 0\} = \{\sqrt{-t} > \varepsilon\}$ is contained in a compact set. □

(2) *Let X be a Hausdorff space and $\mathcal{T}$ a lattice cone of lower semicontinuous functions on X.*

The $\mathcal{S}(X)$-tightness of τ is equivalent to

$$\sup_{K\in\mathfrak{K}(X)} \tau^{\times}(1_{\complement K}\cdot t) = 0$$

for all $t \in \mathcal{T}_-$. If τ is semiregular, then $\tau^{\times}$ may be replaced by τ^*, i.e. for $t \in \mathcal{T}_-$ and $\varepsilon > 0$ there exists $K \in \mathfrak{K}(X)$ such that $-\tau(u) \leq \varepsilon$ for all $u \in \mathcal{T}_-$ with $u \geq t$ and $u = 0$ on K.

This follows from

$$\tau(t) \leq \inf_{s\in\mathcal{S}_-(X)} \tau_{\times}(\max(t,s)) = \inf_{K\in\mathfrak{K}(X)} \tau_{\times}(1_K\cdot t) =$$

$$= \tau(t) - \sup_{K\in\mathfrak{K}(X)} \tau^{\times}(1_{\complement K}\cdot t),$$

where $t \leq 1_K\cdot t \in \mathcal{S}_-(X)$ and $\max(t,s) \geq 1_{\mathrm{supp}(s)}\cdot t$ prove the first equality, and $1_K = 1 - 1_{\complement K}$ proves the second one by Proposition 2.7. If τ is semiregular, one has to use Proposition 6.7. □

7.2 PROPOSITION. *If the restriction of τ to $\mathcal{T}_-$ is represented by an increasing linear functional μ on $\mathcal{S}$, then*

$$\tau_{\times} \leq \mu_{\bullet},$$

and τ is $\mathcal{S}$-tight. If τ is represented by μ, then

$$\tau_{*} \leq \mu_{\bullet}.$$

In the first part, for every function f and $u,v \in \mathcal{T}_-$ with $u - v \leq f$, we have $u,v \in \mathcal{R}^{\bullet}_-(\mu)$, hence

$$\tau(u) - \tau(v) = \mu_{\bullet}(u) - \mu_{\bullet}(v) = \mu_{\bullet}(u - v) \leq \mu_{\bullet}(f),$$

and thus $\tau_{\times}(f) \leq \mu_{\bullet}(f)$.

From the inequalities

$$\tau_{\times} \leq \mu_{\bullet} \leq \mu^{\bullet} \leq \mu^{\times},$$

in view of Theorem 3.6.(ii) we get

$$\tau(t) = \tau_{\times}(t) \leq \inf_{s\in\mathcal{S}_-} \tau_{\times}(\max(t,s)) \leq \inf_{s\in\mathcal{S}_-} \mu^{\times}(\max(t,s)) = \mu^{\bullet}(t) = \tau(t)$$

for $t \in \mathcal{T}_-$, which proves that τ is $\mathcal{S}$-tight.

In the second part, for every function f and $t \in \mathcal{T}$ with $-t \leq f$, we have
$$-\tau(t) = -\mu^\bullet(t) = \mu_\bullet(-t) \leq \mu_\bullet(f),$$
and thus $\tau_*(f) \leq \mu_\bullet(f)$. □

If τ is represented by μ, then all functions in $\{\tau < \infty\}$ are essentially integrable w.r.t. μ. Having applications in mind, one would like to be sure that in fact *all* functions from $\mathcal{T}$ are measurable. The only method we know of proving this is based on the argument in Proposition 4.11 and for that reason requires the hypothesis there, which by Lemma 7.1 is equivalent to μ being $\mathcal{T}$-tight. An additional advantage of this assumption is that it permits the integration theories of the two functionals to be compared.

THEOREM. *If the restriction of* τ *to* $\mathcal{T}_-$ *is represented by a* $\mathcal{T}$*-tight increasing linear functional* μ, *then*
$$\tau_\bullet \leq \mu_\bullet \quad \textit{and} \quad \mathcal{T} \subset \mathcal{R}^0(\tau) \subset \mathcal{R}^0(\mu).$$
Moreover, τ *is represented by* μ *iff* τ *is difference-regular.*

By the proposition, $\tau_\times \leq \mu_\bullet$ and hence $\mathcal{R}_-^\times(\tau) \subset \mathcal{R}_-^\bullet(\mu)$. Since $\mathcal{T}_-$ is coinitial in $\mathcal{R}_-^\times(\tau)$ and, by Lemma 7.1, almost coinitial in $\mathcal{R}_-^\bullet(\mu)$, we infer from Proposition and Corollary 4.3 that $\mathcal{R}^0(\tau) \subset \mathcal{R}^0(\mu)$. Furthermore, considering Theorem and Corollary 3.6, for any function f we get
$$\tau_\bullet(f) = \sup_{v\in\mathcal{T}_-} \inf_{u\in\mathcal{T}_-} \tau_\times(\mathrm{med}(f,-v,u)) \leq \sup_{v\in\mathcal{T}_-} \inf_{u\in\mathcal{T}_-} \mu_\bullet(\mathrm{med}(f,-v,u)) =$$
$$= \mu_{\bullet\bullet}(f) = \mu_\bullet(f),$$
hence $\tau_\bullet \leq \mu_\bullet$.

It remains to prove that τ is represented by μ iff τ is difference-regular. But this is implicit in the inequalities
$$\tau_\times \leq \mu_\bullet \leq \mu^\bullet \leq \tau^\bullet = \tau_\times$$
which by Proposition 6.3 hold on $\mathcal{T}$. □

REMARK. *The functionals* $\tau_\times$ *and* τ_*, *respectively* $\tau_\bullet$, *are lower bounds for every increasing linear functional* μ *on* $\mathcal{S}$ *which represents* τ, *respectively repre-*

sents τ *and is* $\mathcal{T}$*-tight.*

If $\tau_\times$ *is finite on* $\mathcal{S}_-$ *, respectively* τ *is semiregular and* τ_* *is finite on* $\mathcal{S}_-$ *, then* $\tau_\times$ *, respectively* τ_* *, coincides with* $\tau_\bullet$ *on* $\mathcal{S}$ *.*

The first part follows from the proposition and the theorem, the second one from Propositions 6.3 and 6.8. □

Note that whenever $\tau_\times$ is finite on $\mathcal{S}_-$, i.e. whenever any function in $\mathcal{S}$ admits a minorant in $\mathcal{T}$, then every μ is $\mathcal{T}$-tight. Representability by $\tau_\bullet$ will in fact be proved under suitable hypotheses in Corollary 7.4.

7.3 **PROPOSITION.** *Let* μ *be an increasing linear functional on* $\mathcal{S}$ *. Then* τ *and* μ *represent each other iff each of them is difference-regular and* $\tau^\bullet = \mu^\bullet$ *.*

If τ and μ represent each other, then by Proposition 7.2 the functional τ is $\mathcal{S}$-tight and the functional μ is $\mathcal{T}$-tight. The necessity therefore follows from Theorem 7.2. The converse is trivial. □

From this proposition we obtain the following sharper version of the Extension Theorem 6.5 :

COROLLARY. *If* $\mathcal{T}_- \subset \mathcal{S}_- \subset \mathcal{R}^\bullet_-(\tau)$ *, then the functional* $\mu := \tau_\bullet = \tau^\bullet$ *is the smallest and the only difference-regular one among all* $\mathcal{T}$*-tight increasing linear functionals on* $\mathcal{S}$ *which coincide with* τ *on* $\mathcal{T}_-$ *, and we have* $\tau^\bullet = \mu^\bullet$ *.*

In particular, the restriction of μ *to* $\mathcal{S}_-$ *is the only* $\mathcal{T}$*-tight increasing linear functional on* $\mathcal{S}_-$ *which coincides with* τ *on* $\mathcal{T}_-$ *.*

If τ *is semiregular, so is* μ *. If* $\mathcal{T} \subset \mathcal{S}$ *, then* μ *extends* τ *iff* τ *is difference-regular.*

As in the proof of Proposition 6.5 we have $\mathcal{S} \subset \mathcal{R}^0(\tau) \cap \{\tau_\bullet > -\infty\}$, hence $\tau_\bullet = \tau^\bullet$ on $\mathcal{S}$ by Proposition 6.3. Since by definition μ is represented by τ , we infer from Proposition 7.2 that μ is $\mathcal{T}$-tight.

Let κ be a $\mathcal{T}$-tight increasing linear functional on $\mathcal{S}$ which coincides with τ on $\mathcal{T}_-$. By Theorem 7.2, we have $\tau_\bullet \le \kappa_\bullet$. This implies, on the one hand that $\mu \le \kappa$, and on the other hand that $\kappa = \tau^\bullet$ on $\mathcal{S}_-$, since functions in $\mathcal{S}_-$ are essentially integrable w.r.t. τ. Therefore, the restrictions of κ to $\mathcal{S}_-$ and of τ to $\mathcal{T}_-$ represent each other, hence $\kappa^\bullet = \tau^\bullet$ by the proposition. This also shows that κ is difference-regular iff κ coincides with $\tau^\bullet$ on $\mathcal{S}$, i.e iff $\kappa = \mu$.

If τ is semiregular, so is μ by Remark 6.7. Indeed, for $s,t \in \mathcal{S}_-$ with $s \le t$ we have

$$\mu(t) = \tau^\bullet(s) + \tau_\bullet(t-s) = \mu(s) + \tau_*(t-s) \le \mu(s) + \mu_*(t-s)$$

by Proposition 6.8, since $\mathcal{T}_- \subset \mathcal{S}_-$ implies $\tau_* \le \mu_*$ on $\bar{\mathbb{R}}_+^X$. Finally, μ extends τ iff $\tau^\bullet = \tau$ on $\mathcal{T}$, i.e. iff τ is difference-regular. □

REMARK. Let $\mathcal{T} \subset \mathcal{S}$ and τ be semiregular. By Proposition 6.9, the semiregular extension μ of τ is regular whenever it is $\mathcal{S}$–bounded below, e.g. if $\tau_* > -\infty$ on $\mathcal{S}_-$. In view of this result, the above corollary also generalizes the extension theorem in 6.2.

Together with Example 6.5, the corollary yields the following

EXAMPLE. *The unique difference-regular linear extension to $\mathcal{G}(\mathcal{S})$ of a semiregular linear functional on $\mathcal{S}$ is semiregular.*

7.4 **LEMMA.** *Let $\mathcal{S}$ be a lattice cone of negative real-valued functions. Every increasing superlinear functional μ on $\mathcal{S}$ has a unique extension to an increasing functional on $\hat{\mathcal{S}}$. This extension is superlinear and coincides with μ^* on $\hat{\mathcal{S}}$.*

Obviously, the functional $\hat{\mu}$ defined by $\hat{\mu}(f) := \sup_{s\in\mathcal{S},\, s\le f} \mu(s)$ is an increasing superlinear extension of μ to $\mathcal{S}^\uparrow$. If we denote by $\mathcal{S}_\mu$ the set $\{\hat{\mu} = \mu^*\}_-$, then we have $\mathcal{A}(\mathcal{S}_\mu) = \mathcal{S}_\mu$. Indeed, for $f \in \mathcal{A}(\mathcal{S}_\mu)$, a corresponding sequence (f_n) in $\mathcal{S}_\mu$, and a null sequence (ε_n) in $\mathbb{R}_+$ with

$$f_n + \varepsilon_n f_0 \le f \le f_n$$

for $n \ge 1$, we obtain

$$\hat{\mu}(f_n) + \varepsilon_n \hat{\mu}(f_0) \le \hat{\mu}(f) \le \mu^*(f) \le \mu^*(f_n) = \hat{\mu}(f_n),$$

hence $\hat{\mu}(f) = \lim \hat{\mu}(f_n) = \lim \mu^*(f_n) = \mu^*(f)$. Since $\mathcal{S}$ is max–stable and contained in $\mathcal{S}_\mu$, this implies that $\hat{\mathcal{S}} \subset \mathcal{S}_\mu$.

Let now κ be any increasing extension of μ to $\hat{\mathcal{S}}$. Then

$$\hat{\mu}(f) = \sup_{s\in\mathcal{S},\, s\le f} \kappa(s) \le \kappa(f) \le \inf_{s\in\mathcal{S},\, s\ge f} \kappa(s) = \mu^*(f),$$

hence $\hat{\mu}(f) = \kappa(f)$ for every $f \in \hat{\mathcal{S}}$. □

PROPOSITION. *Suppose that all functions in $\mathcal{T}_-$ are $\mathcal{S}$–measurable. Let ν be an upper functional such that the restriction of ν_* to $\mathcal{S}_-$ is linear, and denote by μ its difference–regular linear extension to $\mathcal{S}$.*

If ν is $\mathcal{S}$–tight on $\mathcal{T}_-$ and $\nu_* > -\infty$ on $\mathcal{T}_-$, then the restriction of ν_* to $\mathcal{T}_-$ is linear and represented by μ.

We have $\nu_* \le \mu^\times$ on $\mathbb{R}^X_-$, since for $f \in \mathbb{R}^X_-$ with $\nu_*(f) > -\infty$ and $u,v \in \mathcal{S}_-$ with $f \le u - v$, the superlinearity of ν_* yields

$$\nu_*(f) + \mu(v) = \nu_*(f) + \nu_*(v) \le \nu_*(f + v) \le \nu_*(u) = \mu(u),$$

hence

$$\nu_*(f) \le \mu(u) - \mu(v).$$

Since $\mu^\times$ and ν_* are increasing, and strongly sublinear respectively strongly superlinear on $\mathcal{S}^\wedge_-$, and since each of these functionals coincides with μ on $\mathcal{S}_-$, we infer from Lemmas 1.9.(i) and 7.4 that $\mu^\times$ and ν_* also coincide on $\mathcal{S}^\wedge_-$. By hypothesis, for $t \in \mathcal{T}_-$ and $s \in \mathcal{S}_-$, we have

$$\max(t,s) \in \mathcal{G}_-(\mathcal{S}) = \mathcal{S}^\wedge_- \subset \mathcal{R}^\wedge_-(\mu) = \mathcal{R}^\times_-(\mu)$$

(cf. Example 4.8), which yields

$$\mu_\times(\max(t,s)) = \mu^\times(\max(t,s)) = \nu_*(\max(t,s)).$$

If ν is $\mathcal{S}$–tight on $\mathcal{T}_-$, then for $t \in \mathcal{T}_-$ we have

$$\nu_*(t) = \inf_{s\in\mathcal{S}_-} \nu_*(\max(t,s)) = \inf_{s\in\mathcal{S}_-} \mu_\times(\max(t,s)) = \mu_\bullet(t) \le \mu^\bullet(t) =$$
$$= \inf_{s\in\mathcal{S}_-} \mu^\times(\max(t,s)) = \inf_{s\in\mathcal{S}_-} \nu_*(\max(t,s)) = \nu_*(t)$$

by Theorem 3.6.(i). We have thus proved that the restriction of ν_* to $\mathcal{T}_-$ is linear and represented by μ, if it is finite. □

REMARK. In the proposition we have shown that always

$$\mu_\times(\max(s,t)) = \mu^\times(\max(s,t)) = \nu_*(\max(s,t))$$

holds for all $s \in \mathcal{S}_-$ and $t \in \mathcal{T}_-$.

COROLLARY. *Suppose that all functions in $\mathcal{T}_-$ are $\mathcal{S}$-measurable. If the restriction of $\tau_\bullet$ to $\mathcal{S}_-$ is linear, then its difference-regular linear extension μ to $\mathcal{S}$ is $\mathcal{T}$-tight. The functional τ is represented by μ iff τ is difference-regular and $\mathcal{S}$-tight.*

In fact, μ is $\mathcal{T}$-tight since

$$\mu(s) = \tau_\bullet(s) = \inf_{t\in\mathcal{T}_-} \tau_\times(\max(s,t)) = \inf_{t\in\mathcal{T}_-} \tau_\bullet(\max(s,t)) = \inf_{t\in\mathcal{T}_-} \mu_\times(\max(s,t))$$

for $s \in \mathcal{S}_-$, by the remark applied to $\tau^\bullet$.

The condition is sufficient by the proposition and by Theorem 7.2. The converse follows from Proposition and Theorem 7.2. □

7.5 Uniqueness of the representation is derived from semiregularity in the following *Uniqueness Theorem*:

THEOREM. *Let μ be a difference-regular linear functional on $\mathcal{S}$ which is $\mathcal{T}$-tight, and let κ be a semiregular linear functional on $\mathcal{S}$ with $\kappa \geq \mu$ on $\mathcal{S}_-$. If τ is represented by μ as well as by κ, then κ coincides with μ.*

Indeed, for $s \in \mathcal{S}_-$ and $t \in \mathcal{T}_-$, Proposition 2.7 yields

$$\kappa(s) \leq \kappa_\bullet(\max(s,t)) = \kappa_\bullet(t) + \kappa_\bullet([s-t]^+) \leq \mu_\bullet(t) + \mu_\bullet([s-t]^+) \leq \mu_\bullet(\max(s,t)),$$

since by the Propositions 6.3 and 6.8

$$\mu_\bullet([s-t]^+) \geq \mu_*([s-t]^+) \geq \kappa_*([s-t]^+) = \kappa_\bullet([s-t]^+).$$

Because μ is $\mathcal{T}$-tight, this implies that $\kappa(s) \leq \mu(s)$, hence $\kappa = \mu$ on $\mathcal{S}_-$. Finally, Corollary 6.5 proves that $\kappa = \mu$ on $\mathcal{S}$. □

COROLLARY. *Suppose that $\mathcal{S}_- \subset \mathcal{R}^\bullet_-(\tau)$, and that all functions in $\mathcal{T}_-$ are $\mathcal{S}$-measurable.*

Then the difference-regular linear extension μ to $\mathcal{S}$ of the restriction of $\tau_\bullet$ to $\mathcal{S}_-$ is a $\mathcal{T}$-tight linear functional. It is semiregular if τ is semiregular and $\mathcal{S}$ is a lattice.

The functional μ represents τ iff τ is difference-regular and $\mathcal{S}$-tight. In this case, we have

$$\tau^\bullet = \mu^\bullet \ , \quad \text{and} \quad \mu = \tau_\bullet = \tau^\bullet \ \text{ on } \mathcal{S} .$$

Moreover, if $\mathcal{S}$ is a lattice, τ is semiregular and $\mathcal{S}$-tight, then μ is the only $\mathcal{T}$-tight semiregular linear functional on $\mathcal{S}$ representing τ. If furthermore $\tau_\times$ or τ_* is finite on $\mathcal{S}_-$, then μ coincides with $\tau_\times$, respectively τ_* , on $\mathcal{S}$ and is the only semiregular linear functional on $\mathcal{S}$ representing τ .

The first part is an immediate consequence of Corollary 7.4. For the semiregularity assertion, by Remark 6.7 it is sufficient to prove that μ is semiregular on $\mathcal{S}_-$. Since $\tau^\bullet$ is an increasing extension of μ to $\mathcal{S}^\wedge_-$, Lemma 7.4 shows that $\tau^\bullet = \mu^*$ on $\mathcal{S}^\wedge_-$. Let now $u,v \in \mathcal{S}_-$ with $u \le v$. Then

$$\mu(v) = \tau_\bullet(v) = \tau_\bullet(u) + \tau_\bullet(v-u) = \mu(u) + \tau_*(v-u) \le \mu(u) + \mu_*(v-u) ,$$

since for $t \in \mathcal{T}_-$ with $-t \le v-u$ we have $t \ge u$ and thus $t \in \mathcal{G}_-(\mathcal{S}) = \mathcal{S}^\wedge_-$, hence

$$-\tau(t) = -\tau^\bullet(t) = -\mu^*(t) \le \mu_*(v-u) .$$

The representation assertion follows again from Corollary 7.4. Theorem 7.2 then shows that $\tau^\bullet = \mu^\bullet$, since τ represents the restriction of μ to $\mathcal{S}_-$, and Proposition 6.3 that $\mu = \mu_\times = \mu_\bullet = \mu^\bullet$, hence $\mu = \tau_\bullet = \tau^\bullet$ on $\mathcal{S}$. The last part is a consequence of Remark 7.2 and the Uniqueness Theorem. □

7.6 We now discuss representability by semiregular functionals, weakening the integrability condition.

THEOREM. *Let $\mathcal{S}$ be a lattice cone of functions and suppose that all functions in $\mathcal{T}$ are $\mathcal{S}$-measurable. If τ is regular and if the restriction of $\tau_\bullet$ to $\mathcal{S}_-$ is linear,* *then its difference-regular linear extension* μ *to* $\mathcal{S}$ *is semiregular and* $\mathcal{T}$*-tight.*

Furthermore, τ *is represented by* μ *iff* τ *is* $\mathcal{S}$*-tight. In this case,* μ *is the only* $\mathcal{T}$*-tight semiregular linear functional which represents* τ *.*

For $u,v \in \mathcal{S}_-$ with $u \leq v$, $t \in \mathcal{T}_-$ and $r \in \mathcal{T}$ with $-r \leq \max(v,t)$, we have

$$r + v + \max(u,t) \geq r + \max(u + v, t + u) \geq u ,$$

hence

$$[r + \max(u,t)]_- \geq u - v .$$

Since $r \geq 0$ and t are $\mathcal{S}$-measurable, we have $r \in \mathcal{G}(\mathcal{S})$ and $\max(u,t) \in \hat{\mathcal{S}}_-$, hence $[r + \max(u,t)]_- \in \hat{\mathcal{S}}_-$. As $\tau_\bullet$ is superlinear on $\hat{\mathcal{S}}_-$, Lemma 7.4 yields the coincidence of $\tau_\bullet$ with μ^* and $\mu_\times$ on $\hat{\mathcal{S}}_-$. From Theorem 2.2.(v) we get

$$\tau_\bullet([r + \max(u,t)]_-) \leq \tau_\bullet(r + \max(u,t)) \leq \tau^\bullet(r) + \tau_\bullet(\max(u,t)) ,$$

hence

$$- \tau(r) \leq \tau_\bullet(\max(u,t)) - \tau_\bullet([r + \max(u,t)]_-) =$$

$$= \mu_\times(\max(u,t)) - \mu^*([r + \max(u,t)]_-) \leq \mu_\times(\max(u,t)) + \mu_*(v - u) ,$$

and therefore

$$\tau_*(\max(v,t)) \leq \mu_\times(\max(u,t)) + \mu_*(v - u) .$$

Since μ is $\mathcal{T}$-tight by Corollary 7.4, we obtain

$$\mu(v) = \tau_\bullet(v) = \inf_{t \in \mathcal{T}_-} \tau_*(\max(v,t)) \leq$$

$$\leq \inf_{t \in \mathcal{T}_-} \mu_\times(\max(u,t)) + \mu_*(v - u) = \mu(u) + \mu_*(v - u) .$$

This proves that μ is semiregular by Remark 6.7. The result now follows from Corollary 7.4, Remark 7.2 and Theorem 7.5. □

Remark. Actually, one only has to assume that τ is semiregular, $\tau_\bullet$ is linear on $\mathcal{S}_-$ and $\tau_*(\max(s,t)) > -\infty$ for all $s \in \mathcal{S}_-$, $t \in \mathcal{T}_-$.

This follows from the fact that

$$-\infty < \tau_*(\max(v,t)) = \tau_\bullet(\max(v,t)) = \tau_\times(\max(v,t)) . \quad \square$$

In general, it will be easier to prove the linearity of the functional τ_* than that of the functional $\tau_\bullet$ on $\mathcal{S}_-$, using separation properties (cf. Proposition 14.10). We then must require the finiteness of τ_* on $\mathcal{S}_-$.

ADDENDUM. *If τ is semiregular and if the restriction of τ_* to $\mathcal{S}_-$ is linear, then its difference-regular linear extension μ to $\mathcal{S}$ is semiregular and $\mathcal{T}$-tight.*

Furthermore, τ is represented by μ iff τ is $\mathcal{S}$-tight. In this case, μ is the only semiregular linear functional on $\mathcal{S}$ which represents τ.

One has only to note that τ_* coincides with $\tau_\bullet$ on $\mathcal{S}_-$. The semiregularity of μ is a consequence of the following lemma. □

LEMMA. *Let $\mathcal{S}$ be a lattice cone such that all functions in $\mathcal{T}_+$ are $\mathcal{S}$-measurable. If the restriction of τ_* to $\mathcal{S}_-$ is finite, then it is a semiregular superlinear functional.*

The proof follows the same lines as in the theorem, but is simpler since one has neither to make use of semiregularity nor tightness, replacing the functions $\max(u,t)$ and $\max(v,t)$ by u and v and $\tau_\bullet$ by τ_*. □

7.7 We finish this section with our most general representation theorem, based on the Hahn–Banach–Andenaes theorem for cones of the appendix (cf. § 17 for notations). It will be used to prove the existence of inverse images of Radon integrals (cf. the following Example 7.8 and Theorem 8.18).

Let $\mathcal{K}$ be a lattice cone of negative functions containing $\mathcal{S}_-$ and $\mathcal{T}_-$. We say that an increasing functional ξ on $\mathcal{K}$ has *property* (T) if

$$\xi(k) = \inf_{t\in\mathcal{T}_-} \xi(\max(k,t))$$

holds for all $k \in \mathcal{K}$.

LEMMA.

(i) *If (ξ_i) is a family of increasing functionals on $\mathcal{K}$ having property* (T), *then also $\inf \xi_i$ enjoys this property.*

(ii) *Suppose that $\mathcal{H} \subset \mathcal{T}_-^{\uparrow}$ and $\mathcal{Z}$ are subcones of $\mathcal{K}$. If $\theta : \mathcal{K} \longrightarrow \mathbb{R}_-$ is an increasing sublinear functional having property* (T) *and $\eta : \mathcal{H} \longrightarrow \mathbb{R}_-$ is sublinear, $\zeta : \mathcal{Z} \longrightarrow \mathbb{R}_-$ superlinear, then $(\theta\wedge\eta)\text{⌐}\zeta$ has property* (T).

The first assertion is trivial. For any $l \in \mathcal{K}$ and k , h , z runing respectively through $\mathcal{K}$, $\mathcal{H}$ and $\mathcal{Z}$, we have

$$(\theta \wedge \eta) \frown \zeta(l) = \inf \{\theta(k) + \eta(h) - \zeta(z) : l + z \leq k + h\} =$$

$$= \inf_{t \in \mathcal{T}_-} \inf \{\theta(\max(k,t)) + \eta(h) - \zeta(z) : l + z \leq k + h\} \geq$$

$$\geq \inf_{v \in \mathcal{T}_-} \inf \{\theta(k) + \eta(h) - \zeta(z) : \max(l,v) + z \leq k + h\} =$$

$$= \inf_{v \in \mathcal{T}_-} (\theta \wedge \eta) \frown \zeta(\max(l,v)) \geq (\theta \wedge \eta) \frown \zeta(l) \,,$$

since $h \geq u$ for some $u \in \mathcal{T}_-$ and

$$\max(l, t + u) + z = \max(l + z, t + u + z) \leq \max(k + h, t + h) = \max(k,t) + h \,. \qquad \square$$

THEOREM. *Suppose that all functions in $\mathcal{T}_-$ are $\mathcal{S}$-measurable. If τ is difference-regular, $\mathcal{S}$-tight and $\tau_\bullet > -\infty$ on $\mathcal{S}_-$, then there exists a $\mathcal{T}$-tight difference-regular linear functional μ on $\mathcal{S}$ representing τ .*

Let $\mathcal{K}$ denote the cone $\{\tau_\bullet > -\infty\}_-$ of negative functions, which contains $\mathcal{S}_-$ and $\mathcal{T}_-$. By Theorem 2.2.(iii), $\tau_\bullet$ is superlinear, whereas $\tau^\bullet$ is sublinear on $\mathcal{K}$ with

$$-\infty < \tau_\bullet \leq \tau^\bullet \quad \text{on} \quad \mathcal{K} \,.$$

The preceding lemma shows that

$$\Xi(\tau^\bullet{}_{|\mathcal{K}} \,, \tau_{\bullet|\mathcal{K}} \,, \mathrm{T})$$

enjoys the properties (ii) and (iii) stated in Addendum 17.3. Since

$$\tau^\bullet(k) = \inf_{t \in \mathcal{T}_-} \tau^\times(\max(k,t)) = \inf_{t \in \mathcal{T}_-} \tau^\bullet(\max(k,t))$$

for all $k \in \mathcal{K}$, property (i) is also satisfied. Therefore, there exists an $\mathcal{S}_-$-minimal increasing linear functional κ on $\mathcal{K}$ with property (T) and

$$\tau_\bullet \leq \kappa \leq \tau^\bullet \quad \text{on} \quad \mathcal{K} \,.$$

Let μ denote the difference-regular linear extension of $\kappa_{|\mathcal{S}_-}$ to $\mathcal{S}$. We first prove that μ is $\mathcal{T}$-tight and that $\mathcal{T}_- \subset \mathcal{R}^\bullet_-(\mu)$. Indeed, for all $s \in \mathcal{S}_-$ and $t \in \mathcal{T}_-$ we have

$$\mu_\times(\max(t,s)) = \kappa(\max(t,s))$$

by Lemmas 1.9.(i) and 7.4, since $\mu_\times \leq \kappa$ on $\mathcal{K}$ and $\max(t,s) \in \mathcal{G}_-(\mathcal{S}) = \hat{\mathcal{S}}_-$.
Therefore

$$\mu(s) = \inf_{t\in\mathcal{T}_-} \kappa(\max(s,t)) = \inf_{t\in\mathcal{T}_-} \mu_\times(\max(s,t)) .$$

Furthermore, we have

$$\mu_\bullet(t) = \inf_{s\in\mathcal{S}_-} \mu_\times(\max(t,s)) = \inf_{s\in\mathcal{S}_-} \kappa(\max(t,s)) \geq$$
$$\geq \inf_{s\in\mathcal{S}_-} \tau_\bullet(\max(t,s)) \geq \tau_\bullet(t) = \tau(t) > -\infty ,$$

which by Proposition 6.3 proves that t is essentially integrable w.r.t. μ.

Next we prove that the functional

$$\delta := \tau - \mu^\bullet$$

on $\mathcal{T}_-$ is increasing, linear and satisfies

$$\delta^\times \geq \tau^\times \ , \ \delta_\times \geq \tau_\times \ \text{ and } \ \delta_\bullet \geq \tau_\bullet \ \text{ on } \mathcal{K} .$$

Note first that $\tau_\bullet \leq \mu^\bullet$ on $\mathcal{K}$. In fact,

$$\mu^\bullet(k) = \mu^\times(k) = \inf_{u,v\in\mathcal{S}_-,\ 0\geq u-v\geq k} \kappa(u-v) \geq \kappa(k) \geq \tau_\bullet(k)$$

for $k \in \mathcal{K}$. For $u,t \in \mathcal{T}_-$ with $u \leq t$, this yields

$$\delta(u) - \delta(t) = \tau(u) - \mu^\bullet(u) -(\tau(t) - \mu^\bullet(t)) = \tau_\bullet(u-t) - \mu^\bullet(u-t) \leq 0 .$$

The linearity is a consequence of $\mathcal{T}_- \subset \mathcal{R}^\bullet_-(\mu)$. Finally, we also have

$$\delta(u) - \delta(t) \geq \tau(u) - \tau(t) ,$$

which gives the result.

We now claim that

$$\delta_\times(\max(s,t)) = \delta_\bullet(\max(s,t)) = 0$$

for all $s \in \mathcal{S}_-$ and $t \in \mathcal{T}_-$.

To this end, we first mix Corollary and Addendum 17.3 to get a linear functional ξ on $\mathcal{K}$ with property (T), such that

$$\delta_\bullet \leq \xi \leq \delta^\bullet \ \text{ on } \mathcal{K} \ \text{ and } \ \xi(\max(s,t)) = \delta_\bullet(\max(s,t)) .$$

In fact, note first that $\delta_\bullet \geq \tau_\bullet > -\infty$ on $\mathcal{K}$. By definition, $\delta^\bullet{}_{|\mathcal{K}}$ has property (T), and since $\mathbb{R}_+\cdot\max(s,t) \subset \mathcal{T}_-^\uparrow$, the lemma shows that the conditions of Addendum 17.3 are satisfied for the above sandwich problem considered in Corollary 17.3. So it remains to show that

$$\delta_\bullet(\max(s,t)) \geq \delta_{\bullet|\mathcal{K}} \llcorner \delta^\bullet{}_{|\mathcal{K}}(\max(s,t)) .$$

For $h,k \in \mathcal{K}$ with $h \leq \max(s,t) + k$, we have indeed

$$\delta_\bullet(h) \leq \delta_\bullet(\max(s,t) + k) \leq \delta_\bullet(\max(s,t)) + \delta^\bullet(k) .$$

We next define a functional θ on $\mathcal{K}$ by

$$\theta(k) := \inf_{u\in\mathcal{T}_-} \mu^\bullet(\max(k,u)) .$$

It is sublinear, since $\theta \geq \mu^\bullet \geq \tau_\bullet > -\infty$ on $\mathcal{K}$. For $h,k \in \mathcal{K}$, $u,v \in \mathcal{T}_-$ we have

$$\theta(h+k) \leq \mu^\bullet(\max(h+k,\, u+v)) \leq \mu^\bullet(\max(h,u)) + \mu^\bullet(\max(k,v)) .$$

Furthermore, by the $\mathcal{T}$-tightness of μ, on $\mathcal{S}_-$ it coincides with μ and hence with κ.

To show that $\xi(\max(s,t)) = 0$, it is sufficient to prove the inequalities

$$\tau_\bullet \leq \xi + \theta \leq \tau^\bullet \quad \text{on} \quad \mathcal{K} .$$

Indeed, we then have $\xi + \theta \in \Xi(\tau^\bullet_{|\mathcal{K}}, \tau_{\bullet|\mathcal{K}}, \mathrm{T})$ by property (T) of ξ and θ, and since

$$\xi + \theta = \xi + \kappa \leq \kappa \quad \text{on} \quad \mathcal{S}_- ,$$

the $\mathcal{S}_-$-minimality of κ in $\Xi(\tau^\bullet_{|\mathcal{K}}, \tau_{\bullet|\mathcal{K}}, \mathrm{T})$ implies that $\xi + \kappa = \kappa$ on $\mathcal{S}_-$, hence on $\hat{\mathcal{S}}_-$ by Lemmas 1.9.(ii) and 7.4.

To prove the above inequalities, let $k \in \mathcal{K}$ and $u,v,w \in \mathcal{T}_-$. If

$$v - w \leq \max(k,u) ,$$

then we have

$$\tau(v) - \tau(w) = \delta(v) + \mu^\bullet(v) - [\delta(w) + \mu^\bullet(w)] = \delta(v) - \delta(w) + \mu^\bullet(v-w) \leq$$
$$\leq \delta(v) - \delta(w) + \mu^\bullet(\max(k,u)) ,$$

hence

$$\tau_\times(\max(k,u)) \leq \delta_\times(\max(k,u)) + \mu^\bullet(\max(k,u)) .$$

This yields

$$\tau_\bullet(k) = \inf_{u\in\mathcal{T}_-} \tau_\times(\max(k,u)) \leq \delta_\bullet(k) + \theta(k) \leq \xi(k) + \theta(k) .$$

On the other hand, if $v - w \geq \max(k,u)$, then we get

$$\tau(v) - \tau(w) \geq \delta^\times(\max(k,u)) + \mu^\bullet(\max(k,u)) \geq \xi(k) + \mu^\bullet(\max(k,u)) ,$$

hence

$$\tau^\bullet(k) \geq \xi(k) + \inf_{u\in\mathcal{T}_-} \mu^\bullet(\max(k,u)) = \xi(k) + \theta(k) .$$

Finally, by Theorem 7.2 it is sufficient to prove that μ represents the restriction of τ to $\mathcal{T}_-$, i.e. that $\delta = 0$. For $t \in \mathcal{T}_-$ and $s \in \mathcal{S}_-$, we have indeed

$$0 \geq \delta(t) \geq \delta^\times(t - \max(t,s)) + \delta_\times(\max(t,s)) \geq$$
$$\geq \tau^\times(t - \max(t,s)) = \tau(t) - \tau_\times(\max(t,s)) .$$

Since τ is $\mathcal{S}$-tight, the result follows. $\square$

7.8 COROLLARY. *Suppose that $\mathcal{S}$ is a lattice cone and that all functions in $\mathcal{T}$ are $\mathcal{S}$-measurable. If τ is regular, $\mathcal{S}$-tight and $\tau_{\bullet} > -\infty$ on $\mathcal{S}_-$, then there exists a $\mathcal{T}$-tight semiregular linear functional μ on $\mathcal{S}$ representing τ.*

Define μ as before. By Remark 6.7, it is semiregular if $\tilde{\mu}(d) \geq \mu^*(d)$ for $d \in (\mathcal{S}_- - \mathcal{S}_-)_-$. To this end, it is sufficient to prove the existence of a linear functional $\xi \in \Xi(\tau^{\bullet}_{|\mathcal{K}}, \tau_{\bullet|\mathcal{K}}, \mathrm{T})$ with

$$\xi \leq \kappa \text{ on } \mathcal{S}^{\wedge}_- \cap \mathcal{T}_-^{\uparrow} \quad \text{and} \quad \xi(d) \geq \mu^*(d).$$

Indeed, from

$$\xi(s) \leq \xi(\max(s,t)) \leq \kappa(\max(s,t))$$

for all $s \in \mathcal{S}_-$ and $t \in \mathcal{T}_-$, by property (T) of κ we infer that $\xi \leq \kappa$ on $\mathcal{S}_-$ and therefore $\xi = \kappa = \mu$ on $\mathcal{S}_-$ by the $\mathcal{S}_-$-minimality of κ. This yields

$$\tilde{\mu}(d) = \xi(d) \geq \mu^*(d).$$

By 17.1, Addendum 17.3 and Lemma 7.7, such a functional ξ, i.e. an element of

$$\Xi(\tau^{\bullet}_{|\mathcal{K}} \wedge \kappa_{|\mathcal{S}^{\wedge}_- \cap \mathcal{T}_-^{\uparrow}}, \tau_{\bullet|\mathcal{K}} \vee \mu^*_{|\mathbb{R}_+ \cdot d}, \mathrm{T}),$$

exists iff

$$\tau_{\bullet}(h) + \alpha \cdot \mu^*(d) \leq \tau^{\bullet}(k) + \kappa(u)$$

holds for all $h,k \in \mathcal{K}$, $u \in \mathcal{S}^{\wedge}_- \cap \mathcal{T}_-^{\uparrow}$ and $\alpha \in \mathbb{R}_+$ with $h + \alpha \cdot d \leq k + u$. Since $\tau^{\bullet} = \tau^* < \infty$ on $-\mathcal{T}_-^{\uparrow}$ by regularity, we have

$$\tau^{\bullet}(k) \geq \tau^{\bullet}(h + \alpha \cdot d - u) \geq \tau_{\bullet}(h) + \tau^*(\alpha \cdot d - u).$$

We therefore have to prove that

$$\alpha \cdot \mu^*(d) - \kappa(u) \leq \tau^*(\alpha \cdot d - u).$$

But for $t \in \mathcal{T}$ with $t \geq \alpha \cdot d - u$, we have $t \in \mathcal{S}_-^{\uparrow}$, hence $t \in \mathcal{G}(\mathcal{S})$ and so $(t + u)_- \in \mathcal{S}^{\wedge}_-$ by Proposition 4.9. From Lemma 7.4 we infer that

$$\mu^*([t+u]_-) = \kappa([t+u]_-) = \mu^{\bullet}([t+u]_-) \quad \text{and} \quad \mu^{\bullet}(u) = \kappa(u).$$

Thus

$$\alpha \cdot \mu^*(d) \leq \mu^*([t+u]_-) \leq \mu^{\bullet}(t+u) \leq \tau(t) + \kappa(u),$$

since μ represents τ. □

REMARKS.

(1) *If for* $s \in \mathcal{S}_-$ *there exists* $r \in \mathcal{S}_+$ *with* $-r \leq s$ *and* $\tau^*(r) < \infty$, *then* μ *is regular.*

Note that $\mathcal{T}_+ \subset \mathcal{G}(\mathcal{S})$. So it is not too restrictive to suppose that $\mathcal{T}_+ \subset \mathcal{S}$. Then the above assumption is satisfied if moreover $\tau_* > -\infty$ on $\mathcal{S}_-$.

By Theorem 6.9, we have to prove that μ is $\mathcal{S}$–bounded below. But by the assumption, there exists $t \in \mathcal{T}_+$ with $t \geq r$ and $\tau(t) < \infty$. For $u \in \mathcal{S}_-$ with $-u \leq r$, we therefore have

$$-\mu(u) \leq \tau^\bullet(-u) \leq \tau^\bullet(t) = \tau(t) ,$$

hence $\mu_*(r) \leq \tau(t) < \infty$. □

(2) The corollary still holds if τ is semiregular, $\tau_\bullet > -\infty$ on $\mathcal{S}_-$ and $\tau_*(\max(s,t)) > -\infty$ for all $s \in \mathcal{S}_-$, $t \in \mathcal{T}_-$.

We wonder whether this last restriction can be removed.

EXAMPLE. *Let* X *be a Hausdorff space and* $\mathcal{T}$ *a lattice cone of lower semicontinuous functions on* X *such that for every* $K \in \mathfrak{K}(X)$ *there exists* $t \in \mathcal{T}$ *with* $t \leq -1_K$. *If* τ *is regular and* $\mathcal{S}(X)$*–tight, then there exists a Radon integral representing* τ .

By Example 4.9, the result follows from Theorem 7.7 and Remark 1, since $\mathcal{T}_+ \subset \mathcal{S}(X)$ and $\tau_* > -\infty$ on $\mathcal{S}_-(X)$. □

For a cone $\mathcal{T}$ of arbitrary functions, a similar representation is proved in Theorem 16.9 if $\tau_* > -\infty$ on $\mathcal{S}_-(X)$, τ is regular and satisfies the condition

$$\tau(t) = \inf_{s \in \mathcal{S}(X),\, s \geq t} \tau^*(s)$$

for all $t \in \mathcal{T}$, which by Lemma 7.1 is sharper than $\mathcal{S}(X)$–tightness.

§ 8 RADON INTEGRALS

8.1 To study Radon integrals, the following sharper version of the Daniell property will prove to be decisive. It enables us to extend linear functionals from a lattice cone $\mathcal{S}$ of functions to $\mathcal{S}_\phi$ such that the associated upper functional is an upper integral, hence having the convergence properties discussed in § 5.

DEFINITION. An increasing functional μ on a set $\mathcal{S}$ of functions is said to have the *Bourbaki property* if

$$\mu(\sup s_i) = \sup \mu(s_i)$$

holds for every upward directed family (s_i) in $\mathcal{S}$ with $\mu(s_i) > -\infty$ and $\sup s_i \in \mathcal{S}$.

THEOREM. *For a regular linear functional μ on a lattice cone $\mathcal{S}$ of functions, the following assertions are equivalent:*

(i) *μ has the Bourbaki property.*

(ii) *For every upward directed family (s_i) in $\mathcal{S}_-$ with $\sup s_i = 0$, we have*

$$\sup \mu(s_i) = 0 .$$

(iii) *The restriction μ^ϕ of μ_* to $\mathcal{S}_\phi$ has the Bourbaki property.*

If μ has these properties, then μ^ϕ is a regular linear functional and the only increasing extension of μ to $\mathcal{S}_\phi$ which has the Bourbaki property.

We only have to prove that (ii) implies (iii). To this end, suppose first that (s_i) is an upward directed family in $\mathcal{S}$. Then for every $t \in \mathcal{S}$ with $-t \le \sup s_i$, the family $([s_i + t]_-)$ in $\mathcal{S}_-$ is upward directed and has 0 as its upper envelope. Therefore

$$\sup \mu(s_i) + \mu(t) = \sup \mu(s_i + t) \ge \sup \mu([s_i + t]_-) = 0$$

and hence

$$-\mu(t) \le \sup \mu(s_i) .$$

This implies that

$$\mu_*(\sup s_i) \le \sup \mu(s_i) \le \mu_*(\sup s_i) .$$

Now suppose that (t_j) is an upward directed family in $\mathscr{S}_\phi$, and let (s_i) denote the family of all functions in $\mathscr{S}$ which are majorized by some t_j. Since $\mathscr{S}$ is max-stable, the family (s_i) is upward directed and has $\sup t_j$ as its upper envelope. Consequently,

$$\mu_*(\sup t_j) = \sup \mu(s_i) \leq \sup \mu_*(t_j) \leq \mu_*(\sup t_j) .$$

The linearity and uniqueness of μ^ϕ follow immediately from the Bourbaki property. Since

$$\mu^\phi = \mu_* \leq \mu^\phi{}_* \leq \mu^\phi \quad \text{on } \mathscr{S}_\phi ,$$

μ^ϕ is regular. □

8.2 Definition. Let μ be a regular linear functional defined on a lattice cone. If μ has the Bourbaki property, then μ is called a *Bourbaki integral.*

The upper functionals $\mu^{\phi*}$ *and* $\mu^{\phi\bullet}$, respectively denoted by

$$\int^* \cdot\, d\mu \quad \text{and} \quad \int^\bullet \cdot\, d\mu ,$$

are upper integrals.

This follows from Theorem 8.1 by application of Theorem and Corollary 5.2. □

By definition, the Bourbaki integration theory for μ is the integration theory w.r.t. μ^ϕ and therefore gets incorporated into our general framework. We do not study the Bourbaki theory in detail, since the results will not be needed in this section and are similar to those to be obtained for the Daniell integration theory which will be studied in § 9.

The next corollary shows that every Radon integral is a Bourbaki integral. Note that in this case Bourbaki integration is an abstract Riemann integration since

$$\mathscr{S}(X) = \mathscr{S}(X)_\phi .$$

THROUGHOUT THE REST OF THIS SECTION,
X IS A HAUSDORFF SPACE.

PROPOSITION. *Let $\mathcal{S}$ be a lattice cone of lower semicontinuous functions and μ a regular linear functional on $\mathcal{S}$. If $\mu_\bullet$ is finite on $\mathcal{S}_-(X)$, e.g. if for every $K \in \mathfrak{K}(X)$ there exists $s \in \mathcal{S}$ with $s \leq -1_K$, and if μ is $\mathcal{S}(X)$-tight, then μ is a Bourbaki integral.*

This follows from Theorem 8.1, since $\sup \mu(s_i) = 0$ for every upward directed familiy (s_i) in $\mathcal{S}_-$ with upper envelope 0. Indeed, if s_0 is an element of this family, given $\varepsilon > 0$ there is a compact set K with $\mu^*(1_{\complement K} \cdot s_0) \geq -\varepsilon$, as Example 7.1.2 shows. By hypothesis, there exists $\delta > 0$ with $\delta \cdot \mu_\bullet(-1_K) \geq -\varepsilon$. Using the compactness of K, we infer from Dini's Theorem that $1_K \cdot s_i \geq -\delta \cdot 1_K$ for s_i sufficiently far up. Hence we obtain

$$0 \geq \mu(s_i) \geq \mu_\bullet(1_K \cdot s_i) + \mu^\bullet(1_{\complement K} \cdot s_i) \geq \mu_\bullet(-\delta \cdot 1_K) + \mu^*(1_{\complement K} \cdot s_0) \geq -2\varepsilon . \quad \square$$

COROLLARY. *Every Radon integral is a Bourbaki integral.*

EXAMPLE. *The volume functional v on $\mathcal{T}(\mathbb{R}^n)$ is a Bourbaki integral, and v^ϕ is the unique extension of v to a Radon integral: the Lebesgue integral λ on $\mathbb{R}^n$.*

Recall Example 1.10.2 and note that the $\mathcal{S}(\mathbb{R}^n)$-tightness is trivial since $\mathcal{T}(\mathbb{R}^n) \subset \mathcal{S}(\mathbb{R}^n)$. The first part therefore follows from the proposition. The second part is a consequence of Theorem 8.1 since $\mathcal{T}(\mathbb{R}^n)_\phi = \mathcal{S}(\mathbb{R}^n)$. In fact, for every $s \in \mathcal{S}(\mathbb{R}^n)$ there exists a compact rectangle K and $\alpha \in \mathbb{R}_+$ with $-\alpha \cdot 1_K \leq s$. Given $x \in \mathbb{R}^n$ and $\gamma \in \mathbb{R}_+^*$ such that

$$-\alpha \cdot 1_K(x) + \gamma < s(x) ,$$

there exists an open rectangle G_x with

$$x \in G_x \subset \{s > -\alpha \cdot 1_K(x) + \gamma\} .$$

Hence s is the upper envelope of $-\alpha \cdot 1_K$ and all step functions

$$-\alpha \cdot 1_K + \gamma \cdot 1_{G_x} . \quad \square$$

One can even prove that $\mathcal{T}(\mathbb{R}^n)_\sigma = \mathcal{S}(\mathbb{R}^n)$.

Every Riemann integrable function on $\mathbb{R}$ *is Lebesgue integrable, i.e. integrable w.r.t.* λ^* *, and both integrals coincide.*

Since λ is an extension of v , we have $\lambda^* \le v^* = \iota^*$ by Example 2.5.2. □

THEOREM. *If* X *is locally compact, then every positive linear form* μ *on* $\mathcal{K}(X)$ *is a Bourbaki integral, and the mapping* $\mu \longmapsto \mu^\phi$ *is a bijection between these forms and the Radon integrals on* X .

Since $\mathcal{K}(X)_\phi = \mathcal{S}(X)$, the first assertion follows from Proposition 8.2, and Theorem 8.1 shows that μ^ϕ is a regular linear functional on $\mathcal{S}(X)$, i.e. a Radon integral. Conversely, every Radon integral is a Bourbaki integral, hence the mapping $\mu \longmapsto \mu^\phi$ is bijective. □

8.3 THEOREM. *For every Radon integral* μ *, the functionals* μ^* *and* $\mu^\bullet$ *are upper integrals. Moreover,* $\mu^\bullet$ *has the Bourbaki property on the set of all lower semicontinuous functions* h *with* $\mu_\bullet(h) > -\infty$.

Let (h_i) be an upward directed family of lower semicontinuous functions with $\mu_\bullet(h_i) > -\infty$. The inequality

$$\sup \mu^\bullet(h_i) \le \mu^\bullet(\sup h_i)$$

is obvious. We may therefore assume that $\sup \mu^\bullet(h_i) < \infty$, hence all h_i are essentially integrable by Example 6.1 and the Essential Integrability Criterion 4.4.

For every $s \in \mathcal{S}_-(X)$ we have $\max(h_i,s) \in \mathcal{S}(X)$, and the family

$$(\ \mu(\max(h_i,s)) - \mu^\bullet(h_i)\ )_i$$

is downward directed in $\bar{\mathbb{R}}$. In fact, for essentially integrable functions g and h with $g \le h$ we have

$$\mu^\bullet(g) - \mu^\bullet(\max(g,s)) = \mu^\bullet([g-s]_-) \le \mu^\bullet([h-s]_-) = \mu^\bullet(h) - \mu^\bullet(\max(h,s)) ,$$

since by Corollary 2.11 we may assume g and h to be finite.

By Proposition 6.1, $\mathcal{S}_-(X)$ is almost coinitial in $\bar{\mathbb{R}}^X$ w.r.t. $\mu^\bullet$, hence we have

$$\mu^\bullet(h_i) = \inf_{s\in\mathcal{S}_-(X)} \mu(\max(h_i,s))$$

and thus

$$0 = \inf_i \inf_{s\in\mathcal{S}_-(X)} [\mu(\max(h_i,s)) - \mu^\bullet(h_i)] =$$

$$= \inf_{s\in\mathcal{S}_-(X)} \lim_i [\mu(\max(h_i,s)) - \mu^\bullet(h_i)] =$$

$$= \inf_{s\in\mathcal{S}_-(X)} [\sup_i \mu(\max(h_i,s)) - \sup_i \mu^\bullet(h_i)] =$$

$$= \inf_{s\in\mathcal{S}_-(X)} \mu(\max[\sup_i h_i,s]) - \sup_i \mu^\bullet(h_i) = \mu^\bullet(\sup_i h_i) - \sup_i \mu^\bullet(h_i)$$

by the Bourbaki property of μ. □

8.4 Because of its importance, the proof of the following very general result on representation by Radon integrals is made as self-contained as possible, though we could also have applied Theorem 7.6 to τ or Corollary 7.5 to τ^ϕ (cf. Example 9.13). Note also the general but less informative Example 7.8. For a fully self-contained presentation see *Anger and Portenier* [1991].

THEOREM. *Let $\mathcal{T}$ be a lattice cone of lower semicontinuous functions such that*

$$\mathcal{S}_-(X) \subset \mathcal{T}_\phi .$$

If τ is an $\mathcal{S}(X)$-tight regular linear functional on $\mathcal{T}$, then τ is a Bourbaki integral, and the restriction μ of $\int^\bullet \cdot\, d\tau$ to $\mathcal{S}(X)$ is the only Radon integral representing τ.

Moreover, μ coincides with τ_* on $\mathcal{S}_-(X)$, τ coincides with μ^* on $\mathcal{T}$, and

$$\int^\bullet \cdot\, d\tau = \mu^\bullet .$$

Note first of all that τ is a Bourbaki integral by Proposition 8.2. By hypothesis, $\mathcal{S}_-(X)$ consists of integrable functions w.r.t. τ^ϕ, and μ coincides with $\tau^\phi = \tau_*$ on $\mathcal{S}_-(X)$. Furthermore, the $\mathcal{S}(X)$-tightness of τ gives

$$\tau^\phi(t) = \tau(t) = \inf_{u\in\mathcal{S}_-(X)} \tau_*(\max(t,u)) = \inf_{u\in\mathcal{S}_-(X)} \tau^\phi(\max(t,u))$$

for all $t \in \mathcal{T}_-$. This proves that $\mathcal{S}_-(X)$ is almost coinitial to $\mathcal{T}_-$ w.r.t. $\tau^{\phi *}$, hence in $\bar{\mathbb{R}}^X$ w.r.t. $\tau^{\phi *}$ and $\tau^{\phi \bullet}$ by Propositions 6.1 and 3.5.(i), since $\mathcal{T}_-$ is coinitial in $\mathcal{T}_\phi$. This shows, by Remark 6.1.3, that all functions in $\mathcal{S}(X)$ are measurable w.r.t. τ^ϕ and that

$$\mu = \tau^{\phi \bullet} = \tau^\phi{}_* \quad \text{on } \mathcal{S}(X).$$

Thus, for all $t \in \mathcal{T}_\phi$ we have

$$\tau^\phi(t) = \inf_{u \in \mathcal{S}_-(X)} \tau^\phi(\max(t,u)) = \inf_{u \in \mathcal{S}_-(X)} \mu(\max(t,u)),$$

since $\max(t,u) \in \mathcal{T}_\phi \cap \mathcal{S}(X)$.

For $s \in \mathcal{S}(X)$, this formula now yields

$$\mu(s) = \tau^\phi{}_*(s) = \sup_{t \in \mathcal{T}_\phi,\, -t \le s} - \tau^\phi(t) =$$

$$= \sup_{t \in \mathcal{T}_\phi,\, -t \le s} \sup_{u \in \mathcal{S}_-(X)} - \mu(\max(t,u)) \le \mu_*(s),$$

because

$$- \max(t,u) = \min(-t,-u) \le s.$$

This proves the regularity of μ, and thus, for $t \in \mathcal{T}_\phi$, that

$$\tau^\phi(t) = \inf_{u \in \mathcal{S}_-(X)} \mu(\max(t,u)) = \mu^*(t),$$

since $\mathcal{S}_-(X)$ is almost coinitial in $\bar{\mathbb{R}}^X$ w.r.t. μ^* by Proposition 6.1.

Next we prove that $\int^\bullet \cdot\, d\tau = \mu^\bullet$, which in particular implies that μ represents τ. In fact, for any function f which has an integrable majorant w.r.t. τ^ϕ, we have

$$\int^* f\, d\tau = \inf_{t \in \mathcal{T}_\phi,\, t \ge f} \mu^*(t) \ge \mu^*(f) =$$

$$= \inf_{s \in \mathcal{S}(X),\, s \ge f} \int^\bullet s\, d\tau \ge \int^\bullet f\, d\tau = \int^* f\, d\tau.$$

Application of Theorem 3.6 now yields

$$\int^\bullet f\, d\tau = \inf_{u \in \mathcal{S}_-(X)} \sup_{v \in \mathcal{S}_-(X)} \int^* \operatorname{med}(f,-v,u)\, d\tau =$$

$$= \inf_{u \in \mathcal{S}_-(X)} \sup_{v \in \mathcal{S}_-(X)} \mu^*(\operatorname{med}(f,-v,u)) = \mu^\bullet(f),$$

since $\mathcal{S}\_(X)$ is almost coinitial in $\mathcal{R}\_(\tau^{\phi})$ and $\mathrm{med}(f,-v,u) \leq -v \in \mathcal{R}\_(\tau^{\phi})$.

Finally, if κ is another Radon integral representing τ, we have $\kappa^{\bullet} = \tau^{\bullet}$ on $\mathcal{T}$ and thus on $\mathcal{T}_{\phi}$ by Theorem 8.3. Since $\mathcal{S}\_(X) \subset \mathcal{T}_{\phi}$, Lemma 6.2 proves that $\kappa = \mu$. □

REMARKS.

(1) We only have to asssume that $-1_K \in \mathcal{T}_{\phi}$ for all $K \in \mathfrak{K}(X)$.

This follows from the formula

$$s = \sup_{n \in \mathbb{N}} -\frac{1}{2^n} \cdot \left[1_{\mathrm{supp}(s)} + \sum_{i>0} 1_{\{s \leq -i/2^n\}} \right],$$

the sum having only a finite number of terms since $s \in \mathcal{S}\_(X)$ is bounded below. □

(2) It is possible to weaken the regularity condition to semiregularity if $\mathcal{S}\_(X) \subset \mathcal{T}_{\sigma}$ and τ_* is finite on $\mathcal{S}\_(X)$.

The proof follows the same lines using the essential Daniell upper integral to be introduced in the next section, replacing Proposition 8.2 by Example 9.5.

(3) In Corollary 14.11, the sufficient conditions for representability will be replaced, in the context of Radon measures, by a weak Urysohn separation condition, which is also necessary. Convergence properties will play no role, since the proof is carried out in an abstract Riemann spirit using Addendum 7.6.

(4) For every Radon integral μ with $\mathcal{T}\_ \subset \mathcal{R}^{\bullet}\_(\mu)$, we have $\mu^* = \mu_{\bullet}$ on $\mathcal{T}$.

Indeed, for $t \in \mathcal{T}$ we get

$$\mu_{\bullet}(t) = \mu^{\bullet}(t) = \mu^{\bullet}(t^+) + \mu^{\bullet}(t\_) = \mu(t^+) + \mu^*(t\_) = \mu^*(t)$$

by Example 6.1 and Theorem 2.2.(iv), since $t^+ \in \mathcal{S}(X)$. □

Therefore, the restriction of $\mu^* = \mu_{\bullet}$ to $\mathcal{T}$ defines an increasing linear functional. The only problem to get a bijection between adequate functionals τ on

$\mathcal{T}$ and Radon integrals is to ensure the regularity of τ. This will be done in Corollary 8.6 and Theorem 8.8.

(5) The $\mathcal{S}(X)$-tightness of τ is necessary. This follows from Proposition 7.2. For a direct proof, see *Anger and Portenier* [1991], Theorem 4.2.

8.5 We now give two conditions which enable us to apply the preceding representation theorem.

PROPOSITION. *For a lattice cone $\mathcal{T}$ of lower semicontinuous functions, we have*

$$\mathcal{S}_-(X) \subset \mathcal{T}_\phi$$

if one of the following conditions is satisfied:

(i) $\mathcal{T}_- \neq \{0\}$, *for all* $t \in \mathcal{T}_-$ *we have* $\max(t,-1) \in \mathcal{T}$, *and for any two different points* $x,y \in X$ *there exists a function* $t \in \mathcal{T}$, *continuous at* x, *with*

$$t(x) < 0 \quad \textit{and} \quad t(y) = 0 .$$

(ii) *For any* $x \in X$ *and any neighbourhood* U *of* x, *there exists a function* $t \in \mathcal{T}$, *continuous at* x, *with*

$$x \in \{t < 0\} \subset U .$$

By Remark 8.4.1, one only has to prove that $-1_K \in \mathcal{T}_\phi$ for all $K \in \mathfrak{K}(X)$. This follows by a usual compactness argument, noting in the second case that for any $\varepsilon > 0$, any $x \in X$ and any neighbourhood U of x, there exists a function $t \in \mathcal{T}_-$, continuous at x, with

$$t(x) < -1 , \quad t \geq -1 - \varepsilon \text{ on } U \quad \text{and} \quad t = 0 \text{ on } \complement U :$$

Indeed, choose functions $u,v \in \mathcal{T}_-$, continuous at x, with

$$u(x) = -1 - \tfrac{\varepsilon}{2} , \ v(x) < -1 \quad \text{and} \quad \{v < 0\} \subset U \cap \{u > -1 - \varepsilon\} .$$

Then define $t := \max(u,v)$. □

8.6 The tightness assumption in Theorem 8.4 follows in the case of $\mathcal{W}$–dominated functions from a boundedness condition :

LEMMA. Let $\mathcal{W}$ *be a set of weights and* $\mathcal{T}$ *a lattice cone of* $\mathcal{W}$*-dominated lower semicontinuous functions. Then every* $\mathcal{W}$*-bounded semiregular linear functional* τ *on* $\mathcal{T}$ *is* $\mathcal{S}(X)$*-tight.*

There exists $w \in \mathcal{W}$ with $\tau_*(w) < \infty$. For $t \in \mathcal{T}_-$ and $\varepsilon > 0$, let $\delta > 0$ with $\delta \cdot \tau_*(w) \leq \varepsilon$ and $K := \{t \leq -\delta w\} \in \mathfrak{K}(X)$. For every $u \in \mathcal{T}_-$ with $u \geq t$ and $u = 0$ on K, we have $u \geq -\delta w$ and therefore

$$-\tau(u) \leq \delta \cdot \tau_*(w) \leq \varepsilon .$$

The result now follows from Example 7.1.2. □

COROLLARY. *Let* $\mathcal{T}$ *be a lattice cone of* $\mathcal{W}$*-dominated lower semicontinuous functions with* $\mathcal{S}_-(X) \subset \mathcal{T}_\phi$*. Then every regular linear functional* τ *on* $\mathcal{T}$ *with* $\tau_*(w) < \infty$ *for some* $w \in \mathcal{W}$ *is represented by a unique Radon integral on* X*.*

In particular, there is a bijection between the $\mathcal{W}$*-bounded regular linear functionals* τ *on* $\mathcal{S}^{\mathcal{W}}(X)$ *and the Radon integrals* μ *with* $\mu(w) < \infty$ *for some* $w \in \mathcal{W}$*, given by*

$$\tau^\bullet = \mu^\bullet .$$

The first part is an immediate consequence of Theorem 8.4 and the lemma. Since

$$\mathcal{S}(X) \subset \mathcal{S}^{\mathcal{W}}(X) = \mathcal{S}^{\mathcal{W}}(X)_\phi ,$$

there exists only one Radon integral μ with $\tau^\bullet = \mu^\bullet$. As $\tau_*(w) < \infty$ for some $w \in \mathcal{W}$, we have

$$\mu(w) = \tau(w) = \tau_*(w) < \infty .$$

Conversely, every function in $\mathcal{S}^{\mathcal{W}}_-(X)$ is $\mathcal{W}$-bounded, so it is integrable w.r.t. μ, because $\mu(w) < \infty$ for some $w \in \mathcal{W}$. By Corollary 6.2, the functional $\tau := \mu_\bullet = \mu^\bullet$ on $\mathcal{T}$ is regular and $\mathcal{W}$-bounded. □

8.7 In the case of a vector lattice, the conditions in Proposition 8.5 may be expressed in the following form :

DEFINITION. A vector lattice $\mathcal{T}$ of functions is called *Stonian* if $\min(t,1) \in \mathcal{T}$

for all $t \in \mathcal{T}$, and *rich* if for any $x \in X$ and any neighbourhood U of x there exists $t \in \mathcal{T}$ with

$$x \in \{t > 0\} \subset U .$$

It is said to be *point-separating*, respectively *linearly separating*, if $\mathcal{T} \neq \{0\}$ and if for any two different points $x,y \in X$ there exist $u \in \mathcal{T}$, respectively $u,v \in \mathcal{T}$, with

$$u(x) \neq u(y) \ , \text{ respectively } \ u(x)v(y) \neq u(y)v(x) .$$

REMARK. *This last condition is equivalent to the existence of a function* $t \in \mathcal{T}$ *such that*

$$t(x) = 1 \quad \textit{and} \quad t(y) = 0 .$$

In particular, if $\mathcal{T}$ *is rich, then* $\mathcal{T}$ *is linearly separating. If* $1 \in \mathcal{T}$, *then* $\mathcal{T}$ *is linearly separating iff* $\mathcal{T}$ *is point-separating.*

Indeed, one only has to define

$$t := \frac{v(y)}{u(x)v(y) - u(y)v(x)} \cdot u - \frac{u(y)}{u(x)v(y) - u(y)v(x)} \cdot v . \quad \square$$

EXAMPLES.

(1) $\mathscr{C}(X)$ *and* $\mathscr{C}^b(X)$ *are rich iff* X *is completely regular. If* $\mathscr{C}^0(X)$ *is linearly separating, then* X *is locally compact. Conversely, if* X *is locally compact, then* $\mathscr{C}^0(X)$ *and* $\mathscr{K}(X)$ *are rich.*

(2) *Let* $\mathscr{W}$ *be a set of weights on a completely regular topological space* X *such that every point* $x \in X$ *has a neighbourhood* U *with*

$$\{w \leq \gamma\} \cap U \quad \textit{is compact for all } w \in \mathscr{W} \textit{ and } \gamma \in \mathbb{R}_+ ,$$

e.g. X *is locally compact. Then* $\mathscr{C}^{\mathscr{W}}(X)$ *is rich.*

In fact, if we choose $f \in \mathscr{C}(X)$ with $0 \leq f \leq 1$ and $x \in \{f > 0\} \subset U$, then f is $\mathscr{W}$–dominated, since for $w \in \mathscr{W}$ and $\varepsilon > 0$ the set

$$\{f \geq \varepsilon \cdot w\} \subset \{w \leq \tfrac{1}{\varepsilon}\} \cap U$$

is compact. $\square$

8.8 THEOREM. *If $\mathcal{T}$ is a linearly separating Stonian, or a rich, vector lattice of continuous functions, then there exists a bijection between the $\mathcal{S}(X)$-tight positive linear forms τ on $\mathcal{T}$ and the Radon integrals μ with $\mathcal{T} \subset \mathcal{R}(\mu)$, given by*

$$\int^{\bullet} \cdot \, d\tau = \mu^{\bullet} .$$

By Proposition 8.5, we have $\mathcal{S}_{-}(X) \subset \mathcal{T}_{\phi}$, hence there exists a unique Radon integral μ with $\int^{\bullet} \cdot \, d\tau = \mu^{\bullet}$, as Theorem 8.4 shows. We have $\mathcal{T} \subset \mathcal{R}(\mu)$ since τ is finite on $\mathcal{T}$. Conversely, every Radon integral μ with this property defines an $\mathcal{S}(X)$-tight positive linear form on $\mathcal{T}$ by Proposition 7.2. □

DEFINITION. A Radon integral is said to be *bounded* if 1 is integrable.

We have the following alternative description :

EXAMPLE. *Let X be completely regular. Then there exists a bijection between the $\mathcal{S}(X)$-tight positive linear forms τ on $\mathcal{C}^b(X)$ and the bounded Radon integrals μ, given by*

$$\int^{\bullet} \cdot \, d\tau = \mu^{\bullet} .$$

Using Lemma 8.6 and proceeding as in the proof of Corollary 8.6, we immediately get the following

COROLLARY. *Let $\mathcal{W}$ be a set of weights such that $\mathcal{C}^{\mathcal{W}}(X)$ is linearly separating. Then there is a bijection between the $\mathcal{W}$-bounded positive linear forms on $\mathcal{C}^{\mathcal{W}}(X)$ and the Radon integrals with $\mu(w) < \infty$ for some $w \in \mathcal{W}$.*

8.9 We now give some conditions ensuring $\mathcal{W}$-boundedness.

LEMMA. *Let w be a locally bounded weight. Then every increasing linear functional τ on $\mathcal{S}^{\{w\}}(X)$ or $\mathcal{C}^{\{w\}}(X)$ is w-bounded.*

We first discuss the case $\mathcal{S}^{\{w\}}(X)$. Let us suppose that $\tau_*(w) = \infty$. Then there exists a sequence (t_n) in $\mathcal{S}_-^{\{w\}}(X)$ with

$$-t_n \leq w \quad \text{and} \quad \tau(t_n) \leq -n \cdot 2^n .$$

We prove that $t := \sum \frac{1}{2^n} \cdot t_n$ belongs to $\mathcal{S}^{\{w\}}(X)$. First note that t is lower semicontinuous at $x \in X$. Indeed, for every $\varepsilon > 0$, by the local boundedness of w there exists a neighbourhood U of x and $k \in \mathbb{N}$ such that

$$\sum_{n \geq k} \frac{1}{2^n} \cdot t_n \geq - \sum_{n \geq k} \frac{1}{2^n} \cdot w \geq - \frac{\varepsilon}{2} \quad \text{in } U .$$

By the lower semicontinuity of $\sum_{n<k} \frac{1}{2^n} \cdot t_n$ we can find a neighbourhood $V \subset U$ of x with

$$\sum_{n<k} \frac{1}{2^n} \cdot t_n \geq t(x) - \frac{\varepsilon}{2} \quad \text{in } V ,$$

hence

$$t = \sum_{n<k} \frac{1}{2^n} \cdot t_n + \sum_{n \geq k} \frac{1}{2^n} \cdot t_n \geq t(x) - \varepsilon \quad \text{in } V .$$

Furthermore, t is w–dominated, since for $\varepsilon > 0$ the set $\{t \leq -\varepsilon w\}$ is closed and contained in

$$\{ \sum_{n<k} \frac{1}{2^n} \cdot t_n \leq -\varepsilon w - \sum_{n \geq k} \frac{1}{2^n} \cdot w \} \subset \{ \sum_{n<k} \frac{1}{2^n} \cdot t_n \leq - \frac{\varepsilon}{2} \cdot w \} \in \mathfrak{K}(X)$$

for k sufficiently large.

We now have the contradiction

$$- \infty < \tau(t) \leq \frac{1}{2^n} \cdot \tau(t_n) \leq - n$$

for all $n \in \mathbb{N}$.

In the case $\mathcal{C}^{\{w\}}(X)$, note that t is also upper semicontinuous, and hence continuous, as lower envelope of continuous functions. □

EXAMPLE. *Let X be locally compact. Then all positive linear forms on $\mathcal{C}^0(X)$ are Bourbaki integrals, and are in one–to–one correspondence with the bounded Radon integrals.*

This follows from Lemma 8.6, Proposition 8.2 and Corollary 8.8, since every positive linear form on $\mathscr{C}^0(X) = \mathscr{C}^{\{1\}}(X)$ is 1–bounded by the lemma. □

REMARKS.

(1) The assumption on w may be replaced by the following :

w is bounded on every compact subset, and every $x \in X$ has a neighbourhood U such that $\mathfrak{G}(U) = \mathfrak{G}(\mathfrak{K}(U))$.

Note that t is lower semicontinuous on every compact subset by the same argument as above, and then on U by the assumption on the topology.

(2) More generally, if $\mathcal{W}$ is a set of weights, then $\mathscr{C}^{\mathcal{W}}(X)$ is a locally convex space if endowed with the family of seminorms defined for $w \in \mathcal{W}$ by

$$t \longmapsto \inf \{\alpha \in \mathbb{R}_+ : |t| \leq \alpha w\} ,$$

which is finite, since $-|t|$ is w–bounded.

By the same arguments as in the lemma one can show :

If $\mathscr{C}^{\mathcal{W}}(X)$ is bornological and $\mathscr{C}^{\mathcal{W}}_+(X)$ is sequentially complete, then every positive linear form on $\mathscr{C}^{\mathcal{W}}(X)$ is continuous, i.e. $\mathcal{W}$–bounded.

It is possible to give conditions relating $\mathcal{W}$ to the topology of X and ensuring the above assumptions on $\mathscr{C}^{\mathcal{W}}(X)$.

THROUGHOUT THE REST OF THIS SECTION,
μ IS A RADON INTEGRAL ON X.

8.10 **LEMMA.** *For $A \subset X$ we have*

$$\mu_*(A) = \sup_{K \in \mathfrak{K}(X),\ K \subset A} \mu_*(K)$$

and

$$\mu^*(A) = \inf_{G \in \mathfrak{G}(X),\ G \supset A} \mu(G) .$$

To prove the first formula, note that for $s \in \mathcal{S}_-(X)$ with $-s \leq 1_A$ and $\varepsilon > 0$ we have

$$K := \{s \leq -\varepsilon\} \in \mathfrak{K}(X) \ , \ K \subset A \quad \text{and} \quad -1_K + \max(s,-\varepsilon) \leq s \ ,$$

hence

$$\mu(-1_K) + \mu(\max(s,-\varepsilon)) \leq \mu(s)$$

and therefore

$$-\mu(s) \leq -\sup_{\varepsilon>0} \mu(\max(s,-\varepsilon)) + \sup_{K\in\mathfrak{K}(X),\, K\subset A} \mu_*(K) =$$

$$= \sup_{K\in\mathfrak{K}(X),\, K\subset A} \mu_*(K)$$

by the Bourbaki property of μ. This yields

$$\mu_*(A) \leq \sup_{K\in\mathfrak{K}(X),\, K\subset A} \mu_*(K) \, .$$

On the other hand, for $s \in \mathcal{S}(X)$ with $s \geq 1_A$ and $\alpha > 1$ we have

$$A \subset G := \{\alpha s > 1\} \in \mathfrak{G}(X) \quad \text{and} \quad 1_G \leq \alpha s \, ,$$

hence $\mu(G) \leq \alpha \cdot \mu(s)$. This proves that

$$\inf_{G\in\mathfrak{G}(X),\, G\supset A} \mu(G) \leq \mu^*(A) \, .$$

The reverse inequalities are obvious. □

COROLLARY. (i) $\mu^\bullet$ *is inner* $\mathfrak{K}(X)$*-regular at every measurable set, i.e.*

$$\mu^\bullet(A) = \sup_{K\in\mathfrak{K}(X),\, K\subset A} \mu^\bullet(K)$$

for every set A *with* $1_A \in \mathcal{R}^0(\mu)$.

(ii) $\mu^\bullet$ *is outer* $\mathfrak{G}(X)$*-regular at every integrable set, i.e.*

$$\mu^\bullet(A) = \inf_{G\in\mathfrak{G}(X),\, G\supset A} \mu(G)$$

for every set A *with* $1_A \in \mathcal{R}^*(\mu)$.

One only has to note that $\mu_* = \mu^\bullet$ on $\mathcal{R}^0_+(\mu)$ by Proposition 6.1 and that $\mu^* = \mu^\bullet$ on $\mathcal{R}^*(\mu)$. □

REMARK. Properties of $\mu^\bullet$ on sets could now easily be derived using the corresponding indicator functions. This will be studied in 12.5 when Radon measures are discussed.

8.11 DEFINITION. Let Y be a Hausdorff space. A mapping $f : X \longrightarrow Y$ is called *Lusin measurable* (w.r.t. μ) if for every compact set $K \subset X$ and every $\varepsilon > 0$ there exists a compact subset $L \subset K$ such that

$$\mu^*(K \setminus L) \leq \varepsilon$$

and such that the restriction of f to L is continuous.

LEMMA. *Let (f_n) be a sequence of Lusin measurable mappings on X. Then for every compact set $K \subset X$ and $\varepsilon > 0$, there exists a compact subset $L \subset K$ with $\mu^*(K \setminus L) \leq \varepsilon$ and such that the restriction of each f_n to L is continuous.*

In fact, choose compact sets $L_n \subset K$ with $\mu^*(K \setminus L_n) \leq \frac{\varepsilon}{2^n}$ such that the restriction of f_n to L_n is continuous. Define $L := \bigcap L_n$. Then

$$\mu^*(K \setminus L) \leq \mu^*(\sum 1_{K \setminus L_n}) \leq \sum_n \mu^*(K \setminus L_n) \leq \varepsilon$$

by Lemma 5.3. □

This immediately implies the following

PROPOSITION.

(i) *The set of all Lusin measurable functions is a homogeneous lattice.*

(ii) *The set of all $\overline{\mathbb{R}}$-valued Lusin measurable functions is a lattice cone.*

8.12 THEOREM OF EGOROFF. *Let Y be a metrizable topological space and (f_n) a sequence of Lusin measurable mappings from X to Y, converging pointwise essentially almost everywhere to a mapping $f : X \longrightarrow Y$. Then we have:*

(i) *For every compact set $K \subset X$ and $\varepsilon > 0$, there exists a compact subset $L \subset K$ with $\mu^*(K \setminus L) \leq \varepsilon$ and such that the restrictions of f_n to L are continuous and converge uniformly to f on L.*

(ii) *f is Lusin measurable.*

Let d be a compatible metric on Y. For any compact set K and $\varepsilon > 0$, by Lemma 8.11 there exists a compact subset $H \subset K$ with $\mu^*(K \setminus H) \leq \frac{\varepsilon}{2}$ and such that the restriction of each f_n to H is continuous.

For $k,m \in \mathbb{N}$ with $k \geq 1$ define

$$A_{km} := \bigcup_{i,j \geq m} \{x \in H : d(f_i(x), f_j(x)) \geq \tfrac{1}{k}\},$$

which is a union of countably many compact subsets of H, hence integrable. The sequence $(A_{km})_{m \in \mathbb{N}}$ is decreasing with intersection contained in the essentially null set N where (f_n) does not converge to f, and is therefore a null set. By the Monotone Convergence Theorem 5.4 there exists $m_k \in \mathbb{N}$ such that

$$\mu^*(A_{km_k}) \leq \frac{\varepsilon}{2^{k+2}} .$$

The set $A := \bigcup_{k \geq 1} A_{km_k} \cup (H \cap N)$ is integrable and

$$\mu^*(A) \leq \sum_{k \geq 1} \mu^*(A_{km_k}) \leq \frac{\varepsilon}{4} .$$

By construction, the sequence (f_n) converges uniformly to f on $H \setminus A$. According to Corollary 8.10.(i), we can choose a compact set $L \subset H \setminus A$ with

$$\mu^*(H \setminus A) \leq \mu^*(L) + \frac{\varepsilon}{4} ,$$

hence satisfying (i).

Since the restriction of f to L is continuous, (ii) is also proved. □

COROLLARY.
(i) *For every sequence* (f_n) *of Lusin measurable functions, the functions* $\inf f_n$, $\sup f_n$, $\liminf f_n$ *and* $\limsup f_n$ *are Lusin measurable.*
(ii) *If* f *and* g *are Lusin measurable functions, so is* $f \cdot g$.

The first part is obvious as is (ii) if f and g are finite. The general case then follows from the formula

$$f \cdot g = \lim_n \operatorname{med}(f,n,-n) \cdot \operatorname{med}(g,n,-n) .$$ □

REMARK. The first assertion could also be derived from Proposition 5.10 and the following Theorem of Lusin.

8.13 **LEMMA.** *Every function* $s \in \mathcal{S}(X)$ *is Lusin measurable.*

By Proposition 8.11 and Egoroff's Theorem 8.12.(ii), it is sufficient to prove the lemma for -1_K and 1_G with $K \in \mathfrak{K}(X)$ and $G \in \mathfrak{G}(X)$, since we have

$$s := \sup_n \frac{1}{2^n} \cdot \left[\sum_{i=1}^{n2^n} \left[1_{\{s>i/2^n\}} - 1_{\{s\leq -i/2^n\}} \right] - 1_{\mathrm{supp}(s_-)} \right].$$

For H compact and $\varepsilon > 0$, by Corollary 8.10.(i) there exists a compact set $L \subset H \setminus K$, respectively $L \subset H \cap G$, with

$$\mu^*(H \setminus K) \leq \mu^*(L) + \varepsilon \ , \ \text{respectively} \quad \mu^*(H \cap G) \leq \mu^*(L) + \varepsilon .$$

But the restriction of -1_K to $L \cup (H \cap K)$ and of 1_G to $L \cup (H \setminus G)$ is continuous, and

$$\mu^*(H \setminus [L \cup (H \cap K)]) = \mu^*(H \setminus K) - \mu^*(L) \leq \varepsilon ,$$

respectively

$$\mu^*(H \setminus [L \cup (H \setminus G)]) = \mu^*(H \cap G) - \mu^*(L) \leq \varepsilon . \quad \square$$

Theorem of Lusin. *A function is measurable w.r.t.* μ *iff it is Lusin measurable w.r.t.* μ.

Suppose first that f is an integrable function. Then there exists a decreasing sequence (s_n) in $\mathcal{S}(X)$ with $s_n \geq f$ and

$$\mu^*(f) = \inf \mu(s_n) = \mu^*(\inf s_n)$$

by the Monotone Convergence Theorem 5.4. Since $f \leq \inf s_n$ and both functions are integrable, we infer from Proposition 2.12 that $f = \inf s_n$ a.e. (cf. Remark 5.3). By the lemma and Egoroff's Theorem 8.12.(ii), f is Lusin measurable.

Let now f be measurable w.r.t. μ and K be a compact set. For any $n \in \mathbb{N}$, the function $\mathrm{med}(f, n\cdot 1_K, -n\cdot 1_K)$ is integrable, hence

$$1_K \cdot f = \lim \mathrm{med}(f, n\cdot 1_K, -n\cdot 1_K)$$

is Lusin measurable by Egoroff's Theorem.

Finally, for $\varepsilon > 0$ there exists a compact set $L \subset K$ with $\mu^*(K \setminus L) \leq \varepsilon$ and such that the restriction of $1_K \cdot f$ to L, hence also that of f, is continuous. This finishes the proof of the necessity part.

Conversely, for any compact set K and any $n \in \mathbb{N}$, we have to prove that

$g := \text{med}(f, n \cdot 1_K, -n \cdot 1_K)$ is integrable. Since g is Lusin measurable by Proposition 8.11 and the lemma, for $k \geq 1$ there exists a compact set $L_k \subset K$ with $\mu^*(K \setminus L_k) \leq \frac{1}{k}$ and such that the restriction of g to L_k is continuous. We may suppose that $L_k \subset L_{k+1}$. Since

$$1_{L_k} \cdot g - n \cdot 1_{L_k} \in \mathcal{I}_-(X),$$

the function $1_{L_k} \cdot g$ is integrable, and so is g by Lebesgue's Dominated Convergence Theorem. □

8.14 From 6.10 – 6.13 we infer :

PROPOSITION. *If μ and τ are Radon integrals on X, then $\mu + \tau$ is a Radon integral with*

$$(\mu + \tau)^* = \mu^* + \tau^* \quad \textit{and} \quad (\mu + \tau)^\bullet = \mu^\bullet + \tau^\bullet .$$

A function is integrable, essentially integrable, respectively measurable w.r.t. $\mu + \tau$ iff it is so w.r.t. μ and τ.

The relation $\tau \precsim \mu$ is equivalent to each of the following :

$$\mu \leq \tau \text{ on } \mathcal{I}_-(X) \; , \; \tau \leq \mu \text{ on } \mathcal{I}_+(X) \; , \; \tau^* \leq \mu^* \text{ on } \overline{\mathbb{R}}_+^X \; , \; \tau_* \leq \mu_* \text{ on } \overline{\mathbb{R}}_+^X .$$

In this case, there exists a unique Radon integral ρ with

$$\tau + \rho = \mu .$$

THEOREM. *An upward directed family $(\mu_i)_{i \in I}$ of Radon integrals on X admits a smallest Radon integral $\mu =: \sup \mu_i$ on X with $\mu_i \precsim \mu$ iff $(\mu_i)_{i \in I}$ is locally uniformly bounded, i.e. for any $x \in X$ there exists an open neighbourhood U of x such that*

$$\sup \mu_i(U) < \infty .$$

In this case, for any function $f \geq 0$ we have

$$(\sup \mu_i)_*(f) = \sup\nolimits_{i \in I} \mu_{i*}(f) ,$$

$$(\sup \mu_i)_\bullet(f) = \sup \mu_{i\bullet}(f) \quad \textit{and} \quad (\sup \mu_i)^\bullet(f) = \sup \mu_i^\bullet(f) .$$

In particular,

$$\mathcal{R}^0(\sup \mu_i) = \bigcap_{i \in I} \mathcal{R}^0(\mu_i) .$$

One has just to note that the condition is equivalent to the fact that (μ_i) is uniformly $\mathcal{I}(X)$–bounded below. Indeed, if $t \in \mathcal{I}(X)$ with $-t \leq -1_{\{x\}}$, then one can choose $U := \{t > \frac{1}{2}\}$, since $1_U \leq 2t$. Conversely, for $s \in \mathcal{I}_-(X)$, $\alpha \in \mathbb{R}_+$ and $K \in \mathfrak{K}(X)$ with $s \geq -\alpha \cdot 1_K$, by a usual compactness argument there is $G \in \mathfrak{G}(X)$ with $K \subset G$ and $\sup \mu_i(G) < \infty$. Then one may choose $t := \alpha \cdot 1_G$. □

COROLLARY. *Let* $(\mu_i)_{i \in I}$ *be a family of Radon integrals on* X *. Then*

$$\sum_{i \in I} \mu_i := \sup_J \sum_{j \in J} \mu_j ,$$

where J *runs through all finite subsets of* I *, is a Radon integral iff* $(\mu_i)_{i \in I}$ *is uniformly locally summable, i.e. for any* $x \in X$ *there exists an open neighbourhood* U *of* x *such that*

$$\sum_{i \in I} \mu_i(U) < \infty .$$

In this case, for any function $f \geq 0$ *we have*

$$(\sum_{i \in I} \mu_i)_*(f) = \sum_{i \in I} \mu_{i*}(f) ,$$

$$\sum_{i \in I} \mu_i)_\bullet(f) = \sum_{i \in I} \mu_{i\bullet}(f) \quad \text{and} \quad (\sum_{i \in I} \mu_i)^\bullet(f) = \sum_{i \in I} \mu_i^\bullet(f) .$$

In particular,

$$\mathcal{R}^0(\sup \mu_i) = \bigcap_{i \in I} \mathcal{R}^0(\mu_i) .$$

8.15 Next we discuss Radon integrals with densities.

DEFINITION. A function $w : X \longrightarrow \overline{\mathbb{R}}_+$ is said to be *locally integrable* w.r.t. μ if every point in X has an open neighbourhood U such that

$$w \cdot 1_U \in \mathcal{R}^\bullet(\mu) .$$

REMARK. *If w is locally integrable and $K \subset X$ is compact, then there exists an open neighbourhood U of K such that $w \cdot 1_U$ is essentially integrable. Moreover, w is Lusin measurable and finite essentially a.e.*

In fact, K is covered by finitely many open sets $U_1,\dots,U_n$ such that $w \cdot 1_{U_i} \in \mathcal{R}^\bullet(\mu)$, hence $U := \bigcup U_i$ is an open neighbourhood of K with

$$w \cdot 1_U = \max_i \, w \cdot 1_{U_i} \in \mathcal{R}^\bullet(\mu) .$$

On the one hand, the function $w \cdot 1_U$ is Lusin measurable by Lusin's Theorem 8.13, and thus also w by definition. On the other hand, the condition of Definition 6.14 is satisfied, since sw is measurable for $s \in \mathcal{S}(X)$ by Lemma 8.13 and Corollary 8.12.(ii). So w is finite essentially a.e. by Theorem 6.14. □

THEOREM. *Let $w \geq 0$ be locally integrable w.r.t. μ. Then $w \cdot \mu$ is a Radon integral with*

$$(w \cdot \mu)^\bullet = w \cdot \mu^\bullet ,$$

for which a function f is essentially integrable, resp. measurable iff wf is essentially integrable, resp. measurable w.r.t. μ.

We apply Theorem 6.14. First we have to prove that

$$\mu^\bullet(1_{\{0<w<\infty\}} \cdot s) = \inf_{t \in \mathcal{S}_-(X),\ t \geq s/w} \mu^\bullet(wt)$$

holds for $s \in \mathcal{S}_-(X)$.

Consider the compact set $K := \mathrm{supp}(s)$ and $\alpha \in \mathbb{R}_+^*$ with $s \geq -\alpha \cdot 1_K$. Since $1_{\{0<w<\infty\}} = \sup_n \min(n \cdot w, 1)$ essentially a.e., this function is Lusin measurable by Egoroff's Theorem, and so is $1_{\{0<w<\infty\}} \cdot 1_K$ by Corollary 8.12.(ii). By the inner regularity of $\mu^\bullet$, there exists a compact set $L \subset K \cap \{0 < w < \infty\}$ with

$$\mu^\bullet(K \cap \{0 < w < \infty\} \setminus L) \leq \frac{\varepsilon}{\alpha} ,$$

and we may assume that the restriction of w to L is continuous. The function t defined by $\frac{s}{w}$ on L and by 0 elsewhere on X is in $\mathcal{S}_-(X)$. Since $t \geq \frac{s}{w}$ and

$$w \cdot t = 1_L \cdot s \leq 1_{\{0<w<\infty\}} \cdot s + \alpha \cdot 1_{K \cap \{0<w<\infty\} \setminus L} ,$$

we get

$$\mu^\bullet(wt) \le \mu^\bullet(1_{\{0<w<\infty\}}\cdot s) + \alpha\cdot\mu^\bullet(K \cap \{0 < w < \infty\} \setminus L) \le \mu^\bullet(1_{\{0<w<\infty\}}\cdot s) + \varepsilon ,$$

which proves the formula.

Finally, the regularity of $w\cdot\mu$ follows from the above remark. In fact, we have $-\alpha\cdot 1_U \le s$, so $\mu_*(\alpha w\cdot 1_U) = \mu_\bullet(\alpha w\cdot 1_U) < \infty$ by Proposition 6.8. □

8.16 We now discuss images of Radon integrals.

PROPOSITION. *Let* $p : X \longrightarrow Y$ *be a Lusin measurable mapping from* X *into a Hausdorff space* Y *, and* g *a lower semicontinuous function on* Y *. Then* $g\circ p$ *is measurable.*

By Corollary 8.12, we may assume g to be bounded. In fact,

$$g\circ p = \lim_n \operatorname{med}(g,n,-n)\circ p .$$

For a compact set $K \subset X$ and $\varepsilon > 0$, choose a compact subset $L \subset K$ with $\mu^*(K \setminus L) \le \varepsilon$, and such that the restriction of p to L is continuous. Then, for $\alpha \in \mathbb{R}_+$ with $g \le \alpha$, we have

$$1_L\cdot g\circ p - \alpha\cdot 1_L \in \mathcal{S}_-(X) .$$

This function is Lusin measurable by Lemma 8.13, which by definition proves that $g\circ p$ is also Lusin measurable. □

DEFINITION. A mapping $p : X \longrightarrow Y$ of X into a Hausdorff space Y is said to be *proper* w.r.t. μ if p is Lusin measurable w.r.t. μ and if every point in Y has an open neighbourhood V such that

$$\mu^\bullet(p^{-1}(V)) < \infty .$$

To be able to apply Theorem 6.15, we shall need the following result.

LEMMA. *If* X *is compact and* $p : X \longrightarrow Y$ *is continuous, then* $s_{p-} \in \mathcal{S}_-(Y)$ *for every* $s \in \mathcal{S}_-(X)$.

Since $s_{p-} = 0$ on $\complement p(X)$, $s_p \le 0$ on $p(X)$, which is compact, we only have to prove that s_p is lower semicontinuous on $p(X)$. We may therefore assume that $Y = p(X)$. For $\gamma \in \mathbb{R}$, we obviously have

$$\{s_p \leq \gamma\} \supset p(\{s \leq \gamma\}) .$$

Since $\{s \leq \gamma\}$ is compact, only the dual inclusion remains to be proved. For every $y \in \{s_p \leq \gamma\}$, by the compactness of the non-empty set $p^{-1}(y)$, there exists x in this set with

$$\gamma \geq s_p(y) = \inf s(p^{-1}(y)) = s(x) ,$$

hence

$$y = p(x) \in p(\{s \leq \gamma\}) . \quad \square$$

THEOREM. *If X and Y are Hausdorff spaces and $p : X \longrightarrow Y$ is proper w.r.t. μ, then $p(\mu)$ is a Radon integral on Y with*

$$p(\mu)^\bullet = p(\mu^\bullet) .$$

A function g on Y is essentially integrable, resp. measurable w.r.t. $p(\mu)$ iff $g \circ p$ is essentially integrable, resp. measurable w.r.t. μ.

For all $t \in \mathcal{S}(Y)$, the function $t \circ p$ is measurable by the proposition, and

$$\mu_\bullet(t \circ p) > -\infty .$$

In fact, by the compactness of $T := \operatorname{supp}(t_-)$ there exists an open set $V \supset T$ such that $\mu^\bullet(p^{-1}(V)) < \infty$. For suitable $n \in \mathbb{N}$, we have $-n \cdot 1_V \leq t$ and therefore

$$\mu_\bullet(t \circ p) \geq -n \cdot \mu^\bullet(p^{-1}(V)) > -\infty .$$

Moreover,

$$p(\mu)_*(n \cdot 1_V) \leq p(\mu)(n \cdot 1_V) = n \cdot \mu^\bullet(p^{-1}(V)) < \infty$$

yields that $p(\mu)$ is $\mathcal{S}(Y)$–bounded below.

It only remains to prove that

$$\mu(s) = \inf_{t \in \mathcal{S}_-(Y),\ t \geq s_{p-}} \mu^\bullet(\max(s, t \circ p))$$

for all $s \in \mathcal{S}_-(X)$.

Let $\varepsilon > 0$ be given and choose $\alpha > 0$ with $s \geq -\alpha$. By the compactness of $K := \operatorname{supp}(s)$ and the Lusin measurability of p w.r.t. μ, there exists a compact subset $L \subset K$ with

$$\mu^*(K \setminus L) \leq \frac{\varepsilon}{\alpha}$$

and such that the restriction of p to L is continuous. By the preceding lemma, the function $t := (s|_L)_{p|L-}$ belongs to $\mathcal{S}_-(Y)$. Obviously,

$$t \geq s_{p-} \quad \text{and} \quad s \geq top \text{ on } L \cup \complement K = \complement(K \setminus L),$$

hence

$$\max(s,top) \leq s + \alpha \cdot 1_{K \setminus L}.$$

This yields

$$\mu^{\bullet}(\max(s,top)) \leq \mu(s) + \alpha \cdot \mu^{\bullet}(1_{K \setminus L}) \leq \mu(s) + \varepsilon. \quad \square$$

8.17 Next we discuss *induced Radon integrals.*

Let Y be a subspace of X, measurable w.r.t. μ, and $j : Y \hookrightarrow X$ the canonical embedding.

DEFINITION. For any function g on Y, we denote by g_Y the function on X, equal to g on Y and to 0 on $X \setminus Y$. Let μ_Y be the functional defined on $\mathcal{S}(Y)$ by

$$\mu_Y(t) := \mu^{\bullet}(t_Y).$$

THEOREM. *The functional* μ_Y *is a Radon integral on* Y *with*

$$j(\mu_Y^{\bullet}) = 1_Y \cdot \mu^{\bullet}.$$

For any function f *on* X*, the restriction* $f_{|Y}$ *is essentially integrable, resp. measurable w.r.t.* μ_Y *iff* $1_Y \cdot f$ *is essentially integrable, resp. measurable w.r.t.* μ.

Note first that for $t \in \mathcal{S}(Y)$ the function t_Y is measurable by Theorem 8.13 and Corollary 8.12.(ii), since t has a largest lower semicontinuous extension h to X and

$$t_Y = 1_Y \cdot h.$$

Because t_Y is positive outside and lower bounded on a compact set (contained in Y), we have $\mu_*(t_Y) > -\infty$. Therefore, μ_Y is an increasing linear functional. To prove its regularity, let $\gamma \in \mathbb{R}$ with

$$\gamma < \mu_Y(t) = \mu^{\bullet}(t_Y) = \mu_*(t_Y)$$

be given, and choose $s \in \mathcal{S}(X)$ with $-s \leq t_Y$ and $-\mu(s) > \gamma$. There exist $K \in \mathfrak{K}(X)$ and $\alpha > 0$ such that

$$- \alpha \cdot 1_K \leq s \, , t_Y \, .$$

For any $\varepsilon > 0$ and $G \in \mathfrak{G}(X)$ with $G \supset K$ and $\mu^{\bullet}(G) < \infty$, the function

$$u := \min(s + \varepsilon, \alpha \cdot 1_G)_{|Y}$$

belongs to $\mathscr{S}(Y)$, since $s + \varepsilon$ vanishes outside the compact set $\{s \leq -\varepsilon\} \subset Y$. It satisfies

$$-u \leq t \quad \text{and} \quad - \mu_Y(u) > \gamma$$

for ε sufficiently small. Indeed,

$$u + t = \min(t_Y + s + \varepsilon, t_Y + \alpha \cdot 1_G)_{|Y} \geq 0 \, ,$$

and since

$$1_Y \cdot \min(s + \varepsilon, \alpha \cdot 1_G) \leq s + \varepsilon \cdot 1_G \, ,$$

we get

$$\mu_Y(u) = \mu^{\bullet}(1_Y \cdot \min[s + \varepsilon, \alpha \cdot 1_G]) \leq \mu^{\bullet}(s + \varepsilon \cdot 1_G) = \mu(s) + \varepsilon \cdot \mu^{\bullet}(G) \, .$$

For $s \in \mathscr{S}(X)$ we have

$$j(\mu_Y)(s) = \mu_Y(s \circ j) = \mu^{\bullet}(1_Y \cdot s) = 1_Y \cdot \mu \, (s) \, ,$$

hence $j(\mu_Y) = 1_Y \cdot \mu$. Since j is proper w.r.t. μ_Y and 1_Y is locally integrable w.r.t. μ , the assertions now follow from Theorems 8.15 and 8.16. □

8.18 Finally, we study inverse images of Radon integrals. We restrict our investigations to continuous mappings, in order that Example 7.8 be applicable.

THEOREM. *Let $p : Y \longrightarrow X$ be a continuous mapping from a Hausdorff space Y into X . There is a Radon integral ρ on Y with $p(\rho) = \mu$ such that p is proper w.r.t. ρ iff for any $K \in \mathfrak{K}(X)$ and any $\varepsilon > 0$ there exists $L \in \mathfrak{K}(Y)$ with $\mu^{\bullet}(K \smallsetminus p(L)) \leq \varepsilon$.*

The condition is necessary. Indeed, since p is proper w.r.t. ρ , we have $\rho^{\bullet}((p^{-1}(K)) < \infty$. Thus, by Corollary 8.10.(i), there exists $L \in \mathfrak{K}(Y)$ with

$$L \subset p^{-1}(K) \quad \text{and} \quad \rho^{\bullet}((p^{-1}(K) \smallsetminus L) \leq \varepsilon \, .$$

From Theorem 8.16 we then infer that

$$\mu^{\bullet}(K \smallsetminus p(L)) = \rho^{\bullet}(p^{-1}[K \smallsetminus p(L)]) \leq \rho^{\bullet}((p^{-1}(K) \smallsetminus L) \leq \varepsilon \, .$$

To prove that the condition is sufficient, let $\mathcal{T}$ denote the lattice cone of all functions $s \circ p$ on Y with $s \in \mathcal{S}(X)$. Since $s \circ p$ is lower semicontinuous, every function in $\mathcal{T}$ is $\mathcal{S}(Y)$–measurable.

Note that if $K \in \mathfrak{K}(X)$ is contained in $\complement p(Y)$, then for any $\varepsilon > 0$, there exists $L \in \mathfrak{K}(Y)$ with

$$\mu^{\bullet}(K) = \mu^{\bullet}(K \setminus p(L)) \leq \varepsilon ,$$

hence $\mu^{\bullet}(K) = 0$. For $r,s \in \mathcal{S}(X)$ with $r \circ p = s \circ p$, the set $\{r \neq s\}$ is measurable by Proposition 5.10.(ii) and contained in $\complement p(Y)$, hence $r = s$ essentially a.e. by Corollary 8.10.(i). This yields $\mu(r) = \mu(s)$ by Corollary 2.11, thus

$$\tau : s \circ p \longmapsto \mu(s)$$

is well–defined on $\mathcal{T}$. Obviously, τ is a regular linear functional on $\mathcal{T}$.

For $L \in \mathfrak{K}(Y)$, we have $p(L) \in \mathfrak{K}(X)$ and

$$-1_{p(L)} \circ p \leq -1_{L} .$$

For the existence of ρ, it is now sufficient to prove that τ is $\mathcal{S}(Y)$–tight. Indeed, from Example 7.8 we then infer the existence of a Radon integral ρ on Y which represents τ, i.e. for $s \in \mathcal{S}(X)$ we have

$$p(\rho)(s) = \rho^{\bullet}(s \circ p) = \tau(s \circ p) = \mu(s) .$$

The $\mathcal{S}(Y)$–tightness is by Remark 7.1.2 a consequence of the formula

$$\tau(t) = \inf_{u \in \mathcal{S}_{-}(Y),\ u \geq t} \tau_{*}(u) ,$$

which holds for all $t \in \mathcal{T}_{-}$. In fact, let $t = s \circ p$ with $s \in \mathcal{S}_{-}(X)$ and $\varepsilon > 0$ be given. For $K := \operatorname{supp}(s) \in \mathfrak{K}(X)$ and $\alpha > \|s\|_{\infty}$, there exists $L \in \mathfrak{K}(Y)$ with

$$\mu^{\bullet}(K \setminus p(L)) \leq \frac{\varepsilon}{\alpha} .$$

Then $u := \max(t, -\alpha \cdot 1_{L}) \in \mathcal{S}_{-}(Y)$ and

$$\tau_{*}(u) \leq \tau(t) + \varepsilon .$$

Indeed, for any $r \in \mathcal{S}(X)$ with $-r \circ p \leq u$, we have

$$-r \leq s \quad \text{on } p(L) ,$$

since $u = t$ on L, and hence

$$-r \leq 1_{p(L)} \cdot s \quad \text{on } p(Y) ,$$

since $u \leq 0$. As above, the set $\{-r > 1_{p(L)} \cdot s\}$ is an essentially null set, using Corollary 8.12.(ii). From

$$1_{p(L)} \cdot s \le s + \alpha \cdot 1_{K \setminus p(L)}$$

we infer that

$$-\tau(r \circ p) = -\mu(r) \le \mu^{\bullet}(1_{p(L)} \cdot s) \le \mu(s) + \alpha \cdot \mu^{\bullet}(K \setminus p(L)) \le \tau(t) + \varepsilon ,$$

which gives the result.

Finally, p is proper w.r.t. ρ, since μ is a Radon integral and

$$\rho^{\bullet}((p^{-1}(V)) = \rho^{\bullet}(1_V \circ p) = \mu(V)$$

for every open set V in X. □

An immediate consequence is the following

COROLLARY. *Let* μ *be bounded. There exists a bounded Radon integral* ρ *on* Y *with* $p(\rho) = \mu$ *iff there exists a sequence* (L_n) *in* $\mathfrak{K}(Y)$ *such that*

$$\mu^{*}(X \setminus \bigcup p(L_n)) = 0 .$$

§ 9 INTEGRALS AND DANIELL INTEGRATION

THROUGHOUT THIS SECTION,
μ IS AN INCREASING LINEAR FUNCTIONAL
ON A MIN-STABLE FUNCTION CONE $\mathcal{S}$.

9.1 In the present section, our aim is to construct an upper functional based on μ which has the convergence properties discussed in § 5. To this end, we must impose the Daniell property on μ. Because of Theorem 5.2, we first have to extend μ suitably to a min-stable function cone $\mathcal{T}$ with $\mathcal{S} \subset \mathcal{T} = \mathcal{T}_\sigma$ in order to get an upper integral. The smaller we can choose $\mathcal{T}$, the better the regularity properties, but the fewer the integrable functions. In contrast to the previous section where the stronger Bourbaki property was imposed on μ and led to an extension to $\mathcal{S}_\phi$, we now discuss this extension problem with the smallest such function cone

$$\mathcal{S}^\sigma$$

to obtain the Daniell theory of integration. We have $\mathcal{S}^\sigma \subset \mathcal{S}_{max\ \sigma}$, and if $\mathcal{S}$ is max-stable, then $\mathcal{S}^\sigma = \mathcal{S}_\sigma$.

The function cone $(\mathcal{S}^\sigma)_-$ coincides with $(\mathcal{S}_-)^\sigma$ and will be denoted by

$$\mathcal{S}^\sigma_- .$$

Obviously $(\mathcal{S}^\sigma)_- \supset (\mathcal{S}_-)^\sigma$. On the other hand, by Zorn's Lemma there exists a maximal min-stable function cone $\mathcal{T}$ with $\mathcal{S} \subset \mathcal{T} \subset \mathcal{S}^\sigma$ and such that $t_- \in (\mathcal{S}_-)^\sigma$ for all $t \in \mathcal{T}$. It is easy to see that $\mathcal{T}_\sigma$ has these properties, too. Therefore, $\mathcal{T}_\sigma$ coincides with $\mathcal{T}$, hence $\mathcal{T} = \mathcal{S}^\sigma$, and thus $(\mathcal{S}^\sigma)_- \subset (\mathcal{S}_-)^\sigma$. □

9.2 **THEOREM.** *If μ is regular, then the following assertions are equivalent:*

(i) *μ has the Daniell property.*

(ii) *For every increasing sequence* (s_n) *in* $\mathcal{S}_-$ *with* $\sup s_n = 0$, *we have*

$$\sup \mu(s_n) = 0 .$$

(iii) *The restriction* μ^σ *of* μ_* *to* $\mathcal{S}^\sigma$ *has the Daniell property.*

When these conditions are fulfilled, μ^σ *is the only regular linear extension of* μ *to* $\mathcal{S}^\sigma$ *which has the Daniell property.*

In view of $\mathcal{S}^\sigma \subset \mathcal{S}_{max\,\sigma}$, the only non-trivial implication is a consequence of the lemma below. Since

$$\mu^\sigma = \mu_* \le {\mu^\sigma}_* \le \mu^\sigma \quad \text{on } \mathcal{S}^\sigma ,$$

μ^σ is regular. If κ is any such extension of μ to $\mathcal{S}^\sigma$, then $\kappa^* \ge \kappa_* \ge \mu_*$. By Lemma 1.9.(i), the functionals κ^* and μ_* coincide on $\mathcal{S}_{max}$, and therefore κ and μ^σ coincide, since κ^* is an upper integral by Theorem 5.2. □

Lemma. *If* $\sup \mu(s_n) = 0$ *for every increasing sequence* (s_n) *in* $\mathcal{S}_-$ *with* $\sup s_n = 0$, *then* μ_* *has the Daniell property on* $\mathcal{S}_{max\,\sigma}$ *and is linear there.*

We first prove that

$$\mu_*(\sup t_n) = \sup \mu_*(t_n)$$

holds for every increasing sequence (t_n) in $\mathcal{S}_{max}$. Without loss of generality, we may assume that $\sup \mu_*(t_n) < \infty$. Since μ_* and μ^* coincide on $\mathcal{S}_{max}$ (Corollary 1.9), by Lemma 5.1 for $\varepsilon > 0$ there exists an increasing sequence (s_n) in $\mathcal{S}$ with $s_n \ge t_n$ and $\mu(s_n) \le \mu_*(t_n) + \varepsilon$. For $s \in \mathcal{S}$ with $-s \le \sup t_n$, the sequence $((s_n + s)_-)$ is increasing, contained in $\mathcal{S}_-$ and has 0 as upper envelope. Therefore,

$$0 = \sup \mu((s_n + s)_-) \le \sup \mu(s_n) + \mu(s)$$

and hence

$$-\mu(s) \le \sup \mu(s_n) \le \sup \mu_*(t_n) + \varepsilon ,$$

thus

$$\mu_*(\sup t_n) \le \sup \mu_*(t_n) .$$

Let now (s_n) be an increasing sequence in $\mathcal{S}_{max\,\sigma}$, and let $(s_{nm})_{m\in\mathbb{N}}$, for every $n \in \mathbb{N}$, be an increasing sequence in $\mathcal{S}_{max}$ with $s_n = \sup_m s_{nm}$. Then $t_n := \sup_{i,j\le n} s_{ij}$ defines an increasing sequence in $\mathcal{S}_{max}$ with $t_n \le s_n$ and $\sup t_n = \sup s_n$. Therefore,

$$\mu_*(\sup s_n) = \sup \mu_*(t_n) \le \sup \mu_*(s_n) \le \mu_*(\sup s_n) .$$

The linearity of μ_* on $\mathscr{S}_{max}$ (cf. Corollary 1.9) implies the linearity of μ_* on $\mathscr{S}_{max\,\sigma}$ by the Daniell property. □

REMARKS.

(1) For an increasing linear functional μ with the Daniell property, the classical construction of μ^σ is based on the max-stability of $\mathscr{S}$. In fact, setting

$$\mu^\sigma(\sup s_n) := \sup \mu(s_n)$$

defines unambiguously the only increasing (linear) extension of μ to $\mathscr{S}_\sigma$ with the Daniell property. By Theorem 5.2, $\mu^{\sigma*}$ is an upper integral, for which however, because of the missing regularity property, we cannot expect an effective theory of integration.

Without max-stability, it is just the regularity that allows us to extend μ uniquely to $\mathscr{S}_{max\,\sigma}$ conserving the Daniell property. The extension from $\mathscr{S}_{max}$ to $\mathscr{S}_{max\,\sigma}$ could also be done by the above process.

Without regularity, which is a substitute for the vector lattice situation, we are forced to make the detour through $\mathscr{S}_- - \mathscr{S}_-$ (cf. 9.5).

(2) Not every regular linear functional has the Daniell property. This is illustrated by Example 4.7.1 in which $\mathscr{S}$ is the vector lattice of all continuous real-valued functions on $]0,1]$ having a finite limit at 0, and μ is the positive linear form defined on $\mathscr{S}$ by $\mu(s) = \lim_{x\to 0+} s(x)$. The sequence

$$s_n := (n\cdot id - 1)_-$$

is increasing and has 0 as upper envelope, but $\mu(s_n) = -1$ for all $n \in \mathbb{N}$.

9.3 We are now going to incorporate the Daniell theory for min-stable function cones into our general framework of integration. The preceding theorem and 5.2 justify the following

DEFINITION. A regular linear functional μ having the Daniell property is called a *regular (Daniell) integral*. The upper integrals $\mu^{\sigma*}$ and $\mu^{\sigma\bullet}$ will be denoted respectively by

$$\int^* \cdot d\mu \quad \text{and} \quad \int^\bullet \cdot d\mu ,$$

and are called the *Daniell* and *essential Daniell upper integrals*. The corresponding Daniell lower integrals are denoted analogously.

A function is said to be *μ-integrable, essentially μ-integrable* or *μ-measurable* if it is respectively integrable, essentially integrable or measurable w.r.t. $\int^* \cdot d\mu$, i.e. w.r.t. μ^σ. The corresponding sets of functions are denoted respectively by

$$\mathscr{L}^*(\mu) \ , \ \mathscr{L}^\bullet(\mu) \quad \text{and} \quad \mathscr{L}^0(\mu).$$

If f is μ-integrable or essentially μ-integrable, then $\int^* f\, d\mu$ resp. $\int^\bullet f\, d\mu$ is called the *(Daniell) integral* resp. the *essential (Daniell) integral* of f.

Proposition. *If μ is a regular integral, then $\int^* \cdot d\mu$ is an auto-determined upper integral, and $\mathscr{S}_- \subset \mathscr{L}^*_-(\mu)$ is almost coinitial in $\overline{\mathbb{R}}^X$ w.r.t. $\int^* \cdot d\mu$ and $\int^\bullet \cdot d\mu$. Moreover, every $\mathscr{S}_{max\,\sigma}$-measurable function is μ-measurable. We have*

$$\int_* f\, d\mu = \int_\bullet f\, d\mu \quad \textit{for every function } f \textit{ with } \int_* f\, d\mu > -\infty ,$$

and

$$\int_\bullet f\, d\mu = \int^\bullet f\, d\mu \quad \textit{whenever } f \in \mathscr{L}^0(\mu) \textit{ and } \int_\bullet f\, d\mu > -\infty .$$

This is an immediate consequence of Proposition 6.1 and Remark 6.1.3, applied to the regular integral μ^σ, since $\mathscr{S}_-$ is coinitial in $\mathscr{S}^\sigma_-$. □

Corollary. *If μ is a regular integral, then*

$$\int^* \cdot d\mu \le \mu^* \quad \textit{and} \quad \int^\bullet \cdot d\mu \le \mu^\bullet ,$$

in particular

$$\mathscr{R}^*(\mu) \subset \mathscr{L}^*(\mu) \quad \textit{and} \quad \mathscr{R}^\bullet(\mu) \subset \mathscr{L}^\bullet(\mu) ,$$

and all functions measurable w.r.t. μ are μ-measurable, i.e. $\mathscr{R}^0(\mu) \subset \mathscr{L}^0(\mu)$.

Since μ^σ extends μ, we have $\int^* \cdot d\mu \le \mu^*$. This in turn coupled with Theorem 3.6.(ii) implies the inequality $\int^\bullet \cdot d\mu \le \mu^\bullet$, and with Proposition 4.3 the inclusion $\mathcal{R}^0(\mu) \subset \mathcal{L}^0(\mu)$, since $\mathcal{S}_-$ is almost coinitial in both $\mathcal{R}^*_-(\mu)$ and $\mathcal{L}^*(\mu)$. □

REMARKS.

(1) We use the prefix "μ–" to emphasize that the corresponding concept is used in the *theory of integration in the sense of Daniell* (i.e. w.r.t. $\int^* \cdot d\mu$) and not in the abstract Riemann sense (i.e. w.r.t. μ^*). Accordingly, we use the term *μ–null set* instead of null set w.r.t. $\int^* \cdot d\mu$, etc.

(2) In the Daniell theory, in addition to the general results of § 2 – § 4 , the convergence theorems of § 5 are available.

A μ–measurable function f is μ–integrable, resp. essentially μ–integrable, iff the upper integral $\int^* |f|\, d\mu$, resp. $\int^\bullet |f|\, d\mu$, is finite.

The μ–measurability can be described in the usual set–theoretic terms if 1 is measurable (cf. Theorem 13.6.(iv) and Corollary 14.6).

(3) By definition, the Daniell integration theory for μ is the abstract Riemann integration theory for μ^σ. Here, our concept based on *function cones* instead of *function spaces* as elementary sets of functions (cf. Remark 1.2) proves to be extremely fruitful, since in general $\mathcal{S}^\sigma$ will only be a cone of functions. In contrast to the upper integral $\int^* \cdot d\mu$, the upper functional μ^* will in general fail to have the Daniell property, as the classical Riemann integral ι on $\mathbb{R}$ illustrates:

If (a_n) is an enumeration of $\mathbb{Q}$ and if $A_n := \{a_1,...,a_n\}$, then the increasing sequence (1_{A_n}) has the upper envelope $1_{\mathbb{Q}}$. We have $\sup \iota^*(A_n) = 0$, but $\iota^*(\mathbb{Q}) = \infty$.

That the classical Riemann theory is so impoverished has its reason in the missing stability property of the domain of ι (cf. Theorem 5.2), which however the domain of ι^σ possesses : The function space $\mathcal{E}$ of step functions differs from $\mathcal{E}_\sigma$, whereas

$$\mathcal{E}^\sigma = \mathcal{E}_\sigma = (\mathcal{E}^\sigma)_\sigma .$$

(4) If $\mathcal{S}$ is a lattice cone and μ a Bourbaki integral, then μ^ϕ extends μ^σ , hence we have

$$\mu^{\phi *} \le \mu^{\sigma *} \le \mu^*$$

and, by the same arguments as above,

$$\mu^{\phi \bullet} \le \mu^{\sigma \bullet} \le \mu^\bullet .$$

This shows that every function which is (essentially) integrable in the abstract Riemann or Daniell sense is also (essentially) integrable in the sense of Bourbaki and that the respective integrals coincide.

9.4 The following *Extension Theorem* translates the results of Corollary 6.2 into the Daniell theory. It will be generalized in 9.11.

PROPOSITION. *Let $\mathcal{T}$ be a min-stable cone of functions contained in $\mathcal{S}$, and let τ be a regular integral on $\mathcal{T}$.*

(i) *If $\mathcal{T}_{max} \subset \mathcal{S}$ and if κ is an increasing linear extension of τ to $\mathcal{S}$ with the Daniell property, then*

$$\int_* \cdot \, d\tau \le \kappa \le \int^* \cdot \, d\tau \quad \text{on } \mathcal{S} .$$

(ii) *If $\mathcal{T}^\sigma \subset \mathcal{S}$ and $\mathcal{S}_- \subset \mathcal{L}^*_-(\tau)$, then $\mu := \int_* \cdot \, d\tau = \int^\bullet \cdot \, d\tau$ is the only regular integral on $\mathcal{S}$ which extends τ , and we have*

$$\mu^\bullet = \int^\bullet \cdot \, d\mu = \int^\bullet \cdot \, d\tau .$$

For the proof of (i), note first of all that $\kappa = \tau_*$ on $\mathcal{T}_{max}$ by Lemma 1.9.(ii). Let $s \in \mathcal{S}$, $t \in \mathcal{T}^\sigma$ and (t_n) be an increasing sequence in $\mathcal{T}_{max}$ with

$\sup t_n = t$. If $t \geq s$, then $s = \sup_n (\min(s,t_n))$, and since $\min(s,t_n) \in \mathcal{S}$ we infer from Lemma 9.2 that

$$\kappa(s) = \sup \kappa(\min(s,t_n)) \leq \sup \kappa(t_n) = \sup \tau_*(t_n) = \tau_*(t) = \tau^\sigma(t) ,$$

hence $\kappa(s) \leq \int^* s\, d\tau$. If $-t \leq s$, then $0 = \sup (s + t_n)_-$, hence

$$0 = \sup \kappa((s + t_n)_-) \leq \kappa(s) + \sup \kappa(t_n) = \kappa(s) + \tau^\sigma(t)$$

and therefore $\int_* s\, d\tau \leq \kappa(s)$.

To prove (ii), we can apply Corollary 6.2.(ii) to τ^σ since

$$\mathcal{S}_- \subset \mathcal{L}^*_-(\tau) = \mathcal{R}^*_-(\tau^\sigma).$$

This yields that μ is the only regular linear extension of τ^σ , hence μ extends τ , and that $\mu^\bullet = \tau^{\sigma\bullet} = \int^\bullet \cdot\, d\tau$. Since μ has the Daniell property, μ is a regular integral.

To prove the asserted uniqueness, let κ be a regular integral on $\mathcal{S}$ extending τ . By Corollary 1.9, we have $\kappa_* = \tau_*$ on $\mathcal{T}_{max}$ and hence $\kappa_* = \tau_*$ on $\mathcal{T}_{max\ \sigma}$, as Lemma 9.2 shows. Therefore, κ is an extension of τ^σ . Thus $\kappa = \mu$.

Finally, for the regular extension μ^σ of τ^σ , we get by the same argument

$$\int^\bullet \cdot\, d\mu = \mu^{\sigma\bullet} = \tau^{\sigma\bullet} = \int^\bullet \cdot\, d\tau ,$$

since $\mathcal{S}^\sigma_- \subset \mathcal{R}^*_-(\tau^\sigma)$. □

9.5 We are now going to discuss the Daniell integration theory for non–regular μ . As in § 6, we have to consider the positive linear form $\tilde{\mu}$ on $\mathcal{S}_- - \mathcal{S}_-$. To simplify things, we introduce the following concept; it codifies a property which implies the Daniell property of $\tilde{\mu}$ and which is equivalent to the Daniell property of μ if μ is semiregular :

DEFINITION. The functional μ is said to be a *(Daniell) integral* if for all sequences (s_n) in $\mathcal{S}$ and (t_n) in $\mathcal{S}_-$ for which $(s_n - t_n)$ is increasing with $\sup (s_n - t_n) \in \mathcal{S}$, the equality

$$\mu(\sup\,[s_n - t_n]) = \sup\,[\mu(s_n) - \mu(t_n)]$$

holds.

REMARKS.

(1) *If* μ *is an integral, then* μ *has the Daniell property. If* $\mathcal{S} - \mathcal{S}_- \subset \mathcal{S}$ *, e.g. if* $\mathcal{S}$ *is a function space, then the converse holds.*

The integral is so defined that the positive linear form $\tilde{\mu}$ has the Daniell property on $\mathcal{S}_- - \mathcal{S}_-$, i.e. is a regular integral, iff μ is an integral on $\mathcal{S}_-$.

(2) Our next proposition shows that every regular integral in the sense of Definition 9.3 is an integral.

(3) In Example 13.5 we will show that the integral property cannot be replaced by the following stronger version of the Daniell property : For every increasing sequence (s_n) in $\mathcal{S}$ and every decreasing sequence (t_n) in $\mathcal{S}$ with $\inf t_n \le \sup s_n$, the inequality

$$\inf \mu(t_n) \le \sup \mu(s_n)$$

holds.

PROPOSITION.

(i) *Let* μ *be difference-regular. Then* μ *is an integral iff the restriction of* μ *to* $\mathcal{S}_-$ *is an integral.*

(ii) *Let* μ *be semiregular. Then* μ *is an integral iff* $\sup \mu(s_n) = 0$ *for every increasing sequence* (s_n) *in* $\mathcal{S}_-$ *with* $\sup s_n = 0$.

The conditions are obviously necessary for μ to be an integral. So we examine their sufficiency.

Let μ be a difference-regular integral on $\mathcal{S}_-$. If (s_n) in $\mathcal{S}$ and (t_n) in $\mathcal{S}_-$ are sequences such that $(s_n - t_n)$ is increasing with $s := \sup\,(s_n - t_n) \in \mathcal{S}$, then for all $u,v \in \mathcal{S}_-$ with $u - v \le s$ we have $\min(s_n + v,\, u + t_n) \in \mathcal{S}_-$, and

$$\min(s_n - t_n,\, u - v) + v = \min(s_n + v,\, u + t_n) - t_n$$

defines an increasing sequence with upper envelope u . Consequently,

$$\mu(u) = \sup_n\ [\mu(\min(s_n + v,\, u + t_n)) - \mu(t_n)] \le \sup_n\ [\mu(s_n) + \mu(v) - \mu(t_n)]\,,$$

hence

$$\mu(s) = \mu_{\times}(s) \leq \sup_n [\mu(s_n) - \mu(t_n)] \leq \mu(s) .$$

This concludes the proof of sufficiency in case (i).

In the second case, μ is difference–regular by Proposition 6.7, and μ^* coincides with $\mu^{\times}$ on $\bar{\mathbb{R}}^X_-$. It is therefore sufficient to prove that μ is an integral on $\mathscr{S}_-$. By Proposition 6.3, $\mu^{\times}$ is an upper functional and thus μ^* is submodular on $(\mathscr{S}_- - \mathscr{S}_-)_-$ (cf. Theorem 2.2.(ii)). Let $\varepsilon > 0$ and (s_n), (t_n) be sequences in $\mathscr{S}_-$ such that $(s_n - t_n)$ is increasing. Without loss of generality, we may assume that the upper envelope of this sequence is 0. By Lemma 5.1 there exists an increasing sequence (u_n) in $\mathscr{S}_-$ with $s_n - t_n \leq u_n$ and

$$\mu(u_n) \leq \mu^*(s_n - t_n) + \varepsilon = \mu^{\times}(s_n - t_n) + \varepsilon = \mu(s_n) - \mu(t_n) + \varepsilon .$$

Since

$$0 = \sup (s_n - t_n) \leq \sup u_n \leq 0 ,$$

we have by hypothesis

$$0 = \sup \mu(u_n) \leq \sup [\mu(s_n) - \mu(t_n)] + \varepsilon .$$

This proves the sufficiency part of (ii). □

EXAMPLE. *Let* $\mathscr{S}$ *be a lattice cone of lower semicontinuous functions on a Hausdorff space* X *and* μ *a semiregular linear functional on* $\mathscr{S}$ *. If* $\mu_{\bullet}$ *is finite on* $\mathscr{S}_-(X)$ *, e.g. if for every* $K \in \mathfrak{K}(X)$ *there exists* $s \in \mathscr{S}$ *with* $s \leq -1_K$ *, and if* μ *is* $\mathscr{S}(X)$*–tight, then* μ *is an integral.*

This follows exactly as in Proposition 8.2, replacing μ^* by $\mu^{\times}$ and using (ii) of the proposition. □

9.6 DEFINITION. For an arbitrary integral μ, we define μ–integration to be $\tilde{\mu}$–integration, i.e. Daniell integration w.r.t. the regular integral $\tilde{\mu}$. For the sake of clarity, the upper integral $\int^* \cdot d\tilde{\mu}$ and $\mathscr{L}^*(\tilde{\mu})$ will be denoted by

$$\int^{\times} \cdot d\mu \quad \text{and} \quad \mathscr{L}^{\times}(\mu) .$$

The corresponding lower integral will be denoted analogously. Apart from this notation, we will use the notations from 9.3, but with μ instead of $\tilde{\mu}$ whenever confusion is impossible.

This is justified by the fact that in the regular case there is no significant difference between the Daniell integration theories for μ and $\tilde{\mu}$. We will prove in Theorem 9.7 that the essential upper integrals

$$\left[\int^{*} \cdot d\mu\right]^{\bullet} \quad \text{and} \quad \left[\int^{\times} \cdot d\mu\right]^{\bullet} = \int^{\bullet} \cdot d\tilde{\mu}$$

coincide. We therefore use the notation $\int^{\bullet} \cdot d\mu$ for both of them, and $\int_{\bullet} \cdot d\mu$ for the corresponding essential lower integral.

PROPOSITION. *For every integral μ, the upper integral $\int^{\times} \cdot d\mu$ is auto-determined, and $\mathscr{S}_{-} \subset \mathscr{L}^{\times}_{-}(\mu)$ is almost coinitial in $\overline{\mathbb{R}}^{X}$ w.r.t. $\int^{\times} \cdot d\mu$ and $\int^{\bullet} \cdot d\mu$. Moreover, every $\mathscr{S}_{max\ \sigma}$-measurable function is μ-measurable. We have*

$$\int_{\times} f\, d\mu = \int_{\bullet} f\, d\mu \quad \textit{for every function } f \textit{ with } \int_{\times} f\, d\mu > -\infty\,,$$

and

$$\int_{\bullet} f\, d\mu = \int^{\bullet} f\, d\mu \quad \textit{whenever } f \in \mathscr{L}^{0}(\mu) \textit{ and } \int_{\bullet} f\, d\mu > -\infty\,.$$

Furthermore,

$$\int^{\times} \cdot d\mu \quad \textit{on} \quad \mathscr{L}^{0}(\mu) \cap \left\{\int_{\times} \cdot d\mu > -\infty\right\}^{\sim}$$

and

$$\int^{\bullet} \cdot d\mu \quad \textit{on} \quad \mathscr{L}^{0}(\mu) \cap \left\{\int_{\bullet} \cdot d\mu > -\infty\right\}^{\sim}$$

are integrals.

With the exception of the integral properties, all this follows by application of Proposition 9.3 to $\tilde{\mu}$, upon considering the coinitiality of $\mathscr{S}_{-}$ in $(\mathscr{S}_{-} - \mathscr{S}_{-})_{-}$.

The integral properties are an immediate consequence of Remark 9.5.1, using Corollary and Theorem 4.10 for the linearity. □

COROLLARY. *For every integral* μ *, we have*

$$\int^{\times} \cdot d\mu \le \mu^{\times} \quad \text{and} \quad \int^{\bullet} \cdot d\mu \le \mu^{\bullet} ,$$

as well as

$$\mu_{\times}(s) = \int^{\bullet} s \, d\mu \le \mu(s) \le \int^{\times} s \, d\mu \quad \text{for } s \in \mathcal{S} .$$

In particular,

$$\mathcal{R}^{\times}(\mu) \subset \mathcal{L}^{\times}(\mu) \quad \text{and} \quad \mathcal{R}^{\bullet}(\mu) \subset \mathcal{L}^{\bullet}(\mu) ,$$

and all functions measurable w.r.t. μ *are* μ*-measurable, i.e.*

$$\mathcal{R}^{0}(\mu) \subset \mathcal{L}^{0}(\mu) .$$

For the first and last part, we apply Corollary 9.3 to $\tilde{\mu}$. By Proposition 6.3, for $s \in \mathcal{S}$ we have the inequalities

$$\mu_{\times}(s) \le \int_{\times} s \, d\mu \le \int^{\bullet} s \, d\mu \le \mu^{\bullet}(s) = \mu_{\times}(s) \le \mu(s) .$$

To prove the last inequality claimed in the corollary, we may assume that $\int^{\times} s \, d\mu < \infty$. But then there are sequences (s_n) , (t_n) in $\mathcal{S}_-$ such that $(s_n - t_n)$ is increasing and $s \le \sup (s_n - t_n)$. Since

$$s = \sup_n \min(s_n - t_n, s) = \sup_n \, [\min(s_n, s + t_n) - t_n]$$

and $\min(s_n, s + t_n) \in \mathcal{S}_-$, we have

$$\mu(s) = \sup_n \, [\mu(\min(s_n, s + t_n)) - \mu(t_n)] = \sup_n \int^{\times} \min(s_n - t_n, s) \, d\mu = \int^{\times} s \, d\mu ,$$

μ being an integral and $\int^{\times} \cdot d\mu$ being an upper integral. □

9.7 THEOREM. *If* μ *is a regular integral, then the essential upper integrals* $\left[\int^{*} \cdot d\mu\right]^{\bullet}$ *and* $\left[\int^{\times} \cdot d\mu\right]^{\bullet}$ *coincide. With the common notation* $\int^{\bullet} \cdot d\mu$ *for both of them, the inequalities*

$$\int^{\bullet} \cdot d\mu \le \int^{*} \cdot d\mu \le \int^{\times} \cdot d\mu$$

hold. In particular, the theories of essential μ- *and* $\tilde{\mu}$-*integration and the concepts of measurability coincide, and* $\mathscr{L}^{\times}(\mu) \subset \mathscr{L}^{*}(\mu)$ *with coincidence of the integrals. Furthermore,*

$$\int_{\times} f\,d\mu = \int_{*} f\,d\mu \quad \textit{whenever} \quad \int_{\times} f\,d\mu > -\infty\,,$$

and

$$\int^{*} \cdot\,d\mu \quad \textit{on} \quad \mathscr{L}^{0}(\mu) \cap \{\int_{*} \cdot\,d\mu > -\infty\}^{\sim}$$

is an integral.

To prove that $\int^{*} \cdot\,d\mu \le \int^{\times} \cdot\,d\mu$, one only has to note that both upper integrals coincide on $(\mathscr{S}_{-} - \mathscr{S}_{-})_{\sigma}$. Since $\mathscr{S}_{-}$ is almost coinitial both in $\mathscr{L}^{*}_{-}(\mu)$ and in $\mathscr{L}^{\times}_{-}(\mu) = \mathscr{L}^{*}_{-}(\tilde{\mu})$, Theorem 3.6.(ii) yields $\left[\int^{*} \cdot\,d\mu\right]^{\bullet} \le \left[\int^{\times} \cdot\,d\mu\right]^{\bullet}$.

By Proposition 6.3 and Theorem 6.4, on $\mathscr{S}_{max}$ we have $\mu_{*} = \tilde{\mu}_{\bullet} = \tilde{\mu}^{\bullet}$, hence also $\mu_{*} = \tilde{\mu}^{\sigma\bullet}$ by Proposition 9.6. From Lemma 9.2 we infer that μ^{σ} coincides with $\tilde{\mu}^{\sigma\bullet}$ on $\mathscr{S}^{\sigma}$, which yields $\tilde{\mu}^{\sigma\bullet} \le \mu^{\sigma *}$. Again by Theorem 3.6.(ii), the latter inequality implies the inequality

$$\left[\int^{\times} \cdot\,d\mu\right]^{\bullet} = \tilde{\mu}^{\sigma\bullet\bullet} \le \left[\int^{*} \cdot\,d\mu\right]^{\bullet},$$

since $\mathscr{S}_{-}$ is almost coinitial in $\mathscr{L}^{\bullet}_{-}(\tilde{\mu})$.

The last assertions now follow from Propositions 9.6 and 9.3, and from Remark 9.5.1 and Corollary 4.10. □

9.8 For an arbitrary integral μ and every μ-measurable function f , according to Proposition 9.6, the formula $\int_{\times} f\,d\mu = \int^{\bullet} f\,d\mu$ holds whenever $\int_{\times} f\,d\mu > -\infty$ .This result may be interpreted as a regularity property w.r.t. the function cone $(\mathscr{S}_{-} - \mathscr{S}_{-})_{\sigma}$. To get stronger results, we must postulate as in § 6 that μ is semiregular. If we do so, we can prove regularity properties w.r.t. $\mathscr{S}^{\sigma}$

by extending μ to μ^σ on $\mathcal{S}^\sigma$ in analogy with the regular case and proving that the essential Daniell integration theory for μ is equivalent to the essential abstract Riemann integration theory for μ^σ (cf. Theorem 9.10). In view of Theorem 6.4, the following definition of μ^σ is in harmony with the one given in the regular case :

PROPOSITION. *For every integral μ, we have*

$$\mu_\times = \int^\bullet \cdot \, d\mu \quad \text{on } \mathcal{S}_{max\ \sigma} .$$

The restriction μ^σ of $\mu_\times$ to $\mathcal{S}^\sigma$ is the only difference-regular integral which coincides with μ on $\mathcal{S}_-$. Moreover, μ^σ is an extension of μ iff μ is difference-regular.

By Corollary 9.6, $\mu_\times$ and $\int^\bullet \cdot \, d\mu$ coincide on $\mathcal{S}$ and hence on $\mathcal{S}_{max}$ by Lemma 1.9.(ii). If (t_n) is an increasing sequence in $\mathcal{S}_{max}$, then

$$\int^\bullet \sup t_n \, d\mu = \sup \int^\bullet t_n \, d\mu = \sup \mu_\times(t_n) \leq \mu_\times(\sup t_n) \leq \int^\bullet \sup t_n \, d\mu ,$$

which proves the formula.

Since μ^σ and μ coincide on $\mathcal{S}_-$, we have

$$\mu^\sigma = \mu_\times \leq \mu^\sigma{}_\times \leq \mu^\sigma \quad \text{on } \mathcal{S}^\sigma ,$$

i.e. μ^σ is difference-regular.

Moreover, μ^σ is an integral. Indeed, for sequences (s_n) in $\mathcal{S}^\sigma$ and (t_n) in $\mathcal{S}^\sigma_-$, for which $(s_n - t_n)$ increases to $s \in \mathcal{S}^\sigma$, we have

$$\mu^\sigma(s) = \int^\bullet s \, d\mu = \sup \int^\bullet (s_n - t_n) \, d\mu =$$

$$= \sup \left[\int^\bullet s_n \, d\mu - \int^\bullet t_n \, d\mu \right] = \sup[\mu^\sigma(s_n) - \mu^\sigma(t_n)] .$$

Finally, let κ be another difference-regular integral on $\mathcal{S}^\sigma$ which coincides with μ on $\mathcal{S}_-$. By Lemma 1.9.(ii), $\int^\bullet \cdot \, d\kappa$ coincides with $\int^\bullet \cdot \, d\mu$ on $\mathcal{S}_{-max}$, hence on $\mathcal{S}_{-max\ \sigma} \supset \mathcal{S}^\sigma_-$. Since both functionals are difference-regular, κ coincides with μ^σ on $\mathcal{S}^\sigma$. □

REMARK. *The equality*

$$\int^{\times} \cdot\, d\mu^{\sigma} = \int^{\times} \cdot\, d\mu$$

holds.

Indeed, both upper integrals coincide on $(\mathcal{S}^{\sigma}_{-} - \mathcal{S}^{\sigma}_{-})_{\sigma}$ and are determined by this set. □

9.9 **PROPOSITION.** *If μ is a semiregular integral, then μ^{σ} is semiregular.*

By Remark 6.7 it is sufficient to prove that μ^{σ} is semiregular on $\mathcal{S}^{\sigma}_{-}$. To this end, let $\mathcal{T} \subset \mathcal{S}^{\sigma}_{-}$ be a maximal min-stable cone of functions containing $\mathcal{S}_{-}$ and having the property that the restriction τ of μ^{σ} to $\mathcal{T}$ is semiregular. If we can prove that the restriction τ_{σ} of μ^{σ} to $\mathcal{T}_{\sigma}$ is semiregular, i.e. that the formula

$$\tau_{\sigma}(t) \leq \tau_{\sigma}(s) + \tau_{\sigma*}(t - s)$$

holds for all $s,t \in \mathcal{T}_{\sigma}$ with $s \leq t$, then we will have $\mathcal{T} = \mathcal{S}^{\sigma}_{-}$, and the proof will be complete.

It is sufficient to prove this inequality for $t \in \mathcal{T}$. Indeed, if (t_n) is an increasing sequence in $\mathcal{T}$ with $t = \sup t_n$, then the inequality for t_n implies

$$\begin{aligned}\tau_{\sigma}(t) = \sup \tau_{\sigma}(t_n) \leq \sup\, [\tau_{\sigma}(\min(s,t_n)) + \tau_{\sigma*}((t_n - s)^{+})] \leq \\ \leq \tau_{\sigma}(s) + \tau_{\sigma*}(t - s)\,.\end{aligned}$$

Therefore, let $t \in \mathcal{T}$ and (s_n) be an increasing sequence in $\mathcal{T}$ with upper envelope s. Since τ is semiregular, we have $\tau^{*} = \tau^{\times}$ on $\bar{\mathbb{R}}^{X}_{-}$ by Proposition 6.7. Thus τ^{*} is submodular on $(\mathcal{T} - \mathcal{T})_{-}$, and by Lemma 5.1, for any $\varepsilon > 0$ there exists an increasing sequence (u_n) in $\mathcal{T}$ with $u_n \geq s_n - t$ and

$$\tau(u_n) \leq \tau^{*}(s_n - t) + \varepsilon\,.$$

Then $u := \sup u_n$ belongs to $\mathcal{T}_{\sigma}$, $-u \leq t - s$, and

$$\tau_{\sigma}(u) = \sup \tau(u_n) \leq \sup \tau^{*}(s_n - t) + \varepsilon\,.$$

By the semiregularity of τ this yields

$$\tau(t) = \lim [\tau(s_n) + \tau_*(t - s_n)] = \tau_\sigma(s) - \sup \tau^*(s_n - t) \leq$$
$$\leq \tau_\sigma(s) - \tau_\sigma(u) + \varepsilon \leq \tau_\sigma(s) + \tau_{\sigma*}(t - s) + \varepsilon ,$$

which gives the inequality we seek. □

9.10 We now prove that for semiregular integrals the essential upper integral for μ coincides with the essential upper functional for μ^σ.

LEMMA. *Let μ be a semiregular integral. Then*

$$\int^\times f\, d\mu = \mu^{\sigma\times}(f)$$

for every function f with $\mu^{\sigma\times}(f) < \infty$, i.e. whenever f admits a majorant in $-\mathcal{S}_-$.

By Remark 9.8 and Proposition 9.6, it suffices to prove the inequality

$$\mu^{\sigma\times}(f) \leq \int^\times f\, d\mu$$

for those $f \in (\mathcal{S}_- - \mathcal{S}_-)_\sigma$ satisfying $f \leq -s$ for some $s \in \mathcal{S}_-$. Moreover, we may suppose that $f \leq 0$. Indeed, if the assertion is proved for negative f , by Proposition 2.7 we have

$$\mu^{\sigma\times}(f) + \mu(s) = \mu^{\sigma\times}(f + s) \leq \int^\times (f + s)\, d\mu = \int^\times f\, d\mu + \mu(s) ,$$

which implies the general assertion since $\mu^\sigma(s) = \mu(s) = \int^\times s\, d\mu$ is finite.

Suppose therefore that $f \leq 0$. Then there exists an increasing sequence (t_n) in $(\mathcal{S}_- - \mathcal{S}_-)_-$ with $f = \sup_n t_n$. On this set, μ^* coincides with $\mu^\times$ by Proposition 6.7 and is therefore submodular. For any $\varepsilon > 0$, by Lemma 5.1 there exists an increasing sequence (s_n) in $\mathcal{S}_-$ with $t_n \leq s_n$ and

$$\mu(s_n) \leq \mu^*(t_n) + \varepsilon = \mu^\times(t_n) + \varepsilon = \int^\times t_n\, d\mu + \varepsilon .$$

This gives

$$\int^\times f\, d\mu = \sup \int^\times t_n\, d\mu \geq \sup \mu(s_n) - \varepsilon = \mu^\sigma(\sup s_n) - \varepsilon \geq \mu^{\sigma\times}(f) - \varepsilon ,$$

which is all that remained to be proved. □

THEOREM. *If* μ *is a semiregular integral, then* $\mu^{\sigma\bullet}$ *is an upper integral, more precisely*

$$\mu^{\sigma\bullet} = \int^{\bullet} \cdot \, d\mu \, .$$

Moreover,

$$\mu^{\sigma}_{*}(f) = \int_{\bullet} f \, d\mu$$

for every function $f \in \bar{\mathbb{R}}^{X}$ *with* $\mu^{\sigma}_{*}(f) > -\infty$.

The first assertion follows from the lemma and Theorem 3.6.(ii) with the set $\mathcal{S}_{-}$, which is almost coinitial w.r.t. both upper functionals. The second assertion is a consequence of Proposition 6.8, applied to the functional μ^{σ}, which by Proposition 9.9 is semiregular. □

9.11 Corresponding to Proposition 6.5 and Corollary 7.3, we now consider *extentions of Daniell integrals*. In the following, let $\mathcal{T}$ be a min-stable cone of functions contained in $\mathcal{S}$, and let τ be an integral on $\mathcal{T}$.

PROPOSITION.

(i) *For every integral* κ *on* $\mathcal{S}$ *which coincides with* τ *on* $\mathcal{T}_{-}$, *we have*

$$\int^{\times} \cdot \, d\kappa \le \int^{\times} \cdot \, d\tau \; , \quad \text{and} \quad \int_{\times} \cdot \, d\tau \le \kappa \le \int^{\times} \cdot \, d\tau \;\; \text{on } \mathcal{S} \, .$$

If τ *is difference-regular and* κ *extends* τ, *then* κ^{σ} *extends* τ^{σ}.

(ii) *If* $\mathcal{S}_{-} \subset \mathcal{L}^{\times}_{-}(\tau)$, *then* $\mu := \int_{\times} \cdot \, d\tau = \int^{\bullet} \cdot \, d\tau$ *is the smallest and the only difference-regular integral on* $\mathcal{S}$ *which coincides with* τ *on* $\mathcal{T}_{-}$, *whereas* $\int^{\times} \cdot \, d\tau$ *is the largest one. We have*

$$\int^{\times} \cdot \, d\mu = \int^{\times} \cdot \, d\tau \, .$$

The integrals $\tilde{\tau}^\sigma$ and $\tilde{\kappa}^\sigma$ coincide on $(\mathcal{T}_- - \mathcal{T}_-)_\sigma \subset (\mathcal{S}_- - \mathcal{S}_-)_\sigma$. This immediately implies the first inequality in (i), whereas the others follow from the first one and from Corollary 9.6.

If τ is difference–regular, then by Proposition 6.5.(i) we have

$$\tau = \tau_\times \le \kappa_\times \le \kappa = \tau \quad \text{on } \mathcal{T},$$

hence $\tau_\times = \kappa_\times$ on $\mathcal{T}_{max}$ by Lemma 1.9.(ii), since $\tau_\times$ on $\mathcal{T}_{max}$ and $\kappa_\times$ on $\mathcal{S}_{max} \supset \mathcal{T}_{max}$ are linear by Proposition 6.5.(ii), applied respectively to τ and κ. From Proposition 9.8 we infer that $\tau_\times = \kappa_\times$ on $\mathcal{T}_{max\,\sigma}$. In particular, κ^σ is an extension of τ^σ.

For the second part, we first note that by Corollary 9.6 the upper integral and the essential upper integral coincide with τ on $\mathcal{T}_-$ and that $\mathcal{S}$ consists of τ–measurable functions by Proposition 4.11, since $\mathcal{S}_- \supset \mathcal{T}_-$ is almost coinitial in $\mathcal{L}^\bullet_-(\tau)$.

By Proposition 9.6, μ and $\int^\times \cdot\, d\tau$ are integrals on $\mathcal{S}$. The rest follows from the first part. In fact, for any integral κ on $\mathcal{S}$ which coincides with τ on $\mathcal{T}_-$, we have $\tilde{\kappa}^\sigma = \int^\times \cdot\, d\tau$ on $(\mathcal{S}_- - \mathcal{S}_-)_\sigma$. This proves that $\int^\times \cdot\, d\kappa \ge \int^\times \cdot\, d\tau$ and gives $\int^\times \cdot\, d\kappa = \int^\times \cdot\, d\tau$. Therefore,

$$\mu = \int^\bullet \cdot\, d\tau = \int^\bullet \cdot\, d\kappa = \kappa_\times \quad \text{on } \mathcal{S}$$

by Corollary 9.6, which shows that κ is difference–regular iff $\kappa = \mu$. □

COROLLARY. *If $\mathcal{S}_- \subset \mathcal{L}^\bullet_-(\tau)$, then $\mu := \int_\bullet \cdot\, d\tau = \int^\bullet \cdot\, d\tau$ is the smallest and the only difference–regular $\mathcal{T}$–tight integral on $\mathcal{S}$ which coincides with τ on $\mathcal{T}_-$. We have*

$$\int^\bullet \cdot\, d\mu = \int^\bullet \cdot\, d\tau\,.$$

Furthermore, μ extends τ iff τ is difference–regular.

If τ is semiregular and $\mathcal{T}^\sigma_- \subset \mathcal{S}_-$, then μ is semiregular, and we have

$$\mu^\bullet = \int^\bullet \cdot d\tau .$$

Again, $\mathcal{S}$ consists of τ-measurable functions. Therefore, by Proposition 9.6, μ is an integral on $\mathcal{S}$. Applying Corollary 7.3 to $(\mathcal{T}_- - \mathcal{T}_-)_\sigma \subset (\mathcal{S}_- - \mathcal{S}_-)_\sigma$ and $\tilde{\tau}^\sigma$, we get the $(\mathcal{T}_- - \mathcal{T}_-)_\sigma$-tightness of the restriction of $\int^\bullet \cdot d\tau = \tilde{\tau}^{\sigma\bullet}$ to $(\mathcal{S}_- - \mathcal{S}_-)_\sigma$, hence the $\mathcal{T}$-tightness of μ since $\mathcal{T}_-$ is coinitial in $(\mathcal{T}_- - \mathcal{T}_-)_\sigma$.

Now suppose that κ is a $\mathcal{T}$-tight integral on $\mathcal{S}$ which coincides with τ on $\mathcal{T}_-$. We prove that $\tilde{\kappa}^\sigma$ is $\mathcal{T}$-tight and hence $(\mathcal{T}_- - \mathcal{T}_-)_\sigma$-tight. The $\mathcal{T}$-tightness of κ means that $\mathcal{T}_-$ is almost coinitial in $\mathcal{S}_-$ w.r.t. $\kappa^\times$. By Proposition 9.6, $\mathcal{S}_-$ is almost coinitial in $\mathcal{L}^\times_-(\kappa)$ w.r.t. $\int^\times \cdot d\kappa$ and, by Corollary 9.6, $\int^\times \cdot d\kappa$ coincides with $\kappa^\times$ on $\mathcal{S}_-$. This proves that $\mathcal{T}_-$ is almost coinitial in $\mathcal{L}^\times_-(\kappa)$. Since $\max(s,t) \in (\mathcal{S}_- - \mathcal{S}_-)_{\sigma-}$ for $s \in (\mathcal{S}_- - \mathcal{S}_-)_{\sigma-}$ and $t \in \mathcal{T}_-$, the asserted tightness obtains.

Corollary 7.3, applied to $(\mathcal{T}_- - \mathcal{T}_-)_{\sigma-} \subset (\mathcal{S}_- - \mathcal{S}_-)_{\sigma-}$ and the restriction of $\tilde{\tau}^\sigma$, now shows that

$$\int^\bullet \cdot d\kappa = \tilde{\kappa}^{\sigma\bullet} = \tilde{\tau}^{\sigma\bullet} = \int^\bullet \cdot d\tau ,$$

because $\tilde{\kappa}^\sigma = \tilde{\tau}^\sigma$ on $(\mathcal{T}_- - \mathcal{T}_-)_{\sigma-}$.

From this equality we infer that

$$\kappa \geq \kappa_\times = \int^\bullet \cdot d\kappa = \mu \quad \text{on } \mathcal{S} ,$$

and thus that κ is difference-regular iff $\kappa = \mu$. Furthermore, μ extends τ iff $\int^\bullet \cdot d\tau = \tau$ on $\mathcal{T}$, i.e. iff τ is difference-regular.

If τ is semiregular so is τ^σ by Proposition 9.9, and by Theorem 9.10 we have $\int^\bullet \cdot d\tau = \tau^{\sigma\bullet}$. By hypothesis, Corollary 7.3 can be applied to the cones $\mathcal{T}^\sigma$ and $\mathcal{S}$, since $\mathcal{T}^\sigma_- \subset \mathcal{S}_-$, and to τ^σ. This yields the semiregularity of μ and the equality $\mu^\bullet = \int^\bullet \cdot d\tau$. □

THEOREM. *If* τ *is regular, then for every integral* κ *on* $\mathcal{S}$ *extending* τ *we have*

$$\kappa^{\sigma *} \le \int^* \cdot d\tau \ , \quad \textit{in particular} \quad \int_* \cdot d\tau \le \kappa \le \int^* \cdot d\tau \ .$$

If moreover $\mathcal{S}_- \subset \mathcal{L}^*_-(\tau)$, *then* $\mu := \int_* \cdot d\tau = \int^\bullet \cdot d\tau$ *is the smallest and* $\int^* \cdot d\tau$ *the largest integral on* $\mathcal{S}$ *extending* τ .

If $\int_* \cdot d\tau$ *and* $\int^* \cdot d\tau$ *coincide on* $\mathcal{S}$, *then* μ *is the only integral on* $\mathcal{S}$ *extending* τ . *If moreover* $\mathcal{T}^\sigma \subset \mathcal{S}$, *then* μ *is regular and*

$$\mu^* = \int^* \cdot d\mu = \int^* \cdot d\tau \ .$$

We can apply Corollary 6.2.(i) to the regular functional τ^σ and its extension κ^σ . This immediately yields the first part. From Theorem 9.7 we know that $\int^* \cdot d\tau$ is an integral on $\mathcal{S}$, hence the rest follows from the corollary and the first part with the exception of the formulas. These follow from Corollary 6.2.(ii), applied to μ and μ^σ . In fact, by the corollary,

$$\mu^{\sigma\bullet} = \int^\bullet \cdot d\mu = \int^\bullet \cdot d\tau = \tau^{\sigma\bullet} \quad \text{on } \mathcal{S}^\sigma \ . \ \square$$

9.12 We conclude this section with a discussion of representation by integrals.

DEFINITION. Let τ be an increasing linear functional on a min-stable cone $\mathcal{T}$ of functions. We say that an integral μ on $\mathcal{S}$ is τ*-representing* if

$$\tau(t) = \int_\bullet t \, d\mu = \int^\bullet t \, d\mu$$

holds for all $t \in \mathcal{T}$, i.e. if τ is represented by the functional $\tilde{\mu}^\sigma$.

REMARK. *If an integral* μ *is* τ*-representing, then* τ *is an integral on* $\{\tau < \infty\} \subset \mathcal{L}^\bullet(\mu)$. *If moreover* τ *is difference-regular, then* τ *is an integral on* $\mathcal{T}$, *and* μ *is* τ^σ*-representing as well as* $\tilde{\tau}^\sigma$*-representing.*

A semiregular integral μ *is* τ*-representing iff* τ *is represented by the functional* μ^σ .

If τ *is an integral and* $\mathcal{S}_-$ *is almost coinitial to* $\mathcal{T}_-$ *w.r.t.* $\tau^\times$, *then* τ^σ *and* $\tilde{\tau}^\sigma$ *are* $\mathcal{S}$*-tight.*

The first part follows from Proposition 9.6. The second part is then a consequence of Proposition 9.5.(i). Moreover, $\int_\bullet \cdot d\mu = \int^\bullet \cdot d\mu$ on $\mathcal{T}_{max}$ by Lemma 1.9.(i), and hence on $\mathcal{T}_{max\,\sigma}$, since

$$\int^\bullet \sup t_n \, d\mu = \sup \int^\bullet t_n \, d\mu = \sup \int_\bullet t_n \, d\mu \le \int_\bullet \sup t_n \, d\mu \le \int^\bullet \sup t_n \, d\mu$$

for every increasing sequence (t_n) in $\mathcal{T}_{max}$. Furthermore, $\int^\bullet \cdot d\tau = \int^\bullet \cdot d\mu$ on $\mathcal{T}_{max}$ by Lemma 1.9.(ii), and hence on $\mathcal{T}_{max\,\sigma}$ and on $(\mathcal{T}_- - \mathcal{T}_-)_\sigma$ by the Daniell property.

The assertion concerning semiregularity follows from Theorem 9.10, whereas the last part follows from Remark 7.1.3. □

To get more information, in particular to prove the difference-regularity of a represented τ , we have to ensure that μ is $\mathcal{T}$-tight (cf. Theorem 7.2).

PROPOSITION. *Let* τ *be an integral such that* $\mathcal{S}_- \subset \mathcal{L}^\bullet_-(\tau)$ *and*

$$\max(t - u, s) \in (\mathcal{S}_- - \mathcal{S}_-)_{\sigma-}$$

for all $t,u \in \mathcal{T}_-$ *and* $s \in \mathcal{S}_-$.

Then the restriction of $\int_\bullet \cdot d\tau$ *to* $\mathcal{S}_-$ *is a* $\mathcal{T}$*-tight integral. Its difference-regular linear extension* μ *to* $\mathcal{S}$ *is an integral, which is* τ*-representing iff* τ *is difference-regular and* $\tilde{\tau}^\sigma$ *is* $\mathcal{S}$*-tight. In this case, we have*

$$\int^\bullet \cdot d\tau = \int^\bullet \cdot d\mu \quad \textit{and} \quad \mu = \int_\bullet \cdot d\tau = \int^\bullet \cdot d\tau \quad \textit{on } \mathcal{S} .$$

We can apply Corollary 7.5 to $\tilde{\tau}^\sigma$ and

$$\mathcal{V} := (\mathcal{S}_- - \mathcal{S}_-)_{\sigma-} \subset \mathcal{L}^\bullet_-(\tau) = \mathcal{R}^\bullet_-(\tilde{\tau}^\sigma),$$

since all functions in $(\mathcal{T}_- - \mathcal{T}_-)_{\sigma-}$ are $\mathcal{V}$-measurable. In fact, for all $t,u \in \mathcal{T}_-$ and $v \in \mathcal{V}$, we have

$$\max(t - u, v) = \max[\max(t - u, s), v] \in \mathcal{V} = \hat{\mathcal{V}}$$

for some $s \in \mathcal{S}_-$ with $s \leq v$.

Since $\tilde{\tau}^\sigma{}_\bullet = \int_\bullet \cdot d\tau = \tilde{\mu}^\sigma$ on $\mathcal{V}$, we now infer that $\tilde{\mu}^\sigma$ is $(\mathcal{T}_- - \mathcal{T}_-)_\sigma$-tight, hence $\mathcal{T}$-tight by coinitiality, and therefore that μ is $\mathcal{T}$-tight by Remark 7.1.3. It is an integral by Propositions 9.6 and 9.5.(i). Furthermore, $\tilde{\mu}^\sigma$ represents $\tilde{\tau}^\sigma$ iff $\tilde{\tau}^\sigma$ is $\mathcal{V}$-tight or $\mathcal{S}$-tight. From Theorem 7.2 we infer that $\tilde{\mu}^\sigma$ represents τ iff τ is difference-regular and $\tilde{\mu}^\sigma$ represents the restriction of τ to $\mathcal{T}_-$, i.e. represents $\tilde{\tau}^\sigma$ by the remark.

In this case, we have

$$\int^\bullet \cdot d\tau = \tilde{\tau}^{\sigma\bullet} = \tilde{\mu}^{\sigma\bullet} = \int^\bullet \cdot d\mu,$$

and since

$$\mu = \mu_\times = \int_\bullet \cdot d\mu = \int^\bullet \cdot d\mu \quad \text{on } \mathcal{S}$$

by Corollary 9.6, it follows that

$$\mu = \int_\bullet \cdot d\tau = \int^\bullet \cdot d\tau \quad \text{on } \mathcal{S}. \quad \square$$

9.13 In case of semiregularity, Theorem 9.10 allows us to apply Corollary 7.5 to τ^σ, avoiding differences !

THEOREM. *Let $\mathcal{S}$ be a lattice cone and τ a semiregular integral on $\mathcal{T}$ such that $\mathcal{S}_- \subset \mathcal{L}^\bullet_-(\tau)$. Suppose that all functions in $\mathcal{T}^\sigma_-$ are $\mathcal{S}$-measurable.*

Then the difference-regular linear extension μ to $\mathcal{S}$ of the restriction of $\int_\bullet \cdot d\tau$ to $\mathcal{S}_-$ is a semiregular integral. It is τ-representing iff τ^σ is $\mathcal{S}$-tight.

In this case, μ is the only semiregular τ-representing integral on $\mathcal{S}$ for *which μ^σ is $\mathcal{T}$-tight. We have*

$$\int^\bullet \cdot d\tau = \mu^\bullet = \int^\bullet \cdot d\mu\,, \quad \text{and} \quad \mu^\sigma = \int_\bullet \cdot d\tau = \int^\bullet \cdot d\tau \quad \text{on } \mathcal{S}_\sigma\,.$$

If furthermore $\tau^\sigma{}_*$ is finite on $\mathcal{S}_-$, then μ coincides with $\tau^\sigma{}_*$ on $\mathcal{S}$ and is the only semiregular τ-representing integral on $\mathcal{S}$.

We can apply Corollary 7.5 to τ^σ, which is a semiregular integral by Proposition 9.9 and, by Theorem 9.10, to $\mathcal{S}_-$ as well as to

$$\mathcal{S}_{\sigma-} \subset \mathcal{L}^\bullet_-(\tau) = \mathcal{R}^\bullet_-(\tau^\sigma)\,,$$

since all functions in $\mathcal{T}^\sigma_-$ are also $\mathcal{S}_\sigma$-measurable by Proposition 4.9.

This yields that μ, which coincides with $\tau^\sigma{}_\bullet = \int^\bullet \cdot d\tau$ on $\mathcal{S}_-$, is a semiregular integral by Propositions 9.6 and 9.5.(i). Since μ^σ coincides with $\int^\bullet \cdot d\tau = \tau^\sigma{}_\bullet$ on $\mathcal{S}_{\sigma-}$ by the Daniell property, we infer that μ and μ^σ are $\mathcal{T}^\sigma$-tight, i.e. $\mathcal{T}$-tight by coinitiality. Furthermore, the functional μ respectively μ^σ represents τ^σ iff τ^σ is $\mathcal{S}$-tight. By Remark 9.12, the integral μ is τ-representing iff the functional μ^σ represents τ, i.e. τ^σ. We then have

$$\int^\bullet \cdot d\tau = \tau^{\sigma\bullet} = \mu^\bullet \quad \text{and} \quad \tau^{\sigma\bullet} = \mu^{\sigma\bullet} = \int^\bullet \cdot d\mu$$

by Theorem 9.10, which proves the formulas.

The last assertions are now immediate consequences of Corollary 7.5. □

Remark.

(1) *If $\hat{\mathcal{S}}_- = \mathcal{S}_{\sigma-}$, e.g. $\mathcal{S} = \mathcal{S}_\sigma$, then the $\mathcal{S}$-measurability of all functions in $\mathcal{T}^\sigma_-$ is equivalent to*

$$\max(t,s) \in \mathcal{S}_{\sigma-}$$

for all $t \in \mathcal{T}_-$ and $s \in \mathcal{S}_-$.

The hypotheses of the theorem imply that $\mathcal{S}_-$ has to be close to $\mathcal{S}_{\sigma-}$, since $\mu^\bullet = \mu^{\sigma\bullet}$. Roughly speaking, the condition $\hat{\mathcal{S}}_- = \mathcal{S}_{\sigma-}$ is almost necessary.

EXAMPLE. *Let $\mathcal{T}$ be a lattice cone of lower semicontinuous functions on a Hausdorff space X such that*

$$\mathcal{S}_-(X) \subset \mathcal{T}_\sigma .$$

If τ is an $\mathcal{S}(X)$-tight semiregular linear functional on $\mathcal{T}$, then τ is an integral, and the restriction μ of $\int^\bullet \cdot d\tau$ to $\mathcal{S}(X)$ is the only semiregular τ-representing integral on $\mathcal{S}(X)$. Moreover, we have

$$\int^\bullet \cdot d\tau = \mu^\bullet .$$

Note that τ is an integral by Example 9.5, and that τ^σ is $\mathcal{S}(X)$-tight. Indeed, for $t \in \mathcal{T}_-$ and $s \in \mathcal{S}_-(X)$ we have

$$\tau^{\sigma\times}((t-s)_-) = \tau^{\sigma\times}(t - \max(s,t)) = \tau^\sigma(t) - \tau^\sigma(\max(s,t)) =$$
$$= \tau_\times(t) - \tau_\times(\max(s,t)) = \tau^\times((t-s)_-) ,$$

which proves the assertion by Remark 7.1.1 since $\mathcal{T}_-$ is coinitial in $\mathcal{T}_\sigma$. Furthermore, every integral on $\mathcal{S}(X)$ is $\mathcal{T}$-tight by Example 7.1.1. The assertion therefore follows from the theorem and Remark 1. □

REMARKS.

(2) To ensure that $\mathcal{S}_-(X) \subset \mathcal{T}_\sigma$, it is sufficient to assume that every open subset of X is a Lindelöf space, e.g. X is a Souslin space, and that $\mathcal{T}$ satisfies one of the conditions given in Proposition 8.5.

In fact, the hypothesis on X implies that $\mathcal{T}_\phi = \mathcal{T}_\sigma$ by *Bourbaki* [1974], TG IX, p. 76, Proposition 3, and TG IX, p. 59, Proposition 5. □

(3) The previous result has been proved in the context of Bourbaki integrals in Theorem 8.4. Note that Remark 8.4.2 is a consequence of the above example, since the finiteness of τ_* on $\mathcal{S}_-(X)$ implies that μ is $\mathcal{S}(X)$-bounded below.

9.14 PROPOSITION. *Let $\mathcal{S}$ be a lattice cone and suppose that $\mathcal{T}^\sigma$ consists of $\mathcal{S}$-measurable functions.*

If τ is a semiregular integral, such that the restriction of ${\tau^\sigma}_*$ to $\mathcal{S}_{\sigma-}$ is an integral, then the difference-regular linear extension μ to $\mathcal{S}$ of the restriction of ${\tau^\sigma}_*$ to $\mathcal{S}_-$ is a semiregular $\mathcal{T}$-tight integral. It is τ-representing iff τ^σ is $\mathcal{S}$-tight.

In this case, μ is the only semiregular τ-representing integral on $\mathcal{S}$.

We can apply Addendum 7.6 to the semiregular integral τ^σ, proceeding as in the proof of the preceding theorem. Note that μ^σ and ${\tau^\sigma}_*$ coincide on $\mathcal{S}_{\sigma-}$ by the Daniell property. □

CHAPTER III.

SET–THEORETICAL ASPECTS AND RADON MEASURES

§ 10 LATTICE–MEASURABLE FUNCTIONS

As already announced in Example 1.10.5, we are now going to treat the measure–theoretic standard example in detail. Together with § 11, § 12 and § 13 this section shows how classical abstract measure theory gets incorporated into our functional analytic framework.

10.1 A system $\mathfrak{K}$ of subsets of a set X is called a *lattice* if $\emptyset \in \mathfrak{K}$ and if $\mathfrak{K}$ is stable with respect to finite intersections and finite unions. We denote by $\mathfrak{K}^{\complement}$ the system of all complements $\complement K$ of sets $K \in \mathfrak{K}$. The system

$$\mathfrak{G}(\mathfrak{K})$$

of $\mathfrak{K}$*–open* sets consists of all subsets G of X with the property that $K \setminus G \in \mathfrak{K}$ for all $K \in \mathfrak{K}$.

$\mathfrak{G}(\mathfrak{K})$ is a lattice containing $\mathfrak{K}^{\complement}$. If $X \in \mathfrak{K}$, then $\mathfrak{G}(\mathfrak{K}) = \mathfrak{K}^{\complement}$. A set F belongs to $\mathfrak{G}(\mathfrak{K})^{\complement}$ iff F is *locally in* $\mathfrak{K}$, which means that $F \cap K \in \mathfrak{K}$ for all $K \in \mathfrak{K}$.

A system $\mathfrak{K}$ of subsets of a set X is said to be a *ring* if $\emptyset \in \mathfrak{K}$ and if $\mathfrak{K}$ is stable with respect to set–theoretic differences as well as finite unions. If moreover X belongs to $\mathfrak{K}$, then $\mathfrak{K}$ is said to be an *algebra.*

Every ring is a lattice. Furthermore, an algebra is a lattice, stable with respect to the formation of complements. If $\mathfrak{K}$ is a ring, then $\mathfrak{G}(\mathfrak{K})$ is the algebra (containing $\mathfrak{K}$) of all sets locally in $\mathfrak{K}$. If $\mathfrak{K}$ is an algebra, then $\mathfrak{G}(\mathfrak{K}) = \mathfrak{K}$.

For every lattice $\mathfrak{K}$, the lattices of all unions of (increasing), respectively intersections of (decreasing), sequences of sets from $\mathfrak{K}$ are denoted by $\mathfrak{K}_\sigma$ and $\mathfrak{K}_\delta$.

We always have

$$\mathfrak{G}(\mathfrak{K})_\sigma \subset \mathfrak{G}(\mathfrak{K}_\delta) = \mathfrak{G}(\mathfrak{K}_\delta)_\sigma .$$

If $\mathfrak{K}$ is a lattice (ring, etc.) with $\mathfrak{K} = \mathfrak{K}_\delta$ or $\mathfrak{K} = \mathfrak{K}_\sigma$, then $\mathfrak{K}$ is said to be a *δ-lattice* (*δ-ring*, etc.) or a *σ-lattice* (*σ-ring*, etc.) respectively.

For every δ-lattice $\mathfrak{K}$, the lattice $\mathfrak{G}(\mathfrak{K})$ is a σ-algebra.

Let $\mathfrak{K}$ be a lattice (ring, etc.) and $\mathfrak{E} \subset \mathfrak{K}$. If $\mathfrak{K}$ is the smallest lattice (ring, etc.) containing $\mathfrak{E}$, then $\mathfrak{E}$ is said to be a *generator of the lattice* (*ring*, etc.), and $\mathfrak{K}$ is said to be the *lattice* (*ring*, etc.) *generated by* $\mathfrak{E}$.

10.2 EXAMPLES.

(1) Let X be a topological space, and let $\mathfrak{F}(X)$ denote the lattice of closed subsets of X. Then $\mathfrak{G}(\mathfrak{F}(X))$ is the lattice $\mathfrak{G}(X)$ of open subsets of X.

(2) Let X be a Hausdorff space, and let $\mathfrak{K}(X)$ denote the lattice of compact subsets of X. Then $\mathfrak{G}(X)$ is contained in $\mathfrak{G}(\mathfrak{K}(X))$, and $\mathfrak{G}(\mathfrak{K}(X))$ defines the finest topology on X inducing the given one on every compact subset.

Conversely :

If X is locally compact, or if every point in X has a countable base of neighbourhoods, then every $\mathfrak{K}(X)$-open set is open.

To prove the last assertion, let $G \in \mathfrak{G}(\mathfrak{K}(X))$. If x belongs to the closure of $\complement G$, then in the first case there is a filter containing $\complement G$ and converging to x, i.e. there exists a compact neighbourhood K of x belonging to this filter. In the second case, there exists a sequence (x_n) in $\complement G$ converging to x, hence $K := \{x_n | n \in \mathbb{N}\} \cup \{x\}$ is compact. Since $K \cap \complement G = K \setminus G \in \mathfrak{K}(X)$ is closed, it follows that $x \in K \cap \complement G \subset \complement G$. This proves that $\complement G$ is closed. $\square$

THROUGHOUT THIS SECTION,
$\mathfrak{K}$ IS A LATTICE OF SETS.

10.3 DEFINITION. A function f is said to be *$\mathfrak{K}$-measurable* if, for all $K \in \mathfrak{K}$ and all $\gamma, \delta \in \mathbb{R}$ with $\gamma < \delta$, there exists a set $L \in \mathfrak{K}$ with

$$\{f \leq \gamma\} \cap K \subset L \subset \{f \leq \delta\} \cap K .$$

The set of all $\mathfrak{K}$-measurable functions will be denoted by

$$\mathcal{M}(\mathfrak{K}) .$$

A function f is said to be *$\mathfrak{K}$-semicontinuous* if $\{f > \gamma\} \in \mathfrak{G}(\mathfrak{K})$ for all $\gamma \in \mathbb{R}$, i.e. if

$$\{f \leq \gamma\} \cap K \in \mathfrak{K} \quad \text{for all } \gamma \in \mathbb{R} \text{ and } K \in \mathfrak{K} .$$

REMARKS.

(1) To prove $\mathfrak{K}$-measurability, one has only to consider the two cases

$$\gamma < \delta < 0 \quad \text{and} \quad 0 < \gamma < \delta .$$

(2) *A function f is $\mathfrak{K}$-measurable if f is $\mathfrak{K}$-semicontinuous, or more generally if for all $K \in \mathfrak{K}$, we have*

$$\{f \leq \gamma\} \cap K \in \mathfrak{K} , \quad \textit{respectively} \ \{f < \gamma\} \cap K \in \mathfrak{K} ,$$

for all but at most countably many $\gamma \in \mathbb{R}$.

(3) *If $\mathfrak{L}$ is a lattice with $\mathfrak{K} \subset \mathfrak{G}(\mathfrak{L})^{c}$ such that every set from $\mathfrak{L}$ is contained in some set from $\mathfrak{K}$, e.g. $\mathfrak{L} = \mathfrak{K}_{\delta}$, then $\mathfrak{G}(\mathfrak{K}) \subset \mathfrak{G}(\mathfrak{L})$, and $\mathfrak{K}$-measurability implies $\mathfrak{L}$-measurability.*

(4) *If $\mathfrak{K}$ is an algebra, then a function f is $\mathfrak{K}$-semicontinuous iff*

$$\{f > \gamma\} \in \mathfrak{K} \quad \textit{for all } \gamma \in \mathbb{R} .$$

This follows from $\mathfrak{G}(\mathfrak{K}) = \mathfrak{K}$. □

PROPOSITION. *$\mathfrak{K}$-measurable functions enjoy the following properties:*

(i) *$\mathcal{M}(\mathfrak{K})$ is a positively homogeneous lattice of functions containing the* *constants. A function* f *is measurable iff* f^+ *and* f_- *are each measurable.*

(ii) *If* $\mathfrak{K}$ *is an ring, then* $\mathcal{M}(\mathfrak{K})$ *is homogeneous.*

(iii) *If* $\mathfrak{K}$ *is a* δ*-lattice, then a function is* $\mathfrak{K}$*-measurable iff it is* $\mathfrak{K}$*-semicontinuous. Furthermore, we have*

$$\mathcal{M}(\mathfrak{K}) = \mathcal{M}(\mathfrak{K})_\sigma ,$$

and $\tilde{\mathcal{M}}(\mathfrak{K}) = \tilde{\mathcal{M}}(\mathfrak{K})_\sigma$ *is a lattice cone of functions.*

For any functions f and g and $\alpha,\gamma \in \mathbb{R}$ with $\alpha > 0$, the formulas

$$\{\alpha f \le \gamma\} = \{f \le \tfrac{\gamma}{\alpha}\} ,$$

$$\{\min(f,g) \le \gamma\} = \{f \le \gamma\} \cup \{g \le \gamma\} ,$$

and

$$\{\max(f,g) \le \gamma\} = \{f \le \gamma\} \cap \{g \le \gamma\}$$

prove the first part of (i). For the second part, we just have to note that

$$\{f \le \gamma\} = \{f^+ \le \gamma\} \quad \text{if } \gamma \ge 0$$

and

$$\{f \le \gamma\} = \{f_- \le \gamma\} \quad \text{if } \gamma < 0 .$$

Let now $\mathfrak{K}$ be a ring. If $f \in \mathcal{M}(\mathfrak{K})$, $K \in \mathfrak{K}$ and $\gamma,\delta \in \mathbb{R}$ with $\gamma < \delta$, then for $\gamma < \alpha < \beta < \delta$ we have

$$\{-f \le \gamma\} \subset \{-f < \alpha\} = \complement\{f \le -\alpha\}$$

and

$$\complement\{f \le -\beta\} = \{-f < \beta\} \subset \{-f \le \delta\} .$$

By hypothesis, there exists $L \in \mathfrak{K}$ with

$$\{f \le -\beta\} \cap K \subset L \subset \{f \le -\alpha\} \cap K ,$$

hence

$$\{-f \le \gamma\} \cap K \subset K \setminus \{f \le -\alpha\} \subset K \setminus L \subset K \setminus \{f \le -\beta\} \subset \{-f \le \delta\} \cap K .$$

Since $K \setminus L \in \mathfrak{K}$, this proves (ii).

Let now $\mathfrak{K}$ be a δ-lattice. By Remark 2, the condition in (iii) is always sufficient. Conversely, for any $K \in \mathfrak{K}$, $\gamma \in \mathbb{R}$ and $n \ge 1$ there exists $L_n \in \mathfrak{K}$ with

$$\{f \le \gamma\} \cap K \subset L_n \subset \{f \le \gamma + \tfrac{1}{n}\} \cap K .$$

Hence

$$K \setminus \{f > \gamma\} = K \cap \{f \le \gamma\} = \bigcap_{n \ge 1} K \cap \{f \le \gamma + \tfrac{1}{n}\} = \bigcap_{n \ge 1} L_n \in \mathfrak{K} ,$$

which proves that $\{f > \gamma\} \in \mathfrak{G}(\mathfrak{K})$, i.e. f is $\mathfrak{K}$-semicontinuous.

Furthermore, for any sequence (f_n) in $\mathcal{M}(\mathfrak{K})$ and $\gamma \in \mathbb{R}$ we have

$$\{\sup f_n > \gamma\} = \bigcup_{n \geq 1} \{f_n > \gamma\} \in \mathfrak{G}(\mathfrak{K})_\sigma = \mathfrak{G}(\mathfrak{K})$$

and

$$\{f_1 + f_2 > \gamma\} = \bigcup_{\gamma_1, \gamma_2 \in \mathbb{Q}, \gamma_1 + \gamma_2 > \gamma} \{f_1 > \gamma_1\} \cap \{f_2 > \gamma_2\} \in \mathfrak{G}(\mathfrak{K})$$

if $f_1, f_2 \in \tilde{\mathcal{M}}(\mathfrak{K})$. □

In Theorem and Remark 10.10 we will prove that a substantial subset of $\tilde{\mathcal{M}}(\mathfrak{K})$ is always a function cone.

COROLLARY. *If* $\mathfrak{K}$ *is a* δ*-ring, then together with every sequence* (f_n) *the functions* $\sup f_n$, $\inf f_n$, $\limsup f_n$, $\liminf f_n$ *and, if* (f_n) *converges pointwise in* $\bar{\mathbb{R}}$, *also* $\lim f_n$ *are* $\mathfrak{K}$*-measurable, as is the product of two* $\mathfrak{K}$*-measurable functions.*

If $\mathfrak{K}$ *is a* σ*-algebra, then a function* f *is* $\mathfrak{K}$*-measurable iff*

$$\{f > \gamma\} \in \mathfrak{K} \quad \textit{for all } \gamma \in \mathbb{R} .$$

The assertions for sequences follow immediately, since $\mathcal{M}(\mathfrak{K})$ is homogeneous and stable under the formation of upper envelopes of sequences. For the product of $\mathfrak{K}$-measurable functions f and g , we may therefore assume that f and g are finite, replacing for example f by $\operatorname{med}(f,n,-n)$. Since

$$\{h^2 > \gamma\} = \{|h| > \sqrt{\gamma}\}$$

for $\gamma \in \mathbb{R}_+$, the formula

$$f \cdot g = \tfrac{1}{4} \cdot [(f+g)^2 - (f-g)^2]$$

yields the result. The last assertion is a consequence of Remark 4. □

EXAMPLE. *Let* X *be a Hausdorff space. A function is* $\mathfrak{K}(X)$*-measurable iff it is* $\mathfrak{K}(X)$*-semicontinuous, i.e. lower semicontinuous with respect to the topology* $\mathfrak{G}(\mathfrak{K}(X))$.

10.4 LEMMA. *A function f which takes only a finite number of real values is $\mathfrak{K}$-measurable iff f is $\mathfrak{K}$-semicontinuous. We then have*

$$f = \sum_{j=0}^{p-1} (\omega_{j+1} - \omega_j)\cdot 1_{\{f>\omega_j\}} - \sum_{i=-n}^{-1} (\omega_{i+1} - \omega_i)\cdot 1_{\{f\leq\omega_i\}} ,$$

where

$$\omega_{-n} < \ldots < \omega_{-1} < \omega_0 := 0 < \omega_1 < \ldots < \omega_p$$

denote the finitely many values of f together with the possible value 0.

In particular, for any subset A the function -1_A , respectively 1_A , is $\mathfrak{K}$-measurable iff $A \in \mathfrak{G}(\mathfrak{K})^c$, respectively $A \in \mathfrak{G}(\mathfrak{K})$.

The $\mathfrak{K}$-semicontinuity is sufficient by Remark 10.3. Conversely, for any $\gamma \in \mathbb{R}$ there exists $\delta > \gamma$ with $\{f \leq \gamma\} = \{f \leq \delta\}$. Since for every $K \in \mathfrak{K}$ there exists $L \in \mathfrak{K}$ with

$$\{f \leq \gamma\} \cap K \subset L \subset \{f \leq \delta\} \cap K ,$$

this implies

$$K \setminus \{f > \gamma\} = \{f \leq \gamma\} \cap K = L \in \mathfrak{K}$$

and hence $\{f > \gamma\} \in \mathfrak{G}(\mathfrak{K})$.

The formula is obvious, and the last part follows from

$$\{-1_A \leq \gamma\} = \emptyset , A \text{ or } X$$

and

$$\{1_A > \gamma\} = X , A \text{ or } \emptyset . \quad \square$$

MEASURE-THEORETIC STANDARD EXAMPLE. We denote by

$$\mathscr{E}(\mathfrak{K})$$

the set of all *step functions* , i.e. the set of all $\mathfrak{K}$-measurable or $\mathfrak{K}$-semicontinuous functions which take only a finite number of real values and are positive outside a suitable set from $\mathfrak{K}$.

PROPOSITION. *$\mathscr{E}(\mathfrak{K})$ is the smallest lattice cone of functions containing -1_K and 1_G for $K \in \mathfrak{K}$ and $G \in \mathfrak{G}(\mathfrak{K})$. For any $e \in \mathscr{E}(\mathfrak{K})$ and $\gamma \in \mathbb{R}$, we have*

$$\{e > \gamma\} \in \mathfrak{G}(\mathfrak{K}) \ , \ \text{and} \ \ \{e \leq \gamma\} \in \mathfrak{K} \ \text{if} \ \gamma < 0 .$$

The function cone $\mathscr{E}(\mathfrak{K})$, respectively $\mathscr{E}_-(\mathfrak{K})$, consists of all functions of the type

$$\sum_{j=1}^{p} \beta_j \cdot 1_{G_j} - \sum_{i=1}^{n} \alpha_i \cdot 1_{K_i} \text{ , respectively } -\sum_{i=1}^{n} \alpha_i \cdot 1_{K_i} \text{ ,}$$

where $n,p \in \mathbb{N}$, $\alpha_i,\beta_j \in \mathbb{R}_+$, $K_i \in \mathfrak{K}$ *and* $G_j \in \mathfrak{G}(\mathfrak{K})$.

For $d,e \in \mathcal{E}(\mathfrak{K})$ and $\gamma \in \mathbb{R}$, by the lemma we have

$$\{d + e > \gamma\} = \bigcup_{\alpha,\beta \in \mathbb{R}\,,\,\alpha+\beta>\gamma} \{d > \alpha\} \cap \{e > \beta\} \in \mathfrak{G}(\mathfrak{K}) \,,$$

the union being finite, since d and e take only finitely many values. Obviously, also $\min(d,e)$, $\max(d,e)$ and $d + e$ take only finitely many real values and are positive outside a suitable set from $\mathfrak{K}$. By Proposition 10.3.(i), this proves that $\mathcal{E}(\mathfrak{K})$ is a lattice cone. The remaining assertions follow from the lemma, noting that if e is positive outside $K \in \mathfrak{K}$ and $\gamma < 0$, we have

$$\{e \le \gamma\} = \{e \le \gamma\} \cap K \in \mathfrak{K} \,. \quad \square$$

REMARKS.

(1) For technical reasons (cf. 11.2 and 12.3), we shall need more general representations of $e \in \mathcal{E}_-(\mathfrak{K})$, called *pyramidal decompositions* , than those given in the lemma :

Let $(\gamma_i)_{i\in\mathbb{N}}$ be a strictly decreasing sequence of real numbers with $\gamma_0 = 0$ and containing all elements of $e(X)$. If we define

$$\delta_i := \gamma_{i-1} - \gamma_i > 0 \quad \text{and} \quad K_i := \{e \le \gamma_i\}$$

for $i > 0$, then $(K_i)_{i>0}$ is a decreasing sequence in $\mathfrak{K}$. It is easy to see that

$$K_i = \{e \le \omega_j\} \quad \text{for} \quad \omega_j \le \gamma_i < \omega_{j+1}$$

and that

$$e = -\sum_{i>0} \delta_i \cdot 1_{K_i} \,.$$

One has to note that there is only a finite number of terms since $K_i = \emptyset$ for all sufficiently large i .

(2) *If* $\mathfrak{K}$ *is a ring, then*

$$\mathcal{E}_-(\mathfrak{K}) - \mathcal{E}_-(\mathfrak{K}) \subset \mathcal{E}(\mathfrak{K}) \,.$$

Equality holds iff $\mathfrak{K}$ *is an algebra.*

In fact, the inclusion (resp. the equality) follows from the proposition since $\mathfrak{K} \subset \mathfrak{G}(\mathfrak{K})$ (resp. $\mathfrak{K} = \mathfrak{G}(\mathfrak{K})$). Conversely, if equality holds, then $\mathcal{E}(\mathfrak{K})$ is a vector lattice, hence for any $K \in \mathfrak{K}$ we have

$$-1_{\complement K} = -1 + 1_K \in \mathcal{E}_-(\mathfrak{K})$$

which implies

$$\complement K = \{-1_{\complement K} \leq -1\} \in \mathfrak{K}$$

and proves that $\mathfrak{K}$ is an algebra. $\square$

10.5 **DEFINITION.** Two lattices $\mathfrak{K}$ and $\mathfrak{G}$ are said to be *compatible* if

$$K \setminus G \in \mathfrak{K} \text{ and } G \setminus K \in \mathfrak{G}$$

for all $K \in \mathfrak{K}$ and $G \in \mathfrak{G}$.

This concept enables us to treat set-theoretic regularity conditions by functional analytic ones via the following refinement of the measure-theoretic standard example :

Let $\mathfrak{G}$ be a lattice compatible with $\mathfrak{K}$, and let

$$\mathcal{E}(\mathfrak{K},\mathfrak{G})$$

denote the set of all functions $e \in \mathcal{E}(\mathfrak{K})$ such that

$$\{e > \gamma\} \in \mathfrak{G} \quad \text{for all } \gamma \geq 0 .$$

PROPOSITION. *$\mathcal{E}(\mathfrak{K},\mathfrak{G})$ is the smallest lattice cone of functions containing -1_K and 1_G for $K \in \mathfrak{K}$ and $G \in \mathfrak{G}$.*

For any $d,e \in \mathcal{E}(\mathfrak{K},\mathfrak{G})$ and any $\gamma \geq 0$, $\delta > 0$, the formulas

$$\{\delta d > \gamma\} = \{d > \tfrac{\gamma}{\delta}\},$$

$$\{\min(d,e) > \gamma\} = \{d > \gamma\} \cap \{e > \gamma\},$$

and

$$\{\max(d,e) > \gamma\} = \{d > \gamma\} \cup \{e > \gamma\}$$

prove that $\mathcal{E}(\mathfrak{K},\mathfrak{G})$ is a positively homogeneous lattice. Since

$$\{d + e > \gamma\} = \bigcup_{\alpha,\beta \in \mathbb{R},\, \alpha+\beta>\gamma} \{d > \alpha\} \cap \{e > \beta\}$$

is a finite union, $\mathcal{E}(\mathfrak{K},\mathfrak{G})$ is a lattice cone. In fact, we have just to note that $\alpha > 0$

if $\beta < 0$ and that in this case

$$\{d > \alpha\} \cap \{e > \beta\} = \{d > \alpha\} \setminus \{e \le \beta\} \in \mathfrak{G},$$

since $\mathfrak{G}$ is compatible with $\mathfrak{K}$. □

Remarks.

(1) *A function* e *which takes only a finite number of real values belongs to* $\mathcal{E}(\mathfrak{K},\mathfrak{G})$ *iff*

$$\{e > \gamma\} \in \mathfrak{G} \quad \textit{for all } \gamma \ge 0 \quad \textit{and} \quad \{e \le \gamma\} \in \mathfrak{K} \quad \textit{for all } \gamma < 0 .$$

$\mathcal{E}(\mathfrak{K},\mathfrak{G})$ *consists of all functions of the type*

$$\sum_{j=1}^{p} \beta_j \cdot 1_{G_j} - \sum_{i=1}^{n} \alpha_i \cdot 1_{K_i}$$

where $n,p \in \mathbb{N}$, $\alpha_i,\beta_j \in \mathbb{R}_+$, $K_i \in \mathfrak{K}$ *and* $G_j \in \mathfrak{G}$.

This follows from the proposition and Lemma 10.4. □

(2) *The lattice* $\{\emptyset\}$ *is the smallest and* $\mathfrak{G}(\mathfrak{K})$ *is the largest lattice which is compatible with* $\mathfrak{K}$. *We have*

$$\mathcal{E}(\mathfrak{K},\mathfrak{G}) \subset \mathcal{E}(\mathfrak{K}) = \mathcal{E}(\mathfrak{K},\mathfrak{G}(\mathfrak{K}))$$

and

$$\mathcal{E}\_(\mathfrak{K}) = \mathcal{E}\_(\mathfrak{K},\mathfrak{G}) = \mathcal{E}(\mathfrak{K},\{\emptyset\}).$$

This follows from

$$L \setminus (G \setminus K) = (L \setminus G) \cup (L \cap K) \in \mathfrak{K}$$

for all $G \in \mathfrak{G}(\mathfrak{K})$ and $K,L \in \mathfrak{K}$, from Proposition 10.4 and Remark 1. □

(3) *Every ring* $\mathfrak{K}$ *is compatible with itself, and*

$$\mathcal{E}(\mathfrak{K},\mathfrak{K}) = \mathcal{E}\_(\mathfrak{K}) - \mathcal{E}\_(\mathfrak{K})$$

is the vector lattice of all usual step functions on $\mathfrak{K}$.

(4) *If* $\mathfrak{K}$ *and* $\mathfrak{G}$ *are compatible lattices, so are* $\mathfrak{K}_\delta$ *and* $\mathfrak{G}_\sigma$, *and*

$$\mathcal{E}(\mathfrak{K}_\delta,\mathfrak{G}_\sigma)_\sigma = \mathcal{E}(\mathfrak{K},\mathfrak{G})_\sigma .$$

The formula follows immediately from

$$-1_{\bigcap K_n} = \sup -1_{K_n} \quad \text{and} \quad 1_{\bigcup G_n} = \sup 1_{G_n} . \quad \square$$

10.6 Let $\mathcal{S}$ be a lattice cone of functions. The lattices of the sets

$$\{s \leq -1\} \quad \text{and} \quad \{s > 1\}$$

with $s \in \mathcal{S}$ are respectively denoted by

$$\mathfrak{K}(\mathcal{S}) \quad \text{and} \quad \mathfrak{G}(\mathcal{S}) .$$

The following Stonian condition is sufficient to prove the compatibility of the above lattices :

DEFINITION. A lattice cone $\mathcal{S}$ of functions is said to be *Stonian* if

$$\max(s,-1) , \min(s,1) \in \mathcal{S}$$

for all $s \in \mathcal{S}$.

Obviously, this definition is in harmony with the one given for vector lattices in 8.7.

THEOREM. *If $\mathcal{S}$ is Stonian then $\mathfrak{K}(\mathcal{S})$ and $\mathfrak{G}(\mathcal{S})$ are compatible.*

For all $s,t \in \mathcal{S}$ we have

$$\{s \leq -1\} \setminus \{t > 1\} = \{s \leq -1\} \cap \{t - 2 \leq -1\} =$$
$$= \{s \leq -1\} \cap \{t + 2 \cdot \max(s,-1) \leq -1\} \in \mathfrak{K}(\mathcal{S})$$

and

$$\{s > 1\} \setminus \{t \leq -1\} = \{s > 1\} \cap \{t + 2 \cdot \min(s,1) > 1\} \in \mathfrak{G}(\mathcal{S}) . \quad \square$$

REMARKS.

(1) *We always have*

$$\mathfrak{K}(\mathcal{S}_\sigma) = \mathfrak{K}(\mathcal{S})_\delta \quad \textit{and} \quad \mathfrak{G}(\mathcal{S}_\sigma) = \mathfrak{G}(\mathcal{S})_\sigma .$$

This follows immediately from the formulas

$$\{\sup s_n \leq -1\} = \bigcap \{s_n \leq -1\} \quad \text{and} \quad \{\sup s_n > 1\} = \bigcup \{s_n > 1\}$$

for $s_n \in \mathcal{S}$. □

(2) *If $\mathcal{S}$ is a vector lattice, then the conditions*

$$\max(s,-1) \in \mathcal{S} \ \textit{for all} \ s \in \mathcal{S},$$

$$\min(s,1) \in \mathcal{S} \ \textit{for all} \ s \in \mathcal{S},$$

and

$$(s+1)_- \in \mathcal{S} \ \textit{for all} \ s \in \mathcal{S}$$

are all equivalent.

This follows from the following formulas :

$$\min(s,1) = -\max(-s,-1) \ , \ (s+1)_- = s + \min(-s,1)$$

and

$$\max(s,-1) = s - (s+1)_- . \ \square$$

(3) *If $\mathcal{S}$ is a min-stable cone of negative functions, then $\mathcal{S} - \mathcal{S}$ is Stonian if*

$$(s+1)_- \in \mathcal{S} \ \textit{ for all } \ s \in \mathcal{S} .$$

This follows from the formulas

$$\min(s-t, u-v) = \min(s+v, t+u) - (t+v)$$

and

$$\min(s-t,\ 1) = \min(s, [t+1]_-) - t . \ \square$$

(4) *If $\mathcal{S}$ is Stonian, then $\mathcal{S}_-$, $\hat{\mathcal{S}}_-$, $\mathcal{S}_\sigma$ and $\mathcal{S}_\phi$ are also Stonian.*

The Stonian property is non-trivial only for $\hat{\mathcal{S}}_-$. It is sufficient to prove that the set of all $s \in \hat{\mathcal{S}}_-$ with $\max(s,-1) \in \hat{\mathcal{S}}_-$, containing $\mathcal{S}_-$, is $\mathcal{A}$-closed. In fact, if (s_n) is a sequence in this set and (ε_n) is a null sequence in $\mathbb{R}_+$ with

$$s_n + \varepsilon_n s_0 \le s \le s_n ,$$

then

$$\max(s_n,-1) + \varepsilon_n s_0 \le \max(s_n + \varepsilon_n s_0,\ -1) \le \max(s,-1) \le \max(s_n,-1),$$

which proves that

$$\max(s,-1) \in \mathcal{A}(\hat{\mathcal{S}}_-) = \hat{\mathcal{S}}_- . \ \square$$

(5) *If $\mathscr{S}$ is a lattice cone of negative functions, then $\mathscr{S}$ is Stonian iff*

$$\max(s,-1) \in \mathscr{S} \quad \textit{for all } s \in \mathscr{S} .$$

In this case, -1 is $\mathscr{S}$-measurable.

In the classical theory, where $\mathscr{S}$ is a vector lattice, the Stonian condition is only needed to ensure this measurability. In our framework, where $\mathscr{S}$ is a lattice cone, this is guaranteed by the Stonian condition for $\mathscr{S}_-$. More important however, the Stonian conditions on $\mathscr{S}$, and not as might be supposed the condition of (3), yield compatibility which is indispensable for regularity properties.

Using Proposition 4.11, this last remark yields the following

PROPOSITION. *If ν is an upper functional and $\mathscr{S}_-$ is Stonian and almost coinitial in $\mathfrak{I}_-^\bullet(\nu)$, then all constants are measurable w.r.t. ν. Conversely, if all constants are measurable, then $\tilde{\mathfrak{I}}(\nu)$ is Stonian.*

10.7 EXAMPLES.

(1) For compatible lattices $\mathfrak{K}$ and $\mathfrak{G}$, we have

$$\mathfrak{K}(\mathscr{E}(\mathfrak{K},\mathfrak{G})) = \mathfrak{K} \quad \text{and} \quad \mathfrak{G}(\mathscr{E}(\mathfrak{K},\mathfrak{G})) = \mathfrak{G} ,$$

and $\mathscr{E}(\mathfrak{K},\mathfrak{G})$ is Stonian.

(2) Let X be a Hausdorff space. Then

$$\mathscr{S}(X) = \mathscr{S}(X)_\phi \quad \text{and} \quad \mathscr{S}^{\mathfrak{W}}(X) = \mathscr{S}^{\mathfrak{W}}(X)_\phi$$

are Stonian, and

$$\mathfrak{K}(\mathscr{S}(X)) = \mathfrak{K}(X) , \quad \mathfrak{G}(\mathscr{S}(X)) = \mathfrak{G}(X)$$

are the compatible lattices of respectively compact and open subsets of X.

(3) Let X be a Hausdorff space. The function spaces $\mathscr{C}(X)$, $\mathscr{C}^b(X)$ and $\mathscr{C}^{\mathfrak{W}}(X)$ are Stonian. The lattice

$$\mathfrak{K}(\mathscr{C}(X)) = \mathfrak{K}(\mathscr{C}^b(X))$$

is the lattice $\mathfrak{Z}(X)$ of all sets $\{s = 0\}$ with $s \in \mathscr{C}(X)$, equivalently with $s \in \mathscr{C}^b(X)$.

Moreover,

$$\mathfrak{G}(\mathscr{C}(X)) = \mathfrak{G}(\mathscr{C}^b(X)) = \mathfrak{Z}(X)^{\complement} \quad \text{and} \quad \mathfrak{Z}(X) \subset \mathfrak{F}(X) \cap \mathfrak{G}(X)_\delta .$$

If X is normal, then

$$\mathfrak{Z}(X) = \mathfrak{F}(X) \cap \mathfrak{G}(X)_\delta .$$

Indeed, if (G_n) is a sequence of open sets and if $A := \bigcap G_n$ is closed, then, by Urysohn's lemma, there exist functions $f_n \in \mathscr{C}^b(X)$ with $0 \le f_n \le 1$, $A \subset \{f_n = 0\}$, and $\complement G \subset \{f_n = 1\}$. But then

$$f := \sum \frac{1}{2^n} \cdot f_n \in \mathscr{C}^b(X) \quad \text{and} \quad A = \{f = 0\} \in \mathfrak{Z}(X) . \ \square$$

(4) Let X be a locally compact space. The function spaces $\mathscr{C}^0(X)$ and $\mathscr{K}(X)$ are Stonian.

A proof analogous to the one given for normal spaces in the preceding example shows that

$$\mathfrak{K}(\mathscr{C}^0(X)) = \mathfrak{K}(\mathscr{K}(X)) = \mathfrak{K}(X) \cap \mathfrak{G}(X)_\delta$$

and that

$$\mathfrak{G}(\mathscr{C}^0(X)) = \mathfrak{G}(\mathscr{K}(X))$$

is the lattice of all relatively compact sets in $\mathfrak{G}(X) \cap \mathfrak{K}(X)_\sigma$.

10.8 PROPOSITION. *Let $\mathscr{S}_-$ be Stonian. Then all functions in $\mathscr{S}$ are $\mathfrak{K}(\mathscr{S})$-semicontinuous.*

Let $s \in \mathscr{S}$, $t \in \mathscr{S}_-$ and $\gamma \in \mathbb{R}$. To prove that $\{s \le \gamma\} \cap \{t \le -1\} \in \mathfrak{K}(\mathscr{S})$, we may suppose $\gamma \ge 0$. We have

$$\{s \le \gamma\} \cap \{t \le -1\} = \{s - (\gamma+1) \le -1\} \cap \{t \le -1\} =$$

$$= \{s + (\gamma+1) \cdot \max(t,-1) \le -1\} \cap \{t \le -1\} \in \mathfrak{K}(\mathscr{S}) ,$$

since $[s + (\gamma+1) \cdot \max(t,-1)]_- \in \mathscr{S}_-$. $\square$

COROLLARY. *Every $\mathscr{S}$-measurable function is $\mathfrak{K}(\hat{\mathscr{S}}_-)$-semicontinuous.*

Since $\mathcal{G}_-(\mathcal{S}) = \mathcal{S}^\wedge_-$ is Stonian by Remark 10.6.4, every function in $\mathcal{G}(\mathcal{S})$ is $\mathfrak{K}(\mathcal{S}^\wedge_-)$-semicontinuous. Let $f \in \mathcal{H}(\mathcal{S})$, $s \in \mathcal{S}^\wedge_-$ and $\gamma \in \mathbb{R}$. Then

$$\{f \le \gamma\} \cap \{s \le -1\} = \{f \le \gamma\} \cap \{s \le -1\} \cap \{|\gamma|\cdot s \le \gamma\} =$$
$$= \{\max(f, |\gamma|\cdot s) \le \gamma\} \cap \{s \le -1\} \in \mathfrak{K}(\mathcal{S}^\wedge_-),$$

since $\max(f, |\gamma|\cdot s) \in \mathcal{G}(\mathcal{S})$. □

10.9 The following technical result on the approximation of functions by step functions will be needed to prove that $\mathfrak{K}$-measurable and $\mathcal{E}(\mathfrak{K})$-measurable functions coincide (cf. Theorem 10.10).

Lemma. *Let f be a function and let, for $i = 1,\dots,n\cdot 2^n$, sets G_{in} and K_{in} be given such that*

$$\left\{f > \frac{i+1}{2^n}\right\} \subset G_{in} \subset \left\{f > \frac{i}{2^n}\right\}$$

and

$$\left\{f \le -\frac{i+1}{2^n}\right\} \subset K_{in} \subset \left\{f \le -\frac{i}{2^n}\right\}.$$

Define

$$e_n := \frac{1}{2^n}\cdot \sum_{i=1}^{n2^n} (1_{G_{in}} - 1_{K_{in}}).$$

Then

$$e_n + \max(f_-, -\frac{1}{2^{n-1}}) \le \max(f,-n) \ , \ \min(f,n) \le e_n + \min(f^+, \frac{1}{2^{n-1}}),$$

and

$$f = \lim e_n .$$

If $u \in \mathbb{R}^X_-$ and $u \le f$, then

$$f = \sup_n \,[e_n + \max(u, -\frac{1}{2^{n-1}}) + (u+n)_-],$$

the sequence being not necessarily increasing!

One verifies the assertions pointwise on sets $\{f \in I\}$ for suitable intervals I. □

COROLLARY. *Let $\mathfrak{G}$ be a lattice compatible with $\mathfrak{K}$. If f is $\mathfrak{K}$-semicontinuous, bounded below, positive outside some set from $\mathfrak{K}$, and such that*

$$\{f > \gamma\} \in \mathfrak{G} \quad \textit{for all } \gamma > 0 ,$$

then $f \in \mathscr{E}(\mathfrak{K},\mathfrak{G})_\sigma$.

Let $f \geq -\alpha \cdot 1_K =: u \in \mathscr{E}\_(\mathfrak{K})$ for suitable $\alpha > 0$, $K \in \mathfrak{K}$, and define

$$G_{in} := \left\{ f > \frac{i}{2^n} \right\} \in \mathfrak{G}$$

and

$$K_{in} := \left\{ f \leq -\frac{i}{2^n} \right\} = K \cap \left\{ f \leq -\frac{i}{2^n} \right\} \in \mathfrak{K} .$$

Then

$$e_n + \max(u, -\frac{1}{2^{n-1}}) \in \mathscr{E}(\mathfrak{K},\mathfrak{G}) ,$$

and f is the upper envelope for $n \geq \alpha$ of these functions. Since $\mathscr{E}(\mathfrak{K},\mathfrak{G})$ is a lattice, this proves that $f \in \mathscr{E}(\mathfrak{K},\mathfrak{G})_\sigma$. □

EXAMPLE. *In the topological standard example, we have*

$$\mathscr{I}(X) = \mathscr{E}(\mathfrak{K}(X),\mathfrak{G}(X))_\sigma .$$

Indeed, any function in $\mathscr{I}(X)$ is semicontinuous for the topology $\mathfrak{G}(\mathfrak{K}(X))$, i.e. $\mathfrak{K}(X)$-semicontinuous. □

10.10 PROPOSITION. *The sets $\mathcal{A}(\mathscr{E}\_(\mathfrak{K}))$ and $\hat{\mathscr{E}}\_(\mathfrak{K})$ coincide with the set of all negative $\mathfrak{K}$-measurable functions which are bounded below and vanish outside some set from $\mathfrak{K}$.*

Let $\mathscr{F}$ denote the set of those functions. We first prove that

$$\mathscr{F} \subset \mathcal{A}(\mathscr{E}\_) .$$

Let $f \in \mathscr{F}$ vanish outside $K \in \mathfrak{K}$. For $n \geq 1$ and $i = 1,\ldots,n \cdot 2^n$, choose $K_{in} \in \mathfrak{K}$ such that

$$\left\{ f \leq -\frac{i+1}{2^n} \right\} = K \cap \left\{ f \leq -\frac{i+1}{2^n} \right\} \subset K_{in} \subset K \cap \left\{ f \leq -\frac{i}{2^n} \right\} = \left\{ f \leq -\frac{i}{2^n} \right\} .$$

With the notations of Lemma 10.9 we have

$$e_n = -\frac{1}{2^n} \cdot \sum_{i=1}^{n2^n} 1_{K_{in}}$$

and

$$e_n - \frac{1}{2^{n-1}} \cdot 1_K \leq e_n + \max(f, -\frac{1}{2^{n-1}}) \leq \max(f,-n) = f \leq e_n$$

for sufficiently large n, hence $f \in \mathcal{A}(\mathcal{E}_-)$.

It remains to prove that

$$\mathcal{A}(\mathcal{F}) \subset \mathcal{F} .$$

Let $f \in \mathcal{A}(\mathcal{F})$ and $\gamma,\delta \in \mathbb{R}$ with $\gamma < \delta < 0$ be given. There exists a sequence (f_n) in $\mathcal{F}$ and a null sequence (ε_n) in $\mathbb{R}_+$ such that

$$f_n + \varepsilon_n f_0 \leq f \leq f_n$$

for $n \geq 1$. Choose $\alpha > 0$ and $n \geq 1$ with

$$\gamma + \alpha < \delta \quad \text{and} \quad -\alpha \leq \varepsilon_n f_0 .$$

Since f and hence f_n vanishes outside some $K \in \mathfrak{K}$, there exists a set $L \in \mathfrak{K}$ such that

$$\{f \leq \gamma\} \subset \{f_n + \varepsilon_n f_0 \leq \gamma\} \subset \{f_n \leq \gamma + \alpha\} \subset L \subset \{f_n \leq \delta\} \subset \{f \leq \delta\} .$$

This proves that f is $\mathfrak{K}$-measurable and hence $f \in \mathcal{F}$. □

THEOREM. *A function is $\mathfrak{K}$-measurable iff it is $\mathcal{E}(\mathfrak{K})$-measurable, i.e.*

$$\mathcal{M}(\mathfrak{K}) = \mathcal{H}(\mathcal{E}(\mathfrak{K})) .$$

Furthermore, we have

$$\mathcal{G}(\mathcal{E}(\mathfrak{K})) \subset \mathcal{E}(\mathfrak{K})_\sigma = \mathcal{G}(\mathcal{E}(\mathfrak{K}_\delta)) .$$

In particular,

$$\mathcal{E}_-(\mathfrak{K})_\sigma = \hat{\mathcal{E}}_-(\mathfrak{K}_\delta) ,$$

and a function is $\mathfrak{K}_\delta$-measurable iff it is $\mathcal{E}(\mathfrak{K})_\sigma$-measurable.

First we prove that every $\mathfrak{K}$-measurable function f which is bounded below and positive outside some set from $\mathfrak{K}$ belongs to $\mathcal{G}(\mathcal{E})$. It is sufficient, for every $e \in \mathcal{E}_-$, to prove that the function $f + e$ is $\mathfrak{K}$-measurable, since then by the above proposition $(f + e)_-$ belongs to $\hat{\mathcal{E}}_-$. By induction, we may suppose that $e = -\alpha \cdot 1_K$ with $\alpha > 0$ and $K \in \mathfrak{K}$. For $\gamma < \delta$ and $L \in \mathfrak{K}$ there exist $A, B \in \mathfrak{K}$ such that

$$\{f \le \gamma + \alpha\} \cap L \subset A \subset \{f \le \delta + \alpha\} \cap L$$

and

$$\{f \le \gamma\} \cap L \subset B \subset \{f \le \delta\} \cap L .$$

Then $(A \cap K) \cup B \in \mathfrak{K}$ and

$$\{f - \alpha\cdot 1_K \le \gamma\} \cap L \subset (A \cap K) \cup B \subset \{f - \alpha\cdot 1_K \le \delta\} \cap L .$$

For every $\mathfrak{K}$–measurable function f and all $e \in \mathcal{E}_-$ we therefore have

$$\max(f,e) \in \mathcal{G}(\mathcal{E})$$

by Proposition 10.3.(i), hence $f \in \mathcal{H}(\mathcal{E})$.

Next we prove that every $\mathcal{E}(\mathfrak{K})$–measurable function f is $\mathfrak{K}$–measurable. Suppose first that $f \in \mathcal{G}(\mathcal{E})$. Let $K \in \mathfrak{K}$ and $\gamma,\delta \in \mathbb{R}$ with $\gamma < \delta$ be given. For $n \in \mathbb{N}$ with $n > \delta$, the function $(f - n\cdot 1_K)_- \in \hat{\mathcal{E}}_-$ is $\mathfrak{K}$–measurable by the proposition, hence there exists $L \in \mathfrak{K}$ with

$$\{f \le \gamma\} \cap K =$$

$$= \{[f - n\cdot 1_K]_- \le \gamma - n\} \cap K \subset L \subset \{[f - n\cdot 1_K]_- \le \delta - n\} \cap K =$$

$$= \{f \le \delta\} \cap K .$$

For arbitrary $f \in \mathcal{H}(\mathcal{E})$, let $K \in \mathfrak{K}$ and $\gamma,\delta \in \mathbb{R}$ with $\gamma < \delta$ be given. For $n \in \mathbb{N}$ with $-n < \gamma$, the function $\max(f, -n\cdot 1_K) \in \mathcal{G}(\mathcal{E})$ is $\mathfrak{K}$–measurable, hence there exists $L \in \mathfrak{K}$ such that

$$\{f \le \gamma\} \cap K =$$

$$= \{\max(f, -n\cdot 1_K) \le \gamma\} \cap K \subset L \subset \{\max(f, -n\cdot 1_K) \le \delta\} \cap K =$$

$$= \{f \le \delta\} \cap K .$$

Furthermore, since $\mathcal{G}_-(\hat{\mathcal{E}}_-) = \hat{\mathcal{E}}_-$ is Stonian (cf. Remark 10.6.4), every $f \in \mathcal{G}(\mathcal{E})$ is $\mathfrak{K}(\hat{\mathcal{E}}_-)$–semicontinuous by Proposition 10.8. Therefore, f is $\mathfrak{K}_\delta$–semicontinuous by Remark 10.3.3, since

$$\mathfrak{K} \subset \mathfrak{K}(\hat{\mathcal{E}}_-) \subset \mathfrak{K}(\mathcal{E}_{-\sigma}) = \mathfrak{K}_\delta .$$

Corollary 10.9 then shows that

$$f \in \mathcal{E}(\mathfrak{K}_\delta)_\sigma = \mathcal{E}(\mathfrak{K})_\sigma .$$

Finally, every $f \in \mathcal{E}(\mathfrak{K})_\sigma$ is $\mathfrak{K}_\delta$–semicontinuous, hence $\mathfrak{K}_\delta$–measurable by Proposition 10.3.(iii) and thus $f \in \mathcal{G}(\mathcal{E}(\mathfrak{K}_\delta))$ by the above characterization. For the last assertion, one just has to note that $\mathcal{H}(\mathcal{E}(\mathfrak{K}_\delta)) = \mathcal{H}(\hat{\mathcal{E}}_-(\mathfrak{K}_\delta))$. □

REMARK. The theorem shows in particular that the set of all $\mathfrak{K}$-measurable functions which are positive outside a set from $\mathfrak{K}$ and bounded below is a lattice cone.

EXAMPLE. *If X is a Hausdorff space, then the $\mathscr{E}(\mathfrak{K}(X))$- and the $\mathscr{S}(X)$-measurable functions coincide with those which are $\mathfrak{K}(X)$-semicontinuous, i.e. lower semicontinuous for the topology $\mathfrak{G}(\mathfrak{K}(X))$. Moreover,*

$$\mathscr{S}(X) \subset \mathscr{G}(\mathscr{E}(\mathfrak{K}(X)))$$

and

$$\mathscr{S}\_(X) = \mathscr{E}\_(\mathfrak{K}(X))_\sigma = \mathscr{E}\hat{\_}(\mathfrak{K}(X)) = \mathcal{A}(\mathscr{E}\_(\mathfrak{K}(X))) .$$

If $\mathfrak{G}(X)$ and $\mathfrak{G}(\mathfrak{K}(X))$ coincide, then

$$\mathscr{S}(X) = \mathscr{G}(\mathscr{E}(\mathfrak{K}(X))) .$$

This follows from Propositions 4.9 and 10.3.(iii). □

10.11 **DEFINITION.** A lattice cone $\mathscr{S}$ of functions is said to be *strongly Stonian* if it is Stonian and if

$$(s + 1)\_ , (s - 1)^+ \in \mathscr{S} \quad \text{for all } s \in \mathscr{S} .$$

REMARKS.

(1) Every Stonian vector lattice is strongly Stonian, as Remark 10.6.2 shows.

(2) If $\mathscr{S}$ is a lattice cone of negative functions, then $\mathscr{S}$ is strongly Stonian iff

$$\max(s,-1) , (s + 1)\_ \in \mathscr{S}\_ \quad \text{for all } s \in \mathscr{S} .$$

PROPOSITION. *Let $\mathscr{S}$ be a strongly Stonian lattice cone of functions. Then*

$$\mathscr{S}_\sigma \cap \mathscr{E}\_^{\uparrow}(\mathfrak{K}(\mathscr{S})) = \mathscr{E}(\mathfrak{K}(\mathscr{S}),\mathfrak{G}(\mathscr{S}))_\sigma .$$

For $s \in \mathscr{S}$, the functions

$$-1_{\{s \le -1\}} = \sup_n \max[n \cdot (s + 1 - \tfrac{1}{n})\_ ,-1]$$

and

$$1_{\{s > 1\}} = \sup_n \min[n \cdot (s - 1)^+ ,1]$$

belong to $\mathscr{S}_\sigma$, which proves that $\mathscr{E}_\sigma \subset \mathscr{S}_\sigma \cap \mathscr{E}_-^\dagger$.

The converse follows from Corollary 10.9, since every $s \in \mathscr{S}$ is $\mathfrak{K}(\mathscr{S})$-semicontinuous by Proposition 10.8. □

EXAMPLES.

(1) In the measure-theoretic standard example, $\mathscr{E}(\mathfrak{K},\mathfrak{G})$ is strongly Stonian.

This follows immediately from the formulas

$$\{(s+1)_- > \gamma\} = \{s > \gamma - 1\} \quad \text{for } \gamma < 0$$

and

$$\{(s-1)^+ > \gamma\} = \{s > \gamma + 1\} \quad \text{for } \gamma \geq 0 . \; \square$$

(2) If X is a Hausdorff space, then $\mathscr{S}(X)$, $\mathscr{S}^{\mathscr{W}}(X)$ and all the function spaces $\mathscr{C}(X)$, $\mathscr{C}^{\mathscr{W}}(X)$, $\mathscr{C}^b(X)$, $\mathscr{C}^0(X)$, and $\mathscr{K}(X)$ are strongly Stonian.

§ 11 CONTENTS AND MEASURABLE SETS

In this section, we investigate the relationship between integration theory and abstract finitely additive measure theory by means of the measure-theoretic standard example of § 10 .

THROUGHOUT THIS SECTION,

$\mathfrak{K}$ AND $\mathfrak{G}$ ARE COMPATIBLE LATTICES OF SETS.

11.1 **Definition.** Let m be a positive set function on $\mathfrak{K}$ with $m(\emptyset) = 0$. Then m is said to be

- *finite* if $m(K) < \infty$ for all $K \in \mathfrak{K}$,
- *bounded* if $\sup_{K \in \mathfrak{K}} m(K) < \infty$,
- *increasing* if $m(K) \le m(L)$ for all $K,L \in \mathfrak{K}$ with $K \subset L$,
- *subadditive* if $m(K \cup L) \le m(K) + m(L)$ for all $K,L \in \mathfrak{K}$,
- *additive* if $m(K \cup L) = m(K) + m(L)$ for all $K,L \in \mathfrak{K}$ with $K \cap L = \emptyset$,
- *completely subadditive* if $n \cdot m(K) \le \sum_{i=1}^{p} m(K_i)$

whenever $K \in \mathfrak{K}$ is covered at least $n \in \mathbb{N}$ times by sets $K_1,\dots,K_p \in \mathfrak{K}$, i.e. $K \subset \bigcup_J \bigcap_{j \in J} K_j$, J running through all subsets of $\{1,\dots,p\}$ with n elements,

- *strongly subadditive* (resp. *strongly additive*) , if

$$m(K \cup L) + m(K \cap L) \le m(K) + m(L) \quad (\text{resp.} =)$$

for all $K,L \in \mathfrak{K}$.

A *content* is an increasing, strongly additive set function.

Remarks.

(1) Every content is additive and subadditive. Conversely :

If $\mathfrak{K}$ is a ring and m is additive, then m is a content.

Indeed, we have

$$m(K \cup L) + m(K \cap L) = m(K) + m(L \setminus K) + m(K \cap L) = m(K) + m(L) . \quad \square$$

A further case will be discussed in Remark 12.2.3.

(2) *m is completely subadditive iff $n \cdot 1_K \leq \sum 1_{K_i}$ implies that*

$$n \cdot m(K) \leq \sum m(K_i) ,$$

equivalently, iff $1_K \leq \sum \alpha_i \cdot 1_{K_i}$ with $\alpha_i \in \mathbb{R}_+$ implies that

$$m(K) \leq \sum \alpha_i \cdot m(K_i) .$$

In particular, every completely subadditive set function is increasing and subadditive.

For rational coefficients, this follows immediately from the above formula since repetitions of the K_i are allowed, and then in general by approximation from above. $\square$

11.2 The following corollary shows moreover that every increasing strongly subadditive set function is completely subadditive.

PROPOSITION. *Let m be a finite increasing set function on $\mathfrak{K}$ with $m(\emptyset) = 0$. Then m is completely subadditive iff there exists an increasing and superlinear extension μ of m to $\mathcal{E}_-(\mathfrak{K})$, i.e. a functional μ on $\mathcal{E}_-(\mathfrak{K})$ which satisfies*

$$\mu(-1_K) = -m(K)$$

for all $K \in \mathfrak{K}$. In this case, the smallest such extension is given by

$$m(e) := \sup \; -\sum \alpha_i \cdot m(K_i) ,$$

where $-\sum \alpha_i \cdot 1_{K_i}$ runs through all step functions $\leq e$.

Let m be completely subadditive. Defining m on $\mathscr{E}\_(\mathfrak{K})$ as above, we get $m(K) \leq -m(-1_K)$ by Remark 11.1.2 , hence $m(K) = -m(-1_K)$ for $K \in \mathfrak{K}$. Obviously, m is increasing and superlinear on $\mathscr{E}\_(\mathfrak{K})$. If μ is another functional of this type and if $-\sum \alpha_i \cdot 1_{K_i} \leq e$, then

$$-\sum \alpha_i \cdot m(K_i) = \sum \alpha_i \cdot \mu(-1_{K_i}) \leq \mu(-\sum \alpha_i \cdot 1_{K_i}) \leq \mu(e) ,$$

hence $m(e) \leq \mu(e)$.

Conversely, if $1_K \leq \sum \alpha_i \cdot 1_{K_i}$, then

$$m(K) = -\mu(-1_K) \leq -\mu(-\sum \alpha_i \cdot 1_{K_i}) \leq -\sum \alpha_i \cdot \mu(-1_{K_i}) = \sum \alpha_i \cdot m(K_i) .$$

This proves that m is completely subadditive. □

COROLLARY. *The set function m is strongly subadditive, resp. a content on $\mathfrak{K}$ iff m is strongly superlinear, resp. linear on $\mathscr{E}\_(\mathfrak{K})$. For any pyramidal decomposition of*

$$e = -\sum_{i>0} \delta_i \cdot 1_{K_i} ,$$

we have

$$m(e) = -\sum_{i>0} \delta_i \cdot m(K_i) .$$

Since

$$-1_{K \cup L} - 1_{K \cap L} = -1_K - 1_L$$

for all $K, L \in \mathfrak{K}$, the conditions are obviously sufficient.

To prove the converse, let us first note that with the notations of Remark 10.4.1 we have

$$-\sum_{i>0} \delta_i \cdot m(K_i) = -\sum_{j=-n}^{-1} (\omega_{j+1} - \omega_j) \cdot m(\{e \leq \omega_j\}) .$$

Therefore, independently of the chosen pyramidal decomposition of e , by

$$\mu(e) := -\sum_{i>0} \delta_i \cdot m(K_i)$$

we define a positively homogeneous functional μ on $\mathscr{E}\_(\mathfrak{K})$ with $\mu \leq m$. We will have $\mu = m$ if we can prove that μ is increasing and superlinear.

We first prove that

$$\mu(e - 1_K) \geq \mu(e) - m(K)$$

holds for all $e \in \mathcal{E}_-(\mathfrak{K})$ and $K \in \mathfrak{K}$.

Let $(\gamma_i)_{i\in\mathbb{N}}$ be a strongly decreasing enumeration of $(\mathbb{Z} + [e(X) \cup \{0\}])_-$ with $\gamma_0 = 0$. Then the values taken by $e' := e - 1_K$ are also contained in this set of γ_i. If $p \in \mathbb{N}$ is the index with $\gamma_p = -1$, then $\gamma_i - 1 = \gamma_{i+p}$ for all $i \in \mathbb{N}$.

If we define $K_i := \{e \leq \gamma_i\}$ and $K_i' := \{e' \leq \gamma_i\}$ for $i > 0$, then for $i > p$ we have

$$K_i' = (\{e \leq \gamma_i + 1\} \cap K) \cup \{e \leq \gamma_i\} = (K_{i-p} \cap K) \cup K_i,$$

and for $0 < i \leq p$ we have

$$K_i' = K \cup K_i.$$

Therefore,

$$m(K_i') + m(K_i \cap K) \leq m(K_{i-p} \cap K) + m(K_i)$$

in the first case, and

$$m(K_i') + m(K_i \cap K) \leq m(K) + m(K_i)$$

in the second case.

With the notations of Remark 10.4.1, in view of $\delta_{i+p} = \delta_i$ we get

$$\mu(e') = -\sum_{i>0} \delta_i \cdot m(K_i') \geq$$

$$\geq -\sum_{i>p} \delta_i \cdot [m(K_{i-p} \cap K) + m(K_i) - m(K_i \cap K)]$$

$$-\sum_{i=1}^{p} \delta_i \cdot [m(K) + m(K_i) - m(K_i \cap K)] =$$

$$= \sum_{i>0} \delta_i \cdot m(K_i) - \sum_{i=1}^{p} \delta_i \cdot m(K) = \mu(e) - m(K),$$

since $\sum_{i=1}^{p} \delta_i = \gamma_0 - \gamma_p = 1$. By induction, this implies that μ is superlinear.

Moreover, if m is strongly additive, then μ is linear since in the above inequalities we have in fact equality.

Let now $d,e \in \mathcal{E}_-(\mathfrak{K})$ and (γ_i) be a strictly decreasing sequence of real numbers containing $d(X) \cup e(X)$, with $\gamma_0 := 0$. Then, if $d \leq e$,

$$L_i := \{e \leq \gamma_i\} \subset K_i := \{d \leq \gamma_i\}$$

for $i > 0$ and hence

$$\mu(d) = -\sum_{i>0} \delta_i \cdot m(K_i) \le -\sum_{i>0} \delta_i \cdot m(L_i) = \mu(e) .$$

This proves that μ is increasing.

The supermodularity of μ follows immediately from the formulas

$$\min(e,d) = -\sum_{i>0} \delta_i \cdot 1_{K_i \cup L_i} \quad \text{and} \quad \max(e,d) = -\sum_{i>0} \delta_i \cdot 1_{K_i \cap L_i} . \quad \square$$

11.3 **COROLLARY.** *Difference-regular linear extension of* m *to* μ *, uniquely specified by*

$$\mu(-1_K) = -\, m(K)$$

for $K \in \mathfrak{K}$ *, defines a bijection between the finite contents* m *on* $\mathfrak{K}$ *and the difference-regular linear functionals* μ *on* $\mathscr{E}(\mathfrak{K},\mathfrak{G})$ *.*

If μ is a difference-regular linear functional on $\mathscr{E}(\mathfrak{K},\mathfrak{G})$, then a finite content m is defined on $\mathfrak{K}$ by

$$m(K) = -\, \mu(-1_K) .$$

By Corollary 6.5, there is a unique difference-regular linear extension of m to $\mathscr{E}(\mathfrak{K},\mathfrak{G})$ which therefore coincides with μ . $\square$

TROUGHOUT THE REST OF THIS SECTION,

m IS A FINITE CONTENT ON $\mathfrak{K}$.

NOTATION. When confusion appears unlikely, the functional μ will again be denoted by m , and the corresponding upper functionals μ^*, $\mu^\bullet$, $\mu^\times$ and $\mu^\bullet$ will be denoted respectively by m^*, $m^\bullet$, $m^\times$ and $m^\bullet$. Corresponding notations will be used for the lower functionals.

Note that $m_\times$, $m_\bullet$, $m^\times$ and $m^\bullet$ only depend on the restriction of μ to $\mathscr{E}\_(\mathfrak{K},\mathfrak{G}) = \mathscr{E}\_(\mathfrak{K})$ and therefore do not depend on $\mathfrak{G}$, whereas the functionals m_* , $m_\bullet$, m^* and $m^\bullet$ depend on the chosen compatible lattice $\mathfrak{G}$.

11.4 Definition. If ν is an upper functional, then the systems of all *sets* which are (i.e. whose indicator functions are) *integrable*, *essentially integrable* or *measurable* w.r.t. ν will be denoted respectively by

$$\mathfrak{I}(\nu)\ ,\ \mathfrak{I}^{\bullet}(\nu)\quad\text{and}\quad\mathfrak{I}^{0}(\nu)\,.$$

If μ is an increasing linear functional defined on a min-stable cone of functions, we denote these systems of sets by

$$\mathfrak{R}^{\times}(\mu)\ ,\ \mathfrak{R}^{\bullet}(\mu)\quad\text{and}\quad\mathfrak{R}^{0}(\mu)$$

respectively.

If μ is regular, then we use the notation $\mathfrak{R}^{*}(\mu) := \mathfrak{I}(\mu^{*})$.

Proposition.

(i) *The systems $\mathfrak{I}(\nu)$, $\mathfrak{I}^{\bullet}(\nu)$ and $\mathfrak{I}^{0}(\nu)$ are rings with*

$$\mathfrak{I}(\nu) \subset \mathfrak{I}^{\bullet}(\nu) \subset \mathfrak{I}^{0}(\nu) \subset \mathfrak{S}(\mathfrak{I}^{0}(\nu)) \subset \mathfrak{S}(\mathfrak{I}^{\bullet}(\nu)) \subset \mathfrak{S}(\mathfrak{I}(\nu))\,.$$

$\mathfrak{I}^{0}(\nu)$ is an algebra iff 1 *is measurable.*

(ii) *$\nu^{\bullet}$ is a content on $\mathfrak{I}^{0}(\nu)$ which coincides with ν on $\mathfrak{I}(\nu)$. Furthermore, $\mathfrak{I}^{\bullet}(\nu)$ is the ring of all measurable sets A with $\nu^{\bullet}(A) < \infty$.*

(iii) *If ν is auto-determined, then ν is a content on $\mathfrak{I}^{0}(\nu)$, and $\mathfrak{I}(\nu)$ is the ring of all measurable sets A with $\nu(A) < \infty$.*

Since $\mathcal{I}_{\mathbb{R}}$ and $\mathcal{I}^{\bullet}_{\mathbb{R}}$ are vector lattices and since $\mathcal{I}^{0}$ is a lattice of functions, the formulas

$$1_{A\cup B} = \max(1_A, 1_B)\ ,\quad 1_{A\cap B} = \min(1_A, 1_B)\quad\text{and}\quad 1_{A\smallsetminus B} = 1_A - 1_{A\cap B}$$

prove that $\mathfrak{I}$ and $\mathfrak{I}^{\bullet}$ are rings and that $\mathfrak{I}^{0}$ is a lattice. Since in general $\mathcal{I}^{0}_{\mathbb{R}}$ is not stable w.r.t. differences (cf. Example 4.7.2), we have to prove the measurability of $A \smallsetminus B$ by direct recourse to the definition :

For $u \in \mathcal{I}_{-}$ and $v := \min(1_A, -u) \in \mathcal{I}_{\mathbb{R}}$,

$$\min(1_{A\smallsetminus B}, -u) = \min(v, 1_{\complement B}) = v - \min(v, 1_B) \in \mathcal{I}_{\mathbb{R}}\,,$$

since $0 \le v \le 1$. This proves the first part of (i). Since $\mathfrak{I} \subset \mathfrak{I}^{\bullet} \subset \mathfrak{I}^{0}$, it remains to prove that $\mathfrak{S}(\mathfrak{I}^{0}) \subset \mathfrak{S}(\mathfrak{I}^{\bullet}) \subset \mathfrak{S}(\mathfrak{I})$. To this end, let $A \in \mathfrak{S}(\mathfrak{I}^{0})$ and B be essentially integrable. Then $B \smallsetminus A$ is measurable, hence essentially integrable by Theorem 4.4. If $A \in \mathfrak{S}(\mathfrak{I}^{\bullet})$ and if B is integrable, then $B \smallsetminus A$ is integrable by Proposition 3.4.(ii).

Finally, the assertions (ii) and (iii) follow from Theorem and Corollary 4.10 by means of the Integrability Criteria 4.4 and 4.5, in view of the formula

$$1_{A\cup B} + 1_{A\cap B} = 1_A + 1_B \,. \quad \square$$

COROLLARY. *The upper functional* $m^\times$ *is auto-determined, the set of functions* $-n\cdot 1_K$ *with* $n \in \mathbb{N}$ *and* $K \in \mathfrak{K}$ *is almost coinitial in* $\overline{\mathbb{R}}^X$ *w.r.t.* $m^\times$ *and* $m^\bullet$ *, and every* $\mathfrak{K}$*-measurable function is measurable. For every set* A *, we have*

$$m_*(A) \le m_\times(A) = m_\bullet(A) \le m^\bullet(A) = \sup_{K\in\mathfrak{K}} m^\times(A \cap K) \le m^*(A) \,.$$

The measurable sets form an algebra $\mathfrak{N}^0(m)$ *containing* $\mathfrak{S}(\mathfrak{K})$ *, and in particular* $\mathfrak{K}$ *, on which* $m_\bullet = m^\bullet$ *and* $m^\times$ *are contents extending* m *. Furthermore,* $\mathfrak{N}^\times(m)$ *and* $\mathfrak{N}^\bullet(m)$ *are respectively the rings of measurable sets* A *with* $m^\times(A) < \infty$ *, i.e. contained in some set from* $\mathfrak{K}$ *, and* $m^\bullet(A) < \infty$ *.*

We have

$$\mathfrak{N}^0(m) = \mathfrak{S}(\mathfrak{N}^\times(m)) \,.$$

More precisely, a set A *is measurable iff* $A \cap K$ *is integrable for every* $K \in \mathfrak{K}$ *.*

For the first and last statement, one also has to consider Theorems 10.10 and 3.6.(i), Propositions 6.3 and 4.3, and the formula

$$\min(1_A, n\cdot 1_K) = 1_{A\cap K}$$

for $n \ge 1$. $\square$

REMARK. It may be interesting to note :

A function f *is equal to* 0 mod $m^\bullet$ *iff*

$$m^\bullet(\{|f| > \varepsilon\} = 0$$

for every $\varepsilon > 0$.

This follows from the inequalities

$$1_{\{|f|>\varepsilon\}} \le \frac{1}{\varepsilon}\cdot|f|$$

and

$$\min(|f|, n\cdot 1_K) \le \varepsilon\cdot 1_K + n\cdot 1_{\{|f|>\varepsilon\}\cap K} \,. \quad \square$$

11.5 The last assertion in the preceding corollary is not true in general. More precisely, not every set which is locally essentially integrable w.r.t. an upper functional ν has to be measurable. Likewise, the inequality

$$\nu^{\bullet}(A) \geq \sup_{B \in \mathfrak{I}^{\bullet}(\nu),\, B \subset A} \nu^{\bullet}(B),$$

valid for arbitrary sets A, may fail to be an equality even when A is measurable. The following examples illustrate these phenomena.

EXAMPLES.

(1) Consider again the regular linear functional μ of Example 1.10.6 defined by

$$\mu(s) := \lim_{\mathfrak{F}} \frac{s}{w} .$$

We use the results of Example 4.7.1.

We always have $\mu^{\bullet} = \mu^{*}$.

In fact, for any function f there exists a filter $\mathfrak{U}$ finer than $\mathfrak{F}$ such that

$$\mu^{*}(f) = \lim_{\mathfrak{U}} \frac{f}{w} \quad \text{in } \overline{\mathbb{R}} .$$

By the almost coinitiality of the functions $-nw$ with $n \in \mathbb{N}$, we have

$$\mu^{*}(f) = \operatorname{med}\left[\lim_{\mathfrak{U}} \frac{f}{w}, \lim_{j \in \mathbb{N}} \lim_{\mathfrak{U}} \frac{jw}{w}, \lim_{k \in \mathbb{N}} \lim_{\mathfrak{U}} \frac{-kw}{w}\right] =$$

$$= \lim_{k \in \mathbb{N}} \lim_{j \in \mathbb{N}} \lim_{\mathfrak{U}} \frac{\operatorname{med}(f, jw, -kw)}{w} \leq$$

$$\leq \inf_{k \in \mathbb{N}} \sup_{j \in \mathbb{N}} \limsup_{\mathfrak{F}} \frac{\operatorname{med}(f, jw, -kw)}{w} = \mu^{\bullet}(f) \leq \mu^{*}(f)$$

by Theorem 3.6. □

Suppose now that $\lim_{\mathfrak{F}} w = 0$. *Then a set* A *is integrable iff it is disjoint from some set in* $\mathfrak{F}$. *In this case* $\mu^{*}(A) = 0$. *A set is measurable iff it is integrable or belongs to* $\mathfrak{F}$.

Indeed, if A is disjoint from some set in $\mathfrak{F}$, then $\lim_{\mathfrak{F}} \frac{1}{w} \cdot 1_A = 0$, hence A is integrable with $\mu^{*}(A) = 0$. On the other hand, if $\mathfrak{F} \cap A$ is a filter base, then

$$\lim_{\mathfrak{F} \cap A} \frac{1}{w} \cdot 1_A = \lim_{\mathfrak{F}} \frac{1}{w} = \infty .$$

Therefore, such a set A is not integrable.

Obviously, every set $A \in \mathfrak{F}$ is measurable. Conversely, if A is not integrable and $A \notin \mathfrak{F}$, then both $\mathfrak{F} \cap A$ and $\mathfrak{F} \cap \complement A$ are filter bases with $\lim_{\mathfrak{F} \cap A} \frac{1}{w} \cdot 1_A = \infty$ and $\lim_{\mathfrak{F} \cap \complement A} \frac{1}{w} \cdot 1_A = 0$. Therefore, $\lim_{\mathfrak{F}} \frac{1}{w} \cdot 1_A$ does not exist, i.e A is not measurable. □

By the above characterization, every set is locally essentially integrable. Furthermore, if A is measurable but not integrable, i.e. $A \in \mathfrak{F}$, then

$$\infty = \mu^{\bullet}(A) > \sup_{B \in \mathfrak{R}^{\bullet}(\mu),\, B \subset A} \mu^{\bullet}(B) = 0 .$$

(2) To illustrate the existence of a locally essentially integrable set which is not measurable, let w denote the identity on $]0,1]$ and $\mathfrak{F}$ be the filter generated by the sets $]0,\varepsilon[$ with $\varepsilon > 0$.

By the above characterization, an example of such a set is $\{ \frac{1}{n} \mid n \geq 1 \}$.

REMARK. These pathologies do not occur if $\nu^{\bullet}$ is of the form $m^{\bullet}$ for some finite content m. For an auto-determined upper functional ν, this is characterized by the validity of the following truncation formulas for all negative integrable functions together with the measurability of the constant functions, as we shall see in Corollary 14.6.

The necessity follows from Example 11.6.2, whereas the proof of sufficiency makes essential use of the subsequent Theorem 11.7.

11.6 **DEFINITION.** We say that an upper functional ν satisfies the *truncation formula at* $-\infty$, respectively *at* 0, for a function $f \leq 0$ if

$$\nu(f) = \inf_{n \in \mathbb{N}} \nu(\max(f,-n)) ,$$

respectively

$$\sup_{n \geq 1} \nu(\max(f, -\tfrac{1}{n})) = 0 ,$$

holds.

PROPOSITION. *Let ν be an upper functional and $\mathcal{T}$, $\mathcal{S}$ be downward directed sets of negative functions with $\mathcal{T}$ being almost coinitial to $\mathcal{S}$.*

If ν is finite on $\mathcal{T}$, respectively on $\mathcal{S}$, and satisfies the truncation formula on $\mathcal{T}$ at $-\infty$, respectively at 0, then also on $\mathcal{S}$.

Note first that ν satisfies the truncation formula at $-\infty$ on $\mathcal{T}$ iff the set of functions $-n$ with $n \in \mathbb{N}$ is almost coinitial to $\mathcal{T}$. The corresponding formula on $\mathcal{S}$ therefore follows by the transitivity property of Proposition 3.5.(i).

Suppose now that ν satisfies the truncation formula at 0 on $\mathcal{T}$. From Lemma 3.5, applied to the sequence $(-\frac{1}{n})_{n\geq 1}$, we infer that

$$0 = \inf_{t\in\mathcal{T}} \sup_{n\geq 1} \nu(\max(t, -\frac{1}{n})) \leq \inf_{s\in\mathcal{S}} \sup_{n\geq 1} \nu(\max(s, -\frac{1}{n})) \leq 0 ,$$

which proves the truncation formula at 0 on $\mathcal{S}$. □

Corollary. *Let ν be an upper functional and $\mathcal{T}$ be almost coinitial in $\mathcal{I}_{-}(\nu)$. If ν satisfies one of the truncation formulas on $\mathcal{T}$, then the corresponding formula is satisfied by $\nu^{\bullet}$ on $\mathcal{I}_{-}^{\bullet}(\nu)$, and in particular by ν on $\mathcal{I}_{-}(\nu)$.*

Since $\mathcal{T}$ is almost coinitial in $\mathcal{I}_{-}^{\bullet}(\nu)$ by Corollary 3.5, it suffices to apply the proposition to $\nu^{\bullet}$. □

Examples.

(1) *Let ν be an upper integral for which 1 is measurable. Then ν satisfies both truncation formulas on $\mathcal{I}_{-}(\nu)$.*

This follows immediately from the Monotone Convergence Theorem 5.4. □

(2) *The upper functionals $m^{\times}$ and $m^{\bullet}$ satisfy both truncation formulas on $\mathcal{R}_{-}^{\times}(m)$ and $\mathcal{R}_{-}^{\bullet}(m)$ respectively.*

By the corollary, we just have to note that the set of functions $-n\cdot 1_K$ with $n \in \mathbb{N}$ and $K \in \mathfrak{K}$ is coinitial in $\mathcal{R}_{-}^{\times}(m)$, and that the truncation formulas are trivial for these functions. □

(3) We again discuss Example 1.10.6 where μ is defined by

$$\mu(s) = \lim_{\mathfrak{F}} \frac{s}{w}.$$

We also recall the Examples 4.7.1 and 11.5.1.

The truncation formula at $-\infty$ *is satisfied by* μ^* *on* $\mathcal{R}^{\times}_{-}(\mu)$ *iff*

$$\limsup_{\mathfrak{F}} w < \infty .$$

Note that by transitivity the set of functions $-n$ with $n \in \mathbb{N}$ is almost coinitial to $\mathcal{R}^{\times}_{-}(\mu)$ iff it is almost coinitial to the set of functions $-nw$ with $n \in \mathbb{N}$, which in turn by positive homogenity means that

$$-1 = \mu(-w) = \inf_n \mu^*(\max(-w,-n)) = \inf_n \limsup_{\mathfrak{F}} \max(-1, -\frac{n}{w}) =$$

$$= \inf_n \max(-1, -n\cdot\frac{1}{\limsup_{\mathfrak{F}} w}),$$

i.e.

$$\limsup_{\mathfrak{F}} w < \infty . \quad \square$$

The truncation formula at 0 *is satisfied by* μ^* *on* $\mathcal{R}^{\times}_{-}(\mu)$ *iff*

$$\limsup_{\mathfrak{F}} w > 0 .$$

Indeed, this formula applied to $-w \in \mathcal{R}^{\times}_{-}(\mu)$ means that

$$0 = \sup_{n\geq 1} \mu^*(\max(-w, -\frac{1}{n})) = \sup_{n\geq 1} \limsup_{\mathfrak{F}} \max(-1, -\frac{1}{nw}) =$$

$$= \sup_{n\geq 1} \max(-1, -\frac{1}{n}\cdot\frac{1}{\limsup_{\mathfrak{F}} w}),$$

i.e.

$$\limsup_{\mathfrak{F}} w > 0 .$$

Conversely, for $f \in \mathcal{R}^{\times}_{-}(\mu)$ we have

$$\sup_{n\geq 1} \mu^*(\max(f, -\frac{1}{n})) = \sup_{n\geq 1} \limsup_{\mathfrak{F}} \max(\frac{f}{w}, -\frac{1}{nw}) =$$

$$= \sup_{n\geq 1} \max[\mu^*(f), -\frac{1}{n}\cdot\frac{1}{\limsup_{\mathfrak{F}} w}] = 0 . \quad \square$$

Note that in the pathological Examples 11.5.2 we have $\lim_{\mathfrak{F}} w = 0$, so the truncation formula at 0 is not satisfied.

Finally, using the preceding Example 2 for the necessity and anticipating

Corollary 14.6 for the converse, we have :

$\mu^\bullet$ is of the form $m^\bullet$ for some finite content m iff

$$\lim_{\mathfrak{F}} w \quad \textit{exists in } \mathbb{R}_+^* .$$

11.7 In the following, let ν be an upper functional such that 1 is measurable.

THEOREM. *For every integrable function $f \leq 0$, the sets $\{f < \gamma\}$ and $\{f \leq \gamma\}$ are integrable for all but at most countably many $\gamma < 0$.*

Let Γ be a finite subset of $\mathbb{R}_-$ with $0 \in \Gamma$ and define $\omega := \min \Gamma$. For $\gamma \in \Gamma$ and $\gamma < 0$, respectively $\gamma > \omega$, denote by $\gamma\!\uparrow$ and $\gamma\!\downarrow$ the next element in Γ larger, respectively smaller, than γ. Then, for $\gamma < 0$, the function

$$h_\gamma := \frac{1}{\gamma\!\uparrow - \gamma}[\max(f,\gamma) - \max(f,\gamma\!\uparrow)]$$

is integrable since constants are measurable. Choose sets H_γ with

$$\{f < \gamma\} \subset H_\gamma \subset \{f \leq \gamma\} .$$

We have

$$-1_{H_{\gamma\uparrow}} \leq h_\gamma \leq -1_{H_\gamma} ,$$

hence

$$\nu(H_\gamma) - \nu_*(H_\gamma) = \nu(1_{H_\gamma}) + \nu(-1_{H_\gamma}) \leq \nu(-h_\gamma) + \nu(h_{\gamma\downarrow}) = \nu(h_{\gamma\downarrow}) - \nu(h_\gamma)$$

for $\omega < \gamma < 0$. Therefore,

$$\sum_{\omega<\gamma<0} \nu(H_\gamma) - \nu_*(H_\gamma) \leq \sum_{\omega<\gamma<0} \nu(h_{\gamma\downarrow}) - \nu(h_\gamma) = \nu(h_\omega) - \nu(h_{0\downarrow}) \leq -\nu(h_{0\downarrow}) .$$

This shows that for any integer $n \geq 1$ we have

$$\sum_{\gamma\in\mathbb{R},\gamma\leq-1/n} \nu(H_\gamma) - \nu_*(H_\gamma) \leq -\nu(h_{-1/n}) < \infty ,$$

and so $\nu(H_\gamma) - \nu_*(H_\gamma) > 0$ for at most countably many $\gamma \in \mathbb{R}_-^*$. □

REMARK. We have even shown that for arbitrarily chosen sets H_γ with

$$\{f < \gamma\} \subset H_\gamma \subset \{f \leq \gamma\} ,$$

only a countable number of them may fail to be integrable.

COROLLARY. *If* f *is a measurable function, then for every* $K \in \mathfrak{I}(\nu)$ *the sets*

$$\{f > \gamma\} \cap K \ , \ \{f \geq \gamma\} \cap K \ , \ \{f < \gamma\} \cap K \quad \textit{and} \quad \{f \leq \gamma\} \cap K$$

are integrable for all but at most countably many $\gamma \in \mathbb{R}$.

It is sufficient to prove that f satisfies both conditions defined by strict inequality, since $\mathfrak{I}(\nu)$ is a ring. Further we may suppose $f \leq 0$ since $-f^+$ and f_- are also measurable by Proposition 4.2, and

$$\{f > \gamma\} = \{-f^+ < -\gamma\} \ , \ \{f < \gamma\} = \{-f^+ > -\gamma\} \quad \text{for } \gamma > 0$$

as well as

$$\{f > \gamma\} = \{f_- > \gamma\} \ , \ \{f < \gamma\} = \{f_- < \gamma\} \quad \text{for } \gamma < 0$$

hold.

To this end, let $K \in \mathfrak{I}(\nu)$ and $n \in \mathbb{N}$ be given. Applying the theorem to the integrable function $h := \max(f,-n \cdot 1_K)$, we deduce that for $\gamma \in \,]-n,0[$, with at most countably many exceptions, the sets $\{h < \gamma\}$ and $\{h \leq \gamma\}$ are integrable, hence

$$\{f < \gamma\} \cap K = \{h < \gamma\} \quad \text{and} \quad \{f > \gamma\} \cap K = K \setminus \{h \leq \gamma\}$$

are integrable, too. Taking the union over $n \in \mathbb{N}$ of the exceptional sets of numbers γ , this proves the assertion. □

PROPOSITION. *Let* ν *satisfy both truncation formulas for an integrable function* $f \leq 0$ *and choose for every* $\gamma < 0$ *a set* H_γ *with*

$$\{f < \gamma\} \subset H_\gamma \subset \{f \leq \gamma\} .$$

Furthermore, define

$$f^*(\gamma) := - \nu_*(H_\gamma) \quad \textit{and} \quad f_*(\gamma) := - \nu(H_\gamma) \quad \textit{for } \gamma < 0$$

and

$$f^*(\gamma) := f_*(\gamma) := 0 \quad \textit{for } \gamma \geq 0 .$$

Then these functions are decreasing on $\mathbb{R}_-^*$ *with* $f_* \leq f^*$ *and essentially Riemann integrable on* $\mathbb{R}$ *with*

$$\nu(f) = \iota^\bullet(f_*) = \iota^\bullet(f^*) .$$

For $n \in \mathbb{N}$ and $\Gamma := \{-\frac{i}{2^n} : i = 0,\ldots,n\cdot 2^n\}$, with the functions h_γ as in the proof of the theorem, we have

$$\sum_{\omega\leq\gamma<0} h_\gamma = 2^n\cdot\max(f,-n)\,.$$

This implies that f^* and f_* are finite and

$$\sum_{\omega\leq\gamma<0} f^*(\gamma) + 2^n\cdot\nu(\max(f,-\frac{1}{2^n})) = -\sum_{\omega\leq\gamma<0}\nu_*(H_\gamma) + \nu(h_{0\downarrow}) \leq \sum_{\omega\leq\gamma<0}\nu(h_\gamma) =$$

$$= 2^n\cdot\nu(\max(f,-n)) = \sum_{\omega\leq\gamma<0}\nu_*(h_\gamma) \leq -\sum_{\omega\leq\gamma<0}\nu(H_\gamma) \leq \sum_{\omega<\gamma<0} f_*(\gamma)\,.$$

On the other hand, the inequalities

$$f^* \leq \sum_{\omega\leq\gamma<0} f^*(\gamma)\cdot 1_{[\gamma,\gamma\uparrow[}$$

and

$$\max(f_*,-n\cdot 1_{[-n,n]}) \geq \sum_{\omega<\gamma<0} f_*(\gamma)\cdot 1_{[\gamma\downarrow,\gamma[} - n\cdot 1_{[0\downarrow,0[}$$

give

$$\iota^\bullet(f^*) + \nu(\max(f,-\frac{1}{2^n})) \leq \nu(\max(f,-n)) \leq \iota_*(\max(f_*,-n\cdot 1_{[-n,n]})) + \frac{n}{2^n}\,,$$

from which we deduce by using the truncation formulas that

$$\iota^\bullet(f^*) \leq \nu(f) \leq \iota_\bullet(f_*) \leq \left\{\begin{matrix}\iota^\bullet(f_*)\\ \iota_\bullet(f^*)\end{matrix}\right\} \leq \iota^\bullet(f^*)\,,$$

since $f_* \leq f^*$.

Thus we have proved that f_* and f^* are essentially Riemann integrable (cf. Example 3.8.1) and that both essential integrals coincide with $\nu(f)$. □

EXAMPLE. In general, the exceptional set of numbers γ in the above proposition cannot be avoided.

In fact, Example 4.7.2 shows that for the integrable function f defined on $[0,1]$ by 0 at 0 and by

$$x\cdot\sin\frac{1}{x} - 1\,,$$

for $x \neq 0$, any set H with

$$\{f < -1\} \subset H \subset \{f \le -1\}$$

is not integrable, even not measurable.

11.8   Now we are in a position to give a complete characterization of measurability w.r.t. a finite content m on $\mathfrak{K}$:

THEOREM. *For any function f , the following assertions are equivalent* :

(i) f *is measurable w.r.t.* m .

(ii) f *is* $\mathcal{E}(\mathfrak{R}^{\times}(m))$–*measurable.*

(iii) f *is* $\mathfrak{R}^{\times}(m)$–*measurable.*

(iv)   *For any* $K \in \mathfrak{K}$, *the sets*

$$\{f > \gamma\} \cap K \quad , \textit{ equivalently } \{f \ge \gamma\} \cap K \ , \ \{f < \gamma\} \cap K \textit{ or } \{f \le \gamma\} \cap K \ ,$$

are integrable for all but at most countably many $\gamma \in \mathbb{R}$.

Moreover, the set of all measurable functions which are bounded below and positive outside an integrable set is the function cone $\mathcal{G}(\mathcal{E}(\mathfrak{R}^{\times}(m)))$.

By Remark 10.3.2, every function f with one of the properties stated in (iv) is $\mathfrak{R}^{\times}(m)$–measurable, since every integrable set is contained in some set $K \in \mathfrak{K}$ and

$$\{f \le \gamma\} \cap K = K \smallsetminus (\{f > \gamma\} \cap K) \ , \ \{f < \gamma\} \cap K = K \smallsetminus (\{f \ge \gamma\} \cap K) \ .$$

Furthermore, (iii) implies (ii) by Theorem 10.10. From Proposition 4.11 we infer that (ii) implies (i), since $\mathcal{E}_-(\mathfrak{R}^{\times}(m))$ is almost coinitial in $\mathscr{R}_-^{\bullet}(m)$ by Corollary 11.4.

The remaining implication is a consequence of Corollary 11.7. Finally, the last assertion follows from the formula

$$\mathcal{G}(\mathcal{E}(\mathfrak{R}^{\times}(m))) = \mathcal{H}(\mathcal{E}(\mathfrak{R}^{\times}(m))) \cap \mathcal{E}_-^{\uparrow}(\mathfrak{R}^{\times}(m))$$

(cf. Proposition 4.9). $\square$

REMARKS.

(1)   Example 4.7.2 together with the characterization given at the end of Example 11.6.3 shows that in general $\tilde{\mathscr{R}}^0(m)$ will not be a function cone.

(2) The theorem shows that set–theoretic measurability has to be stated in terms of $\mathfrak{M}^{\times}(m)$ instead of $\mathfrak{M}^{0}(m)$. However :

If there exists a sequence (K_n) *in* $\mathfrak{K}$ *such that every set* $K \in \mathfrak{K}$ *is contained in some* K_n, *e.g. if* $X \in \mathfrak{K}$, *then a function* f *is measurable iff, with the exception of at most countably many* $\gamma \in \mathbb{R}$, *the sets*

$$\{f > \gamma\} \;,\text{or}\; \{f < \gamma\} ,$$

are measurable.

This follows from Corollary 11.4 since, if f is measurable, then there is a universal at most countable set of exceptional numbers γ such that for all other $\gamma \in \mathbb{R}$ the sets

$$\{f < \gamma\} \cap K \;,\text{or}\; \{f < \gamma\} \cap K ,$$

are integrable for every $K \in \mathfrak{K}$. □

In Theorem 13.6.(iv) we shall see that in case m is a measure, a function f is measurable iff for every $\gamma \in \mathbb{R}$ the sets

$$\{f > \gamma\} \;,\text{or}\; \{f < \gamma\} ,$$

are measurable.

11.9 The following proposition shows that for a finite content m on $\mathfrak{K}$ the algebra $\mathfrak{M}^{0}(m)$ is the system of Carathéodory measurable sets :

PROPOSITION. *For every set* A, *the following assertions are equivalent :*

(i) A *is measurable w.r.t.* m.

(ii) *For every set* B, *equivalently every* $B \in \mathfrak{K}$, *we have*

$$m_{\times}(B) = m_{\times}(B \cap A) + m_{\times}(B \setminus A) .$$

(iii) *For every set* B, *equivalently every* $B \in \mathfrak{K}$, *we have*

$$m^{\times}(B) = m^{\times}(B \cap A) + m^{\times}(B \setminus A) .$$

Suppose first that A is measurable. In (ii), it is sufficient to prove the inequality

$$m_{\times}(B) \leq m_{\times}(B \cap A) + m_{\times}(B \setminus A) .$$

Let $f \in \mathcal{E}\_(\mathfrak{R}) - \mathcal{E}\_(\mathfrak{R}) \subset \mathcal{R}^{\times}(m)$ with $0 \leq f \leq 1_B$ be given. Then

$$\tilde{m}(f) = m_{\times}(f) = m_{\times}(\min(f, 1_A)) + m_{\times}(\min(f, 1_{\complement A})) \leq m_{\times}(B \cap A) + m_{\times}(B \setminus A),$$

since $\min(f, 1_A), \min(f, 1_{\complement A}) \in \mathcal{R}^{\times}(m)$ by the measurability of 1_A. This implies the inequality.

Conversely, if A satisfies condition (ii) for all sets $B \in \mathfrak{R}$, then

$$m_{\times}(B \cap A) \leq m^{\times}(B \cap A) = m^{\times}(1_B - 1_{B \setminus A}) \leq m^{\times}(1_B) + m^{\times}(-1_{B \setminus A}) =$$
$$= m_{\times}(B) - m_{\times}(B \setminus A) = m_{\times}(B \cap A) < \infty .$$

This proves that $A \cap B$ is integrable, hence A is measurable by Corollary 11.4.

The proof for the equivalence of (i) and (iii) is similar. □

§ 12 REGULARITY OF CONTENTS

THROUGHOUT THIS SECTION,

$\mathfrak{K}$ AND $\mathfrak{G}$ ARE COMPATIBLE LATTICES AND m IS A FINITE CONTENT ON $\mathfrak{K}$.

12.1 We first study the set functions m^* and m_*.

LEMMA. *For every increasing positively homogeneous functional* μ *on* $\mathcal{E}(\mathfrak{K},\mathfrak{G})$ *and every set* A, *we have*

$$\mu^*(A) = \inf_{G\in\mathfrak{G},\, G\supset A} \mu^*(G)$$

and

$$\mu_*(A) = \sup_{K\in\mathfrak{K},\, K\subset A} \mu_*(K) .$$

For $e \in \mathcal{E}(\mathfrak{K},\mathfrak{G})$ with $e \geq 1_A$ and $\alpha > 1$, we have

$$G := \{\alpha e > 1\} \in \mathfrak{G} \quad \text{and} \quad 1_G \leq \alpha e ,$$

hence $\mu(G) \leq \alpha\cdot\mu(e)$. This proves that

$$\inf_{G\in\mathfrak{G},\, G\supset A} \mu(G) \leq \mu^*(A) .$$

On the other hand, if $e \in \mathcal{E}_-$ with $-e \leq 1_A$, we have

$$-e \leq 1_{\{e<0\}} = 1_{\{e\leq-\gamma\}}$$

for some $\gamma > 0$, and

$$A \supset \{e \leq -\gamma\} \in \mathfrak{K} .$$

Therefore,

$$-\mu(e) \leq -\mu(-1_{\{e\leq-\gamma\}}) = \mu_*(\{e \leq -\gamma\}) ,$$

which proves that

$$\mu_*(A) \leq \sup_{K\in\mathfrak{K},\, K\subset A} \mu_*(K) .$$

The reverse inequalities are obvious. □

A special case of the following similar result for general function cones was already proved in Lemma 8.10 :

ADDENDUM. *Let* $\mathcal{S}$ *be a function cone and* $A \subset X$. *If* μ *is an increasing posi-*

tively homogeneous functional on $\mathcal{S}$ *then*

$$\mu^*(A) = \inf_{G \in \mathfrak{G}(\mathcal{S}),\, G \supset A} \mu^*(G)\,.$$

If $\mathcal{S}_-$ *is strongly Stonian, and if* μ *is linear and satisfies the truncation formula at* 0 *on* $\mathcal{S}_-$ *, then*

$$\mu_*(A) = \sup_{K \in \mathfrak{K}(\mathcal{S}),\, K \subset A} \mu_*(K)\,.$$

The first assertion is proved as in the lemma. To prove the second one, note that for $s \in \mathcal{S}_-$ with $-s \le 1_A$ and $n \ge 1$ we always have

$$(s + \tfrac{1}{n})_- \ge -1_{\{s \le -1/n\}}\,,$$

hence

$$- \mu(s) = - \mu(\max(s, -\tfrac{1}{n})) - \mu((s + \tfrac{1}{n})_-) \le$$

$$\le - \mu(\max(s, -\tfrac{1}{n})) - \mu^*(-1_{\{s \le -1/n\}}) = - \mu(\max(s, -\tfrac{1}{n})) + \mu_*(\{s \le -\tfrac{1}{n}\})\,,$$

and therefore $- \mu(s) \le \sup_{n \in \mathbb{N}} \mu_*(\{s \le -\frac{1}{n}\})$. Since

$$\{s \le -\tfrac{1}{n}\} = \{ns \le -1\} \in \mathfrak{K}(\mathcal{S})\,,$$

the result follows. □

For contents, the lemma yields the following description of the set functions defined by the corresponding upper and lower functionals :

COROLLARY.

(i) *For every set* A *, we have*

$$m_*(A) = \sup_{K \in \mathfrak{K},\, K \subset A} m(K)$$

and

$$m^*(A) = \inf_{G \in \mathfrak{G},\, G \supset A} m(G)\,.$$

In particular, the set function m_* *is independent of* $\mathfrak{G}$.

(ii) *For* $G \in \mathfrak{G}(\mathfrak{K})$ *and* $F \in \mathfrak{G}(\mathfrak{K})^{\complement}$ *, we have*

$$m(G) = m_\times(G) = \sup_{K \in \mathfrak{K}} m(K) - m(K \setminus G)$$

and

$$m_\times(F) = m_*(F) = \sup_{K \in \mathfrak{K}} m(F \cap K)\,.$$

Since $m_*(K) = -m(-1_K) = m(K)$ for $K \in \mathfrak{K}$ and $m^*(G) = m(G)$ for $G \in \mathfrak{G}$, assertion (i) follows directly from the lemma.

The formulas in (ii) are consequences of Corollary 11.4. In fact, we have

$$m(G) = m_\times(G) = \sup_{K\in\mathfrak{K}} m_\times(G \cap K) = \sup_{K\in\mathfrak{K}} m(K) - m(K \setminus G),$$

whereas the right hand side of the second formula is equal to $m_\times(F)$ and, by (i), also to $m_*(F)$. □

REMARK. The above formulas show how $m_*(A)$ and $m^*(A)$ for arbitrary sets A can be calculated from the values of m on $\mathfrak{K}$, although $m^*(A)$ depends on the choice of $\mathfrak{G}$. Note that $m_*(K) = m(K)$ and $m^*(G) = m(G)$ for $K \in \mathfrak{K}$ and $G \in \mathfrak{G}$, whereas in general $m^*(K)$ differs from $m(K)$ and $m_*(G)$ differs from $m(G)$.

12.2 We now discuss how regularity properties of the functional m on $\mathscr{E}(\mathfrak{K},\mathfrak{G})$ can be translated into properties of the content m. To this end, using Corollary 12.1, we give the following

DEFINITION. If m coincides with m^* on $\mathfrak{K}$, i.e. if

$$m(K) = \inf_{G\in\mathfrak{G},\, G\supset K} \sup_{L\in\mathfrak{K}} m(L) - m(L \setminus G)$$

holds for all $K \in \mathfrak{K}$, then m is said to be *$\mathfrak{G}$-regular*.

If

$$m(L) \le m(K) + m_*(L \setminus K)$$

holds for all $K,L \in \mathfrak{K}$ with $K \subset L$, then m is said to be *semiregular*.

This terminology will sometimes also be used for arbitrary increasing, not necessarily finite set functions m, with the convention

$$m_*(A) := \sup_{K\in\mathfrak{K},\, K\subset A} m(K)$$

for every set A. In the sequel, a semiregular content however is always assumed to be finite, unless otherwise stated.

If for every $K \in \mathfrak{K}$ there is a set $G \in \mathfrak{G}$ with $G \supset K$ and $m_*(G) < \infty$, then m is said to be *$\mathfrak{G}$-bounded*, and if

$$m(K) = \inf_{G\in\mathfrak{G},\, G\supset K} m_*(G) ,$$

then m is said to be *continuous on the right w.r.t.* $\mathfrak{G}$.

Remarks.

(1) *Every* $\mathfrak{G}$*-regular content* m *is semiregular and continuous on the right w.r.t.* $\mathfrak{G}$.

Indeed, for $K \subset L$ we have

$$m(L) - m(K) = m_*(L) - m^*(K) = m_*(1_L) + m_*(-1_K) \leq$$
$$\leq m_*(1_L - 1_K) = m_*(L \setminus K) ,$$

and

$$m(K) = m^*(K) = \inf_{G\in\mathfrak{G},\, G\supset K} m(G) \geq \inf_{G\in\mathfrak{G},\, G\supset K} m_*(G) \geq m(K)$$

by Corollary 12.1.(i). □

Continuity on the right w.r.t. $\mathfrak{G}$ means that for every $K \in \mathfrak{K}$ and every $\varepsilon > 0$ there exists a set $G \in \mathfrak{G}$ with $G \supset K$ such that

$$m(L) - m(K) \leq \varepsilon$$

for all $L \in \mathfrak{K}$ with $K \subset L \subset G$. In this case, m is obviously $\mathfrak{G}$–bounded.

In the following theorem we will prove that, conversely, semiregularity and $\mathfrak{G}$–boundedness imply $\mathfrak{G}$–regularity.

(2) *If* $\mathfrak{K}$ *is a ring, then every finite content on* $\mathfrak{K}$ *is semiregular, and even* $\mathfrak{G}$*-regular if in addition* $\mathfrak{K} \subset \mathfrak{G}$.

(3) *Let* m *be an increasing set function on* $\mathfrak{K}$ *with* $m(\emptyset) = 0$. *If* m *is additive and semiregular, then*

$$m\,(L) = m\,(K) + m_*(L \setminus K)$$

for $K,L \in \mathfrak{K}$ *with* $K \subset L$, *and* m *is a semiregular, not necessarily finite content.*

Indeed, for $H \in \mathfrak{K}$ with $H \subset L \setminus K$ we obtain

$$m(K) + m(H) = m(K \cup H) \leq m(L) ,$$

which proves that equality holds. From this, for arbitrary $K,L \in \mathfrak{K}$, we deduce

$$m(K \cup L) + m(K \cap L) = m(K) + m_*(K{\cup}L \setminus K) + m(K \cap L) =$$
$$= m(K) + m_*(L \setminus K{\cap}L) + m(K \cap L) = m(K) + m(L) .$$

This proves that m is strongly additive. □

(4) *For every upward directed family (m_i) of semiregular, possibly non-finite contents on $\mathfrak{K}$, the set function $m := \sup m_i$ is a semiregular, possibly non-finite content.*

In fact,

$$m(L) = \sup_i m_i(L) \le \sup_i m_i(K) + \sup_i \sup_{H\in\mathfrak{K},\, H\subset L\setminus K} m_i(H) =$$
$$= m(K) + \sup_{H\in\mathfrak{K},\, H\subset L\setminus K} m(H) = m(K) + m_*(L \setminus K)$$

holds for $K,L \in \mathfrak{K}$ with $K \subset L$. Finally, m is strongly additive since (m_i) is upward directed. □

12.3 THEOREM. *The functional m on $\mathcal{E}(\mathfrak{K},\mathfrak{G})$ is semiregular, respectively regular, iff the content m on $\mathfrak{K}$ is semiregular, respectively $\mathfrak{G}$-regular.*

Moreover, $\mathfrak{G}$-regularity is equivalent to semiregularity and $\mathfrak{G}$-boundedness. In particular, every bounded semiregular content is $\mathfrak{G}(\mathfrak{K})$-regular.

If m is semiregular on $\mathcal{E}(\mathfrak{K},\mathfrak{G})$, then for all $K,L \in \mathfrak{K}$ with $K \subset L$ we have

$$m(L) = -\, m(-1_L) \le -\, m(-1_K) + m_*(-1_K - (-1_L)) = m(K) + m_*(L \setminus K) .$$

Therefore, m is semiregular on $\mathfrak{K}$.

Conversely, since m is difference-regular by definition, in view of Remark 6.7 we have to prove that

$$m(e) \le m(d) + m_*(e - d)$$

for all $d,e \in \mathcal{E}_-(\mathfrak{K})$ with $d \le e$. To this end, let (γ_i) be a strictly decreasing sequence of real numbers with $\gamma_0 = 0$ and containing all elements of $d(X) \cup e(X)$. If we define

$$L_i := \{d \le \gamma_i\} \quad \text{and} \quad K_i := \{e \le \gamma_i\} ,$$

then with the notations of Remark 10.4.1 there exists $n \in \mathbb{N}$ such that

$$d = -\sum_{i=1}^{n} \delta_i \cdot 1_{L_i} \quad \text{and} \quad e = -\sum_{i=1}^{n} \delta_i \cdot 1_{K_i} .$$

Let $\varepsilon > 0$ be given and define $\delta := \max_{i=1,\ldots,n} \delta_i$. Since $K_i \subset L_i$, there exist $M_i \in \mathfrak{K}$ with $M_i \subset L_i \setminus K_i$ and

$$m(L_i) \leq m(K_i) + m(M_i) + \tfrac{\varepsilon}{n\delta} .$$

Then $c := -\sum_{i=1}^{n} \delta_i \cdot 1_{M_i}$ belongs to $\mathcal{E}_-(\mathfrak{K})$, we have $-c \leq e - d$ and

$$m(d) + m_*(e - d) \geq m(d) - m(c) = -\sum_{i=1}^{n} \delta_i \cdot [m(L_i) - m(M_i)] \geq$$

$$\geq -\sum_{i=1}^{n} \delta_i \cdot [m(K_i) + \tfrac{\varepsilon}{n\delta}] \geq m(e) - \varepsilon .$$

This proves that m is semiregular on $\mathcal{E}(\mathfrak{K},\mathfrak{G})$.

If m is regular on $\mathcal{E}(\mathfrak{K},\mathfrak{G})$, then for $K \in \mathfrak{K}$

$$m(K) = -m(-1_K) = -m_*(-1_K) = m^*(K) ,$$

which proves that m is $\mathfrak{G}$-regular. The latter in turn implies that m is semiregular and $\mathfrak{G}$-bounded (cf. Remark 12.2.1). From this we infer that m is regular on $\mathcal{E}(\mathfrak{K},\mathfrak{G})$. In fact, by Theorem 6.9.(iii) it suffices to prove that m is $\mathcal{E}(\mathfrak{K},\mathfrak{G})$-bounded below. This boundedness is proved as follows: For $e \in \mathcal{E}_-(\mathfrak{K})$ there exist $\alpha > 0$ and $K \in \mathfrak{K}$ with $e \geq -\alpha \cdot 1_K$ and then, because m is $\mathfrak{G}$-bounded, there exists $G \in \mathfrak{G}$ with $G \supset K$ and $m_*(G) < \infty$. This shows that $e \geq -\alpha \cdot 1_G$ and $m_*(\alpha \cdot 1_G) = \alpha \cdot m_*(G) < \infty$. □

COROLLARY. *There is a bijection between the semiregular, respectively $\mathfrak{G}$-regular, contents m on $\mathfrak{K}$ and the semiregular, respectively regular, linear functionals μ on $\mathcal{E}(\mathfrak{K},\mathfrak{G})$, given by*

$$\mu(-1_K) = -m(K) \quad \textit{and} \quad \mu(1_G) = m_*(G)$$

for $K \in \mathfrak{K}$ and $G \in \mathfrak{G}$.

This follows immediately from Corollaries 11.3 and 12.1.(i) and from Proposition 6.7. □

12.4 In classical measure theory, the following regularity properties play a vital part :

DEFINITION. A set function l is said to be *inner* $\mathfrak{K}$*-regular*, respectively *outer* $\mathfrak{G}$*-regular, at a set* A if the formula

$$l(A) = \sup_{K\in\mathfrak{K},\ K\subset A} l(K)\ , \quad \text{respectively} \quad l(A) = \inf_{G\in\mathfrak{G},\ G\supset A} l(G)\ ,$$

is meaningful and holds.

m_* and m^* are inner $\mathfrak{K}$-regular, respectively outer $\mathfrak{G}$-regular at every set.

This follows from Corollary 12.1.(i). □

PROPOSITION. *If m is semiregular, then*

$$m_*(A) = m_\bullet(A)$$

for any set A. If A is measurable, then moreover $m_*(A) = m^\bullet(A)$, i.e. $m^\bullet$ is an inner $\mathfrak{K}$-regular content on $\mathfrak{R}^0(m)$.

This follows immediately from Proposition 6.8 and Corollary 11.4. □

THEOREM. *If m is $\mathfrak{G}$-regular, then m^* is an auto-determined upper functional with*

$$m^\bullet \le m^* \le m^\times .$$

The ring $\mathfrak{R}^*(m)$ consists of all measurable sets A with $m^*(A) < \infty$. Furthermore, m^* is an outer $\mathfrak{G}$-regular content on $\mathfrak{R}^0(m) = \mathfrak{G}(\mathfrak{R}^*(m))$,

$$m^*(A) = m^\times(A) \quad \text{if } m^\times(A) < \infty ,$$

and

$$m^\bullet(A) = m^*(A) \quad \text{if } m^*(A) < \infty ,$$

i.e. $m^\bullet$ is outer $\mathfrak{G}$-regular on $\{m^* < \infty\}$.

This follows immediately from Proposition 6.1, Theorem 6.4 and Proposition 11.4.(iii). □

COROLLARY. *If m is semiregular, then a set A is measurable w.r.t. m iff*

$$m_*(B) = m_*(B \cap A) + m_*(B \setminus A)$$

for every set B, or equivalently for every $B \in \mathfrak{K}$.

If m is $\mathfrak{G}$-regular, then A is measurable w.r.t. m iff

$$m^*(B) = m^*(B \cap A) + m^*(B \setminus A)$$

for every set B, equivalently for every $B \in \mathfrak{G}$ or every $B \in \mathfrak{K}$.

For the proof we make use of Proposition 11.9. The first assertion follows immediately from the coincidence of the set functions m_* and $m_\times$. In the regular case, since m^* is a content on $\mathfrak{M}^0(m) \supset \mathfrak{G}$, the second formula is necessarily valid for $B \in \mathfrak{G}$ and therefore, by Corollary 12.1.(i), for every set B. Conversely, this implies the formula 11.9.(iii) for $B \in \mathfrak{K}$, since $m^\times = m^*$ on $\{m^\times < \infty\}$. □

REMARK. In the case of a ring $\mathfrak{K}$, the classical concept of measurability in the sense of Carathéodory coincides with the one of measurability w.r.t. m, since m is $\mathfrak{K}$-regular.

12.5 **DEFINITION.** Let X be a Hausdorff space with $\mathfrak{K}(X)$ and $\mathfrak{G}(X)$ denoting the systems of compact and open subsets respectively. A *Radon measure* on X is a $\mathfrak{G}(X)$-regular content on $\mathfrak{K}(X)$.

Thus, Radon measures are exactly the semiregular contents on $\mathfrak{K}(X)$ which are $\mathfrak{G}(X)$-bounded. By Proposition 12.4 and a usual compactness argument, $\mathfrak{G}(X)$-boundedness is equivalent to the fact that m is *locally bounded*. By this is meant that every point in X has a neighbourhood U with $m_*(U) < \infty$. If X is locally compact, then obviously every semiregular content on $\mathfrak{K}(X)$ is a Radon measure.

In Example 13.4.2 we prove that every Radon measure is a measure in the sense of Definition 13.1.

THEOREM. *There is a bijection between the Radon integrals μ and the Radon* *measures* m *on* X *, given by*

$$\mu(s) = m^{\bullet}(s) \quad \textit{and} \quad m(K) = -\mu(-1_K)$$

for $s \in \mathcal{S}(X)$ *and* $K \in \mathfrak{K}(X)$ *. We have* $\mu^{\bullet} = m^{\bullet}$ *.*

Let μ be a Radon integral. Then μ defines a content m on $\mathfrak{K}(X)$ with $m_* = \mu_*$ on sets by the first formula of Lemma 8.10. This shows that m is semiregular. It is $\mathfrak{G}(X)$-bounded by the second formula, hence regular.

Let m be a Radon measure. Then m is a regular linear functional on $\mathcal{E} := \mathcal{E}(\mathfrak{K}(X),\mathfrak{G}(X))$, which is obviously $\mathcal{S}(X)$-tight. Therefore, m has the Bourbaki property by Proposition 8.2. From Theorem 8.1 and Corollary 8.2 we infer that $\mu := m^{\phi}$ is the only Radon integral extending m , since $\mathcal{S}(X) = \mathcal{E}_{\phi}$ by Example 10.9.

By Example 10.10, every function in $\mathcal{S}(X)$ is $\mathcal{E}$-measurable, hence

$$\mu = m^{\phi} = m_* = m^{\bullet} \quad \text{on } \mathcal{S}(X) .$$

This yields

$$m^*(f) \geq \mu^*(f) = \inf_{s\in\mathcal{S}(X),\, s\geq f} m^{\bullet}(s) \geq m^{\bullet}(f) = m^*(f)$$

for every function f which has an integrable majorant w.r.t. m . For any f we therefore have

$$\mu^{\bullet}(f) = \inf_{u\in\mathcal{E}_-} \sup_{v\in\mathcal{E}_-} \mu^*(\mathrm{med}(f,-v,u)) =$$
$$= \inf_{u\in\mathcal{E}_-} \sup_{v\in\mathcal{E}_-} m^*(\mathrm{med}(f,-v,u)) = m^{\bullet}(f) ,$$

since $\mathcal{E}_-$ is coinitial in $\mathcal{S}_-(X)$. □

REMARKS.

(1) Note that for the proof of the equality $\mu^{\bullet} = m^{\bullet}$, only abstract Riemann concepts are needed. In fact, a proof avoiding the Bourbaki property, and extending the theorem to semiregular linear functionals on $\mathcal{S}(X)$ and semiregular contents on $\mathfrak{K}(X)$, can be based, using Example 10.10, on essential integration and the representation theorem Corollary 7.5 (cf. Remark 14.7).

This "simplicity" is due to the generalization of the concept of Radon integrals as regular linear functionals on $\mathcal{S}(X)$ for arbitrary Hausdorff spaces X rather than positive linear forms on $\mathcal{K}(X)$ for locally compact X , but is paid for by more complicated construction methods for such integrals. Actually, we understand the Bourbaki and the Daniell theory more as a method of extension of inte-

grals, from $\mathcal{S}$ to $\mathcal{S}_\phi$ respectively to $\mathcal{S}^\sigma$, than as an integration theory.

In the locally compact case, the correspondence between positive linear forms on $\mathcal{K}(X)$ and Radon measures on X is more subtile. In fact, one has either to extend the linear forms to $\mathcal{S}(X)$ via the Bourbaki property (cf. Corollary 8.3) and then use the correspondence between Radon integrals and measures outlined above, or to use the Representation Theorem 7.6 (cf. Example 14.14.3).

(2) An elementary way to define μ from m may use the one-sided relatively uniform approximation

$$\mathcal{S}_-(X) = \mathcal{A}(\mathcal{E}_-(\mathfrak{K}(X))$$

(cf. Lemma 7.4 and Example 10.10), and the version of the Extension Theorem outlined in Addendum 6.5. In this case, the regularity of μ has to be proved by arguments similar to those given in § 6.

The following Example 1, due to *D.H. Fremlin*, shows that semiregularity is strictly weaker than regularity.

EXAMPLES.

(1) (cf. *Fremlin* [1975], p. 104 et seq.) A set G in $X := \mathbb{R}^2$ is defined to be open if the intersection of G with every horizontal and every vertical line is open in Euclidean topology. Obviously, $\mathfrak{G}(X)$ induces the Euclidean topology on each such line. Since $\mathfrak{G}(X)$ is finer than the product topology on $\mathbb{R}^2$, the projections

$$p_i : (x_1, x_2) \longmapsto x_i$$

are continuous for $i = 1,2$, and X is Hausdorff.

We claim that a set $K \subset X$ is compact iff it is the union of finitely many compact subsets of horizontal or vertical lines. By continuity of the projections, it is sufficient to prove that there exist finite sets $K_1, K_2 \subset \mathbb{R}$ such that

$$K \subset p_1^{-1}(K_1) \cup p_2^{-1}(K_2).$$

In fact, otherwise we could choose a sequence (x^n) in K inductively, such that each of the sequences $(p_i(x^n))$ has distinct elements. Let $x \in K$ be a cluster point of (x^n). Removing x from the sequence if necessary, we may assume that $x \in X \setminus \{x^n : n \in \mathbb{N}\}$. But this set being obviously open, we have a contradiction.

Define a semiregular content m on $\mathfrak{K}(X)$ by

$$m(K) = \sum_{x_1 \in \mathbb{R}} \lambda(K \cap p_1^{-1}(x_1)) + \sum_{x_2 \in \mathbb{R}} \lambda(K \cap p_2^{-1}(x_2)) .$$

Note that for $i = 1,2$ there are only finitely many x_i such that $K \cap p_i^{-1}(x_i)$ has strictly positive one-dimensional Lebesgue measure.

We claim that m is not $\mathfrak{G}$-regular for any lattice $\mathfrak{G}$ compatible with $\mathfrak{K}(X)$. Since $\mathfrak{G} \subset \mathfrak{G}(\mathfrak{K}(X)) = \mathfrak{G}(X)$, it is sufficient to prove that m is not locally bounded. Indeed, for any open neighbourhood G of an arbitrary point, the projection $p_1(G)$ contains a non-degenerate interval G_1, and $G \cap p_1^{-1}(x_1)$ contains a non-degenerate compact interval for every $x_1 \in G_1$. Hence $m_*(G) = \infty$.

The example shows moreover that a semiregular linear functional on a min-stable function cone $\mathcal{S}$ need not be regular, not even $\mathcal{G}(\mathcal{S})$-bounded below, using Remark 1.

(2) The Lebesgue measure on $\mathbb{R}^n$ can now be constructed from the Lebesgue integral of Example 8.2. For a set-theoretical construction, we refer to Example 14.13.1.

The following proposition shows that our definition of a Radon measure is equivalent to that of *Choquet*, stated as (ii):

PROPOSITION. *Let X be a Hausdorff space and m a finite content on $\mathfrak{K}(X)$. The following assertions are equivalent:*

(i) *m is a Radon measure.*

(ii) *m is continuous on the right w.r.t. $\mathfrak{G}(X)$.*

(iii) *m is locally bounded and has the Bourbaki property, i.e.*

$$m(\bigcap K_i) = \inf m(K_i)$$

holds for every downward directed family (K_i) in $\mathfrak{K}(X)$.

By Remark 12.2.1, assertion (i) implies (ii). If (ii) holds, then m is locally bounded since every one-point set is compact. Let $G \in \mathfrak{G}(X)$ be a neighbourhood of $K := \bigcap K_i$. So $K_i \subset G$ for some i and therefore

$$m(K) \le m(K_i) \le m_*(G) .$$

Taking the infimum on G yields the Bourbaki property.

It remains to prove that (iii) implies (i), i.e. the semiregularity of m. For $K,L \in \mathfrak{K}(X)$ with $K \subset L$, consider the downward directed family of all compact neighbourhoods H of K in the subspace L, whose intersection is obviously K. Thus, for any $\varepsilon > 0$, there exists some H with

$$m(H) \le m(K) + \varepsilon .$$

If A denotes the closure of $L \setminus H$ in L, then A is compact, contained in $L \setminus K$, and $L = H \cup A$. This gives

$$m(L) = m(H \cup A) \le m(H) + m(A) \le m(K) + m(A) + \varepsilon \le$$
$$\le m(K) + m_*(L \setminus K) + \varepsilon$$

and proves (i). □

12.6 Let l be a (not necessarily finite) content defined on a lattice $\mathfrak{L}$. By l^∇ we denote the *finite part* of l, i.e. the restriction of l to the lattice

$$\mathfrak{L}^\nabla := \{l < \infty\}$$

of all sets $L \in \mathfrak{L}$ with $l(L) < \infty$.

Lemma. *We have*

$$l^\nabla_* = l^\nabla_\times = l^{\nabla\bullet} \le l = l^{\nabla\times} \quad on\ \mathfrak{L} ,$$

with equality holding on $\mathfrak{L}^\nabla$.

By Corollary 11.4, we have $\mathfrak{L}^c \subset \mathfrak{G}(\mathfrak{L}) \subset \mathfrak{G}(\mathfrak{L}^\nabla) \subset \mathfrak{N}^0(l^\nabla)$, which by Corollary 12.1.(ii) and Proposition 6.3 proves that $l^\nabla_* = l^\nabla_\times = l^{\nabla\bullet} \le l^{\nabla\times}$ on $\mathfrak{L}$ and $l^{\nabla\bullet} = l = l^{\nabla\times}$ on $\mathfrak{L}^\nabla$. It therefore remains to prove that $l(L) < \infty$ whenever $L \in \mathfrak{L}$ and $l^{\nabla\times}(L) < \infty$. But in this case there are functions $d,e \in \mathcal{E}_-(\mathfrak{L}^\nabla)$ with $1_L \le d - e \le -e \le \alpha \cdot 1_K$ for suitable $\alpha > 0$ and $K \in \mathfrak{L}^\nabla$, hence $L \subset K$ and therefore $l(L) \le l(K) < \infty$. □

In our set-up, we are forced to consider $\mathcal{E}_-(\mathfrak{L}^\nabla)$ instead of $\mathcal{E}_-(\mathfrak{L})$ to avoid the functional value $-\infty$. However, we still get all informations on l as the lemma shows. Note that classical integration theory is based on the functional $l^{\nabla\times}$. More important however is the theory of essential integration, based on the functional $l^{\nabla\bullet}$. Therefore, the coincidence of this functional with l on $\mathfrak{L}$ delimits an important class of contents :

DEFINITION. The content l is said to be *semifinite at* $L \in \mathfrak{L}$ if $l(L)$ coincides with $l_*^\nabla(L) = l_\times^\nabla(L) = l^{\nabla\bullet}(L)$, i.e. if

$$l(L) = \sup_{K \in \mathfrak{L},\ K \subset L,\ l(K) < \infty} l(K) .$$

If this holds for all $L \in \mathfrak{L}$, then l is said to be *semifinite*.

REMARKS.

(1) If $\mathfrak{L}$ is a ring, then $\mathfrak{L}$ is compatible with the ring $\mathfrak{L}^\nabla$, and the finite part l^∇ of l is an $\mathfrak{L}$-regular content on $\mathfrak{L}^\nabla$. The classical integration theory (in the sense of Riemann) is based on the increasing linear functional κ on $\mathfrak{F}(\mathfrak{L}^\nabla,\mathfrak{L})$, specified by $\kappa(L) = l(L)$ for $L \in \mathfrak{L}$.

κ is the restriction of $l^{\nabla\times}$, and we have $\kappa^* = \kappa^\times = l^{\nabla\times}$. In particular, κ is regular, semiregular or difference-regular iff the content l is semifinite, equivalently iff κ coincides with the functional l^∇.

This follows from the above lemma. □

(2) *The content $m^\bullet$ on $\mathfrak{R}^0(m)$ is semifinite.*

This follows immediately from Corollary 11.4 since

$$\mathfrak{R}^\times(m) \subset \mathfrak{R}^\bullet(m) = \mathfrak{R}^0(m)^\nabla . \quad \square$$

(3) *l is semifinite iff*

$$l_*(A) = l_*^\nabla(A)$$

for every subset A. In this case, l is semiregular iff its finite part l^∇ is semiregular.

In fact, the first part and the necessity of the second part being obvious, from the semiregularity of l^{∇} we infer that

$$l(L) = \sup_{H \in \mathfrak{L}^{\nabla},\, H \subset L} l(H) \leq$$

$$\leq \sup_{H \in \mathfrak{L}^{\nabla},\, H \subset L} l^{\nabla}(H \cap K) + \sup_{H \in \mathfrak{L}^{\nabla},\, H \subset L} l^{\nabla}_{*}(H \setminus H \cap K) \leq$$

$$\leq l(K) + l_{*}(L \setminus K)$$

for $K, L \in \mathfrak{K}$ with $K \subset L$. $\square$

NOTATION. By integration theory w.r.t. the content l we always understand the integration theory w.r.t. its finite part l^{∇}. When confusion appears unlikely, we denote the functional l^{∇} again by l.

PROPOSITION. *A set function l on $\mathfrak{L}$ is a semiregular semifinite content iff there exists an upward directed family (l_i) of $\mathfrak{S}(\mathfrak{L})$-regular bounded contents on $\mathfrak{L}$ with*

$$l = \sup l_i \,.$$

By Remark 12.2.4, the condition is sufficient. Conversely, define l_K on $\mathfrak{L}$ by

$$l_K(L) := l(L \cap K)$$

for $K \in \mathfrak{L}^{\nabla}$. Then (l_K) is an upward filtering family of bounded contents with $l = \sup l_K$, since l is semifinite. By Theorem 12.3 it remains to prove that each l_K is semiregular, which is almost obvious. $\square$

12.7 In the following, let $\mathfrak{L}$ be a lattice containing $\mathfrak{K}$. We describe precisely the smallest and largest extensions of contents from $\mathfrak{K}$ to $\mathfrak{L}$. Based on the functional analytic extension theorems, some of the results seem to be new even in the classical setting.

PROPOSITION.

(i) *For every extension l of m to a content on $\mathfrak{L}$, we have*

$$l^\times \leq m^\times \;\; ; \;\; \textit{in particular, } m_\times \leq l \leq m^\times \;\; \textit{on } \mathfrak{L}.$$

If all sets in $\mathfrak{L}$ are measurable w.r.t. m, e.g. if $\mathfrak{L}$ is contained in the algebra generated by $\mathfrak{S}(\mathfrak{K})$, then the following assertions hold:

(ii) *Among all extensions l of m to a content on $\mathfrak{L}$, the set function*

$$q := m_\times = m^\bullet$$

is the smallest, and the only one with

$$l(L) = \sup_{K\in\mathfrak{K}} \; l(L \cap K) \quad \textit{for all } L \in \mathfrak{L},$$

whereas $r := m^\times$ is the largest.

Moreover, q is semifinite, and we have $q^\bullet = m^\bullet$ and $r^\times = m^\times$.

(iii) *If m is semiregular, then q is semiregular and the only extension of m to a content on $\mathfrak{L}$ which is inner $\mathfrak{K}$-regular.*

Since the finite part of l extends m to $\mathscr{E}_-(\mathfrak{L}^\nabla) \supset \mathscr{E}_-(\mathfrak{K})$, the inequality $l^\times \leq m^\times$ follows from Proposition 6.5.(i). This implies (i) since by Lemma 12.6 we have

$$m_\times \leq l_\times \leq l^\times = l \leq m^\times \quad \text{on } \mathfrak{L}.$$

If all sets in $\mathfrak{L}$ are measurable w.r.t. m, then (i) and Corollary 11.4 show that $q = m_\times = m^\bullet$ is the smallest and $r = m^\times$ is the largest extension of m to a content on $\mathfrak{L}$. Moreover, q satisfies the formula on $\mathfrak{L}$ and thus is semifinite. Conversely, since $L \cap K \in \mathfrak{M}^\times(m)$ for all $L \in \mathfrak{L}$ and $K \in \mathfrak{K}$, we have

$$l(L \cap K) = m_\times(L \cap K),$$

hence $l = q$. Since $\mathscr{E}_-(\{q < \infty\}) \subset \mathscr{R}^\bullet_-(m)$, the equality $q^\bullet = m^\bullet$ is a consequence of Corollary 7.3, as well as the semiregularity of q^∇ and hence, by Remark 12.6.3, that of q, if m is semiregular. The last assertion in (ii) results from Proposition 6.5.(ii), since $\mathscr{E}_-(\{r < \infty\}) \subset \mathscr{R}^\times_-(m)$, whereas the inner $\mathfrak{K}$-regularity of q in (iii) follows from Proposition 12.4. □

THEOREM. *If all sets in $\mathfrak{L}$ are measurable w.r.t. m, e.g. if $\mathfrak{L}$ is contained in the algebra generated by $\mathfrak{S}(\mathfrak{K})$, and if m is $\mathfrak{S}$-regular, then $r := m^*$ is an extension of m to a content on $\mathfrak{L}$ with $r^\bullet = m^\bullet$. Its finite part is semiregular, and regular w.r.t. every lattice compatible with $\mathfrak{L}$ and containing $\mathfrak{S}$.*

If moreover $\mathfrak{G} \subset \mathfrak{L}$, *then among all extensions of* m *to a content on* $\mathfrak{L}$, *the set function* r *is the largest one which is inner* $\mathfrak{K}$*-regular on* $\mathfrak{G}$, *and the smallest one which is outer* $\mathfrak{G}$*-regular. In particular, it is the only one which is inner* $\mathfrak{K}$*-regular on* $\mathfrak{G}$ *and outer* $\mathfrak{G}$*-regular.*

Obviously, r is an extension of m to a content on $\mathfrak{L}$. Since

$$\mathcal{E}_-(\{r < \infty\}) \subset \mathfrak{R}^*_-(m) \subset \mathfrak{R}^{\bullet}_-(m) ,$$

and $r = m^{\bullet}$ on $\mathcal{E}_-(\{r < \infty\})$, the equality $r^{\bullet} = m^{\bullet}$ and the semiregularity of $r^{\triangledown}$ result from Corollary 7.3, whereas regularity follows from Corollary 6.2.(ii).

Finally, if $\mathfrak{G} \subset \mathfrak{L}$, we infer from Theorem 12.4 that r is outer $\mathfrak{G}$-regular and inner $\mathfrak{K}$-regular on $\mathfrak{G}$. Let now l be an extension of m to a content on $\mathfrak{L}$. For any $L \in \mathfrak{L}$ we then have

$$r(L) = \inf_G r(G) = \inf_G m^{\bullet}(G) \le \inf_G l(G) \ge l(L) ,$$

where G runs through all sets in $\mathfrak{G}$ with $G \supset L$. Since equality holds at the first place iff l is inner $\mathfrak{K}$-regular on $\mathfrak{G}$, respectively at the second place iff l is outer $\mathfrak{G}$-regular, this proves the characterizations of r . $\square$

In contrast to the Extension Theorems 6.5 and 7.3, we could not work with a universal cone, since $\mathcal{E}_-(\mathfrak{L}^{\triangledown})$ depends on the extension l .

REMARKS.

(1)   The formula for l characterizing the content q is actually equivalent to the semifiniteness and the $\mathcal{E}_-(\mathfrak{K})$-tightness of l on $\mathcal{E}_-(\mathfrak{L}^{\triangledown})$.

Indeed, l satisfies the formula iff $l = q$. Corollary 7.3 therefore proves the necessity, and conversely, applied to $\mathcal{E}_-(\mathfrak{K}) \subset \mathcal{E}_-(\mathfrak{L}^{\triangledown})$, shows that $l^{\triangledown} = m_{\bullet}$ on $\mathcal{E}_-(\mathfrak{L}^{\triangledown})$, i.e. $l = q$ on $\mathfrak{L}^{\triangledown}$. This yields the formula on $\mathfrak{L}$ by the semifiniteness. $\square$

(2)   The integration theory for an arbitrary content l extending m may be quite different from that for q , in contrast to that for r , both in the proposition and the theorem. Indeed, the finite part of r is a restriction of that of q , whereas l and q may differ on $\{l < \infty\} \subset \{q < \infty\}$.

If m is semiregular, then tightness and semifiniteness of l is equivalent to inner $\mathfrak{K}$-regularity. This is the reason for the often misleading use of the term "tight" instead of what we call semiregular.

12.8 COROLLARY 1. *Every finite content on $\mathfrak{K}$ has a unique extension to a (finite) content on the ring $\mathfrak{R}$ generated by $\mathfrak{K}$.*

The upper functionals associated to the content and its extension coincide.

Since all sets in $\mathfrak{R}$ are integrable, the assertion follows from (ii). $\square$

REMARK. As already mentioned in Remark 6.3.2, less regularity would be available if instead of contents on lattices only the extensions of these contents to the generated rings were considered.

EXAMPLE. Let $\mathfrak{K}$ be the lattice of all compact intervals in $\mathbb{R}$, containing 0 if non-empty. Let m be the content on $\mathfrak{K}$ which assigns to each interval its length. It is obviously not semiregular. There exists a unique extension of m to a finite content on the ring generated by $\mathfrak{K}$. This is one way to construct the Lebesgue measure on $\mathbb{R}$ (cf. Example 13.2).

Stieltjes measures can be introduced analogously.

COROLLARY 2. *Let $\mathfrak{R}$ be a ring which contains $\mathfrak{K}$ and which is contained in the algebra generated by $\mathfrak{G}(\mathfrak{K})$. Then restriction to $\mathfrak{K}$ defines a bijection between the inner $\mathfrak{K}$-regular contents on $\mathfrak{R}$ which are finite on $\mathfrak{K}$, and the semiregular contents on $\mathfrak{K}$.*

The essential upper functionals of each content and of its restriction coincide.

This follows from Proposition 12.7, (iii) and (ii), since $\mathfrak{R}$ consists of measurable sets by Corollary 11.4, and in view of the fact that $L \setminus K \in \mathfrak{R}$ for $K,L \in \mathfrak{K}$, which shows that semiregularity follows from inner $\mathfrak{K}$-regularity. $\square$

To be able to handle semifinite contents, one has only to note that every extension of the finite part is also an extension, since by Proposition 12.7.(i) it is infinite on sets of infinite content.

COROLLARY 3. *Every semifinite content has a smallest extension, which is semifinite, and a largest extension to a content on any larger lattice consisting of measurable sets.*

COROLLARY 4. *Every content can be additively decomposed into a semifinite content and a content taking only the values* 0 *and* ∞ *, such that semiregularity is inherited by each summand.*

Let l be a content on a lattice $\mathfrak{L}$. Since all sets in $\mathfrak{L}$ are locally in $\mathfrak{L}^\nabla$, the set function q , defined on $\mathfrak{L}$ by

$$q(L) := \sup_{K\in\mathfrak{L}^\nabla} l(L\cap K) ,$$

is a semifinite content on $\mathfrak{L}$, which is semiregular if l and hence its finite part l^∇ is.

Define r on $\mathfrak{L}$ by

$$r(L) := 0 \quad \text{if} \quad q(H) = l(H) \quad \text{for all } H \in \mathfrak{L} \text{ with } H \subset L$$

and else by $r(L) := \infty$. Then $l = q + r$, since $r(L) = \infty$ implies $l(L) = \infty$. Obviously, r is increasing and $r(\emptyset) = 0$. To prove that r is a content is therefore equivalent to show that

$$r(K\cup L) = 0 \quad \text{for } K, L \in \mathfrak{L} \text{ with } r(K) = r(L) = 0 .$$

Let $H \in \mathfrak{L}$ with $H \subset K \cup L$. If $l(H)$ is finite, then $q(H) = l(H)$, and otherwise

$$\infty = l(H) \le l(H\cap K) + l(H\cap L) = q(H\cap K) + q(H\cap L) \le 2\cdot q(H) ,$$

hence also $q(H) = l(H)$.

It remains to show that r is semiregular if l is. To prove that

$$r(L) \le r(K) + r_*(L\setminus K)$$

for $K,L \in \mathfrak{L}$ with $K \subset L$, it is sufficient to show that $r(K) = r_*(L\setminus K) = 0$ implies $r(L) = 0$. Since for all $A \in \mathfrak{L}$ with $A \subset L\setminus K$ we have $q(A) = l(A)$, for all $H \in \mathfrak{L}$ with $H \subset L$ we get $q_*(H\setminus K) = l_*(H\setminus K)$. Therefore, by Remark 12.2.3 we infer that

$$q(H) = q(H \cap K) + q_*(H \setminus H \cap K) = l(H \cap K) + l_*(H \setminus H \cap K) = l(H). \quad \square$$

12.9 **LEMMA.** *Let $\mathfrak{R}$ be a ring generated by a system $\mathfrak{E}$ which is stable with respect to finite unions or finite intersections. Then the function space $\mathcal{E}_-(\mathfrak{R}) - \mathcal{E}_-(\mathfrak{R})$ coincides with the subspace $\mathcal{E}$ generated by the indicator functions of sets from $\mathfrak{E}$.*

Since

$$1_{E \cap F} = 1_E + 1_F - 1_{E \cup F},$$

we have under either stability hypothesis

$$1_E \cdot 1_F = 1_{E \cap F} \in \mathcal{E}$$

for all $E,F \in \mathfrak{E}$. Consequently, $\mathcal{E}$ is an algebra. The formulas

$$1_{A \cup B} = 1_A + 1_B - 1_A \cdot 1_B \quad \text{and} \quad 1_{A \setminus B} = 1_A - 1_A \cdot 1_B$$

show that the system $\mathfrak{R}'$ of all sets whose indicator functions belong to $\mathcal{E}$ is a ring containing $\mathfrak{E}$ and therefore containing $\mathfrak{R}$. This implies that

$$\mathcal{E}_-(\mathfrak{R}) - \mathcal{E}_-(\mathfrak{R}) \subset \mathcal{E}_-(\mathfrak{R}') - \mathcal{E}_-(\mathfrak{R}') \subset \mathcal{E}. \quad \square$$

This lemma immediately yields the following *Uniqueness Theorem*:

PROPOSITION. *A finite content on a ring is uniquely determined by its values on any generator of the ring which is stable with respect to finite unions or finite intersections.*

EXAMPLE. The above uniqueness theorem generally does not hold for semifinite contents, even if the generator is a lattice on which both contents are semifinite :

Let X be an uncountable set and let Y be a countable subset. The system $\mathfrak{K}$ consisting of all subsets of X having a finite complement together with all finite subsets of Y is a lattice. The ring generated by $\mathfrak{K}$ is the algebra $\mathfrak{A}$ of all subsets of X which are finite or have a finite complement.

For $i = 1,2$ we define two different semifinite contents on $\mathfrak{A}$ whose restrictions to $\mathfrak{K}$ are semifinite and coincide; namely

$$l_i(A) := i \cdot \mathfrak{card}(A \setminus Y) + \mathfrak{card}(A \cap Y) .$$

REMARK. We have shown how classical abstract measure theory has to be incorporated into our framework of functional analytic integration theory (cf. Example 12.6.1). Thus, *finite* (which does not mean bounded) and not *semifinite* contents (which arise naturally only in extension problems) should be the starting point. This is also indicated by the preceding example.

Moreover, our concept makes clear that inner and not outer regularity plays the major role.

§ 13 MEASURES AND MEASURABILITY

In this section we continue the treatment of the measure-theoretic standard example by studying the Daniell property for contents, which should be equivalent to the Daniell property of the corresponding linear functional.

THROUGHOUT THIS SECTION,
$\mathfrak{K}$ AND $\mathfrak{G}$ ARE COMPATIBLE LATTICES OF SETS.

13.1 DEFINITION. An increasing set function m on $\mathfrak{K}$ with $m(\emptyset) = 0$ has the *Daniell property*, respectively the *Daniell property at the empty set*, if

$$m(\bigcap K_n) = \inf m(K_n)$$

for every decreasing sequence (K_n) in $\mathfrak{K}$ with $m(K_n) < \infty$ and $\bigcap K_n \in \mathfrak{K}$, respectively $\bigcap K_n = \emptyset$.

A content m on $\mathfrak{K}$ is a *measure* if

$$m(\bigcup_{n=1}^{\infty} (K_n \setminus L_n)) = \sum_{n=1}^{\infty} [m(K_n) - m(L_n)]$$

holds for all sequences of sets $K_n, L_n \in \mathfrak{K}$ with $L_n \subset K_n$ and $m(L_n) < \infty$, for which $(K_n \setminus L_n)$ is a sequence of pairwise disjoint sets with $\bigcup_{n=1}^{\infty} (K_n \setminus L_n) \in \mathfrak{K}$.

PROPOSITION. *Every measure m on $\mathfrak{K}$ has the following properties:*

(i) *m is σ-additive, i.e.*

$$m(\bigcup K_n) = \sum m(K_n)$$

holds for every sequence (K_n) of pairwise disjoint sets in $\mathfrak{K}$ with $\bigcup K_n \in \mathfrak{K}$.

(ii) *For every increasing sequence (K_n) in $\mathfrak{K}$ with $\bigcup K_n \in \mathfrak{K}$,*

$$m(K_n) = \sup m(K_n) .$$

(iii) *m has the Daniell property.*

Conversely, let $\mathfrak{K}$ be a ring and m a content on $\mathfrak{K}$. If m is σ-additive or has the second property, or is a finite content with the Daniell property at the empty set, then m is a measure.

Suppose that m is a measure. It is obviously σ-additive. If (K_n) is an increasing sequence in $\mathfrak{K}$ with $\bigcup K_n =: K \in \mathfrak{K}$, then $(K_n \setminus K_{n-1})$ with $K_0 := \emptyset$ is a sequence of pairwise disjoint sets such that $\bigcup (K_n \setminus K_{n-1}) = K$. We may assume that $m(K_n) < \infty$ for all $n \in \mathbb{N}$. Then

$$m(K) = \sum_{n=1}^{\infty} [m(K_n) - m(K_{n-1})] = \sup m(K_n) .$$

If (K_n) is a decreasing sequence in $\mathfrak{K}$ with $\bigcap K_n =: K \in \mathfrak{K}$ and $m(K_n) < \infty$, then K and all $K_n \setminus K_{n+1}$ define a sequence of pairwise disjoint sets with $K \cup \bigcup (K_n \setminus K_{n+1}) = K_1$. Thus

$$m(K) + \sum_{n=1}^{\infty} [m(K_n) - m(K_{n+1})] = m(K_1)$$

and therefore

$$m(K) = m(K_{n+1}) - \sum_{i=n+1}^{\infty} [m(K_i) - m(K_{i+1})] = \inf m(K_n) .$$

If $\mathfrak{K}$ is a ring, the proof of the converse assertions is even more elementary. □

REMARK. A finite content m on $\mathfrak{K}$ with the property (ii) is semiregular if

$$L \setminus K \in \mathfrak{K}_\sigma \quad \text{for } K,L \in \mathfrak{K} \text{ with } K \subset L .$$

In fact, if the sets $K_n \in \mathfrak{K}$ increase to $L \setminus K$, then $L = K \cup \bigcup K_n$, and since K and K_n are disjoint, we have

$$m(L) = \sup m(K \cup K_n) = m(K) + \sup m(K_n) \leq m(K) + m_*(L \setminus K) . \quad □$$

In Corollary 13.4 we will prove that m is a measure if furthermore m has the Daniell property at the empty set.

13.2 REMARK. The concept of measure is chosen in such a way that:

A finite content m *on* $\mathfrak{K}$ *is a measure iff its unique extension to a finite content* r *on the ring* $\mathfrak{R}$ *generated by* $\mathfrak{K}$ *is a measure.*

Indeed, if $\mathfrak{D}$ denotes the system of all sets $K \setminus L$ with $K,L \in \mathfrak{K}$, then $\mathfrak{D}$ is stable w.r.t. finite intersections, and the system of all finite unions of pairwise disjoint sets from $\mathfrak{D}$ is a ring which contains $\mathfrak{K}$, hence coincides with $\mathfrak{R}$.

If m is a measure, then, by this observation, proving the σ-additivity of r is reduced to proving the formula

$$r(R) = \sum r(D_n),$$

where (D_n) runs through all sequences of pairwise disjoint sets from $\mathfrak{D}$ with $\bigcup D_n = R$. By definition, this formula holds for $R \in \mathfrak{K}$ and therefore also for $R \in \mathfrak{D}$, hence for all $R \in \mathfrak{R}$ by the above description of these sets.

The converse is trivial. □

As already announced in Example 12.8, the Lebesgue measure on $\mathbb{R}$ can be constructed by the following

EXAMPLE. *Let* $\mathfrak{K}$ *be the lattice of all compact intervals* $K \subset \mathbb{R}$ *which contain* 0 *if non-empty. Let* m *be the content on* $\mathfrak{K}$ *assigning to each interval its length. Then* m *is a measure.*

Indeed, let $K_n \supset L_n$ be sets from $\mathfrak{K}$ such that $(K_n \setminus L_n)$ is a sequence of pairwise disjoint sets with $\bigcup (K_n \setminus L_n) = [a,b]$ for $a \le 0 \le b$. Without loss of generality, we may assume that $0 \in K_0 \setminus L_0$, i.e. $L_0 = \emptyset$ and $K_0 = [a_0,b_0]$ with $a_0 \le 0 \le b_0$, and that $0 \notin K_n \setminus L_n$, i.e. $K_n = [a_n,b_n]$ and $L_n = [c_n,d_n]$ with $a_n \le c_n \le 0 \le d_n \le b_n$ for all $n \ge 1$. We have to prove that

$$b - a = b_0 - a_0 + \sum_{n=1}^{\infty} (b_n - a_n - (d_n - c_n)).$$

To this end, let C denote the set of all $c \in [a,0]$ with $[a,c[= \bigcup [a_n,c_n[$ and $c - a = \sum (c_n - a_n)$, where the union and the sum extend over all $n \ge 1$

with $[a_n,c_n[\subset [a,c[$. Obviously, $a \in C$. Having in mind that the sets $[a_0,0]$ and $[a_n,c_n[$ for $n \geq 1$ form a decomposition of $[a,0]$, it is easy to see that $\sup C \in C$ and then, by contradiction, that $\sup C = a_0$. These properties of the set C yield

$$-a = -a_0 + \sum_{n=1}^{\infty} (c_n - a_n) \quad \text{and analogously} \quad b = b_0 + \sum_{n=1}^{\infty} (b_n - d_n) . \quad \square$$

Stieltjes measures can be treated the same way.

13.3 Proposition. *Let m be a finite, increasing and strongly subadditive set function on $\mathfrak{K}$. Then m has the Daniell property iff the smallest increasing superlinear extension of m to $\mathcal{E}_-(\mathfrak{K})$ has this property.*

The condition is obviously sufficient. Conversely, assume that m has the Daniell property on $\mathcal{E}_-(\mathfrak{K})$. Let (e_i) be an increasing sequence in $\mathcal{E}_-(\mathfrak{K})$ having $e \in \mathcal{E}_-(\mathfrak{K})$ as upper envelope. If $e_0 \geq -\alpha\cdot 1_K$ for some $\alpha > 0$ and some $K \in \mathfrak{K}$, and if

$$e = -\sum_{j=-n}^{-1} (\omega_{j+1} - \omega_j)\cdot 1_{\{e \leq \omega_j\}}$$

is a pyramidal decomposition of e (cf. Remark 10.4.1), then for every $\varepsilon > 0$ and $j = -n,\ldots,0$ the sets $K_i^j := \{e_i \leq \omega_j - \varepsilon\}$ are decreasing in $\mathfrak{K}$. For a sufficiently small $\varepsilon > 0$ and $j \neq -n$ we have

$$\bigcap_i K_i^j = \{e \leq \omega_j - \varepsilon\} = \{e \leq \omega_{j-1}\}$$

and

$$\bigcap_i K_i^{-n} = \{e \leq \omega_{-n} - \varepsilon\} = \emptyset .$$

It is easy to see that

$$e_i \geq -\alpha\cdot 1_{K_i^{-n}} - \sum_{j=-n}^{-1} (\omega_{j+1} - \omega_j)\cdot 1_{K_i^{j+1}} - \varepsilon\cdot 1_K .$$

From this inequality we infer, in view of the monotonicity, the superlinearity and the definition of m on $\mathcal{E}_-(\mathfrak{K})$ in Corollary 11.2, that

$$\sup_i m(e_i) \geq -\alpha\cdot\inf_i m(K_i^{-n}) - \sum_{j=-n}^{-1} (\omega_{j+1} - \omega_j)\cdot\inf_i m(K_i^{j+1}) - \varepsilon\cdot m(K) =$$

$$= -\alpha\cdot m(\emptyset) - \sum_{j=-n}^{-1} (\omega_{j+1} - \omega_j)\cdot m(\{e \leq \omega_j\}) - \varepsilon\cdot m(K) =$$

$$= m(e) - \varepsilon\cdot m(K),$$

which completes the proof. □

REMARK. *The equality* $\sup m(e_i) = 0$ *holds for every increasing sequence* (e_i) *in* $\mathcal{E}_-(\mathfrak{K})$ *having* 0 *as upper envelope iff* m *has the Daniell property at the empty set.*

This is implicit in the proof because here $e = 0$. □

13.4 THEOREM. *A finite content* m *on* $\mathfrak{K}$ *is a measure iff the functional* m *on* $\mathcal{E}(\mathfrak{K},\mathfrak{S})$*, or equivalently on* $\mathcal{E}_-(\mathfrak{K})$*), is an integral.*

By Proposition 9.5.(i) and Remark 9.5.1, the functional m on $\mathcal{E}(\mathfrak{K},\mathfrak{S})$, or on $\mathcal{E}_-(\mathfrak{K})$, is an integral iff $\tilde{m}$ has the Daniell property on $\mathcal{E}_-(\mathfrak{K}) - \mathcal{E}_-(\mathfrak{K})$. If r denotes the unique extension of m to a content on the ring $\mathfrak{R}$ generated by $\mathfrak{K}$, then $\tilde{r} = \tilde{m}$, since by Lemma 12.9 the function space $\mathcal{E}_-(\mathfrak{K}) - \mathcal{E}_-(\mathfrak{K})$ coincides with $\mathcal{E}_-(\mathfrak{R}) - \mathcal{E}_-(\mathfrak{R})$. The above properties therefore hold iff the semiregular functional r is an integral on $\mathcal{E}_-(\mathfrak{R})$, i.e. has the Daniell property (cf. Proposition 9.5.(ii)). In view of Propositions 13.3 and 13.1, this is equivalent to r being a measure. The result now follows from Remark 13.2. □

REMARKS.

(1) *Difference-regular linear extension defines a bijection between the finite measures on* $\mathfrak{K}$ *and the difference-regular integrals on* $\mathcal{E}(\mathfrak{K},\mathfrak{S})$.

This follows from Corollary 11.3 and the theorem. □

(2) If m is semiregular, it is not necessary to make the detour through the generated ring, i.e. to make use of Remark 13.2 which was proved by purely

set–theoretical methods. Indeed, from the Daniell property of $\tilde{m}$ we infer directly that m is a measure, and in view of Proposition 9.5.(ii) and Remark 13.3, the integral property is already a consequence of the Daniell property at the empty set. This proves the following

COROLLARY. *A semiregular finite content is a measure iff it has the Daniell property at the empty set.*

EXAMPLES.

(1) *The Riemann integral ι on $\mathbb{R}$ is a regular integral.*

Indeed, if m denotes the measure from Example 13.2, then

$$\iota(e) = \int^{\times} e \, dm$$

for all $e \in \mathcal{E}$. □

An analogous result holds for Stieltjes integrals.

(2) *Every Radon measure is a measure.*

This is an immediate consequence of the preceding corollary, since the Daniell property at the empty set is trivially satisfied on $\mathfrak{K}(X)$. □

13.5 EXAMPLE. The following example shows that *it is not possible to introduce the concepts of measure and integral by any of the stronger versions of the Daniell property* which are near at hand, e.g. in the case of measure by the requirement

$$\inf m(K_n) \leq \sup m(L_n)$$

for every decreasing sequence (K_n) and every increasing sequence (L_n) in $\mathfrak{K}$ with

$$\bigcap K_n \subset \bigcup L_n ,$$

or by the requirement

$$m(\bigcup_{n=1}^{\infty}(K_n \setminus L_n)) = \sup\,[m(K_n) - m(L_n)]$$

for every increasing sequence $(K_n \setminus L_n)$ of differences of sets $K_n, L_n \in \mathfrak{K}$ with $L_n \subset K_n$, $m(L_n) < \infty$ and $\bigcup (K_n \setminus L_n) \in \mathfrak{K}$.

Similarly, for the concept of integral the condition given in Remark 9.5.3 analogous to the first one above is not sufficient.

To demonstrate these claims, we construct from Example 13.2 the lattice $\mathfrak{K}'$ of all sets $K' = K \cap \mathbb{Q}$ with $K \in \mathfrak{K}$. The mapping $e \longmapsto e' := e \cdot 1_{\mathbb{Q}}$ of $\mathcal{E}_-(\mathfrak{K})$ to $\mathcal{E}_-(\mathfrak{K}')$ is an order-preserving cone homomorphism which is injective and therefore bijective, since for all $x \in \mathbb{R}$

$$e(x) = \liminf_{y \to x} e'(y)\,.$$

Consequently,

$$m'(K') := m(\overline{K'})$$

defines a finite content m' on $\mathfrak{K}'$ such that

$$m'(e') = m(e) = \int^{\times} e\,dm$$

holds for $e' \in \mathcal{E}_-(\mathfrak{K}')$.

This content cannot be a measure and therefore m' cannot be an integral on $\mathcal{E}_-(\mathfrak{K}')$, since the extension r' of m' to the ring $\mathfrak{R}'$ generated by $\mathfrak{K}'$ does not have the Daniell property :

Indeed, choose $\varepsilon > 0$ and let (q_n) be an enumeration of $[0,1] \cap \mathbb{Q}$. Then the sets

$$A_k' := [0,1]' \setminus \bigcup_{n=0}^{k} \left[-\frac{\varepsilon}{2^n} + q_n\,,\, q_n + \frac{\varepsilon}{2^n}\right]'$$

form a decreasing sequence in $\mathfrak{R}'$ with empty intersection. Since

$$r'(\bigcup_{n=0}^{k} \left[-\frac{\varepsilon}{2^n} + q_n\,,\, q_n + \frac{\varepsilon}{2^n}\right]') \leq \sum_{n=0}^{k} \frac{2\varepsilon}{2^n} \leq 4\varepsilon\,,$$

we have however

$$r'(A_k') \geq 1 - 4\varepsilon$$

for all k.

We shall prove below that for every increasing sequence (s_n') and every decreasing sequence (t'_n) in $\mathcal{E}_-(\mathfrak{K}')$ with $\inf t'_n \le \sup s'_n$ the condition

$$\inf m'(t'_n) \le \sup m'(s'_n)$$

from Remark 9.5.3 is satisfied. In particular, m' then has the first of the stronger versions of the Daniell property on $\mathfrak{K}'$, and also the second one : In fact, if $(K'_n \setminus L'_n)$ is an increasing sequence of differences of sets $K'_n, L'_n \in \mathfrak{K}'$ with $L'_n \subset K'_n$ and $\bigcup (K'_n \setminus L'_n) \in \mathfrak{K}'$, then either all these differences are empty or $0 \in (K'_n \setminus L'_n)$ for all sufficiently large n, hence $L'_n = \emptyset$ and therefore

$$\sup [m'(K'_n) - m'(L'_n)] = \sup m'(K'_n) = m'(\bigcup K'_n) = m'(\bigcup (K'_n \setminus L'_n)) .$$

To prove the above inequality first note that all s_n and t_n and hence also $\sup s_n$ and $\inf t_n$ are monotone decreasing on $\mathbb{R}_-$ and monotone increasing on $\mathbb{R}_+$. Therefore, the functions $\sup s_n$ and $\inf t_n$ are continuous at all but at most countably many points. Since the inequality $\inf t_n \le \sup s_n$ holds on $\mathbb{Q}$, it also holds at every common point of continuity, hence m-almost everywhere. We may suppose that

$$\inf m(t_n) = \inf m'(t'_n) > -\infty .$$

By the Monotone Convergence Theorem 5.4, we then have

$$\inf m'(t'_n) = \inf \int^{\times} t_n \, dm = \int^{\times} \inf t_n \, dm \le$$

$$\le \int^{\times} \sup s_n \, dm = \sup \int^{\times} s_n \, dm = \sup m'(s'_n) . \quad \square$$

13.6 **Proposition.** *Let ν be an upper integral.*

(i) *$\mathfrak{I}(\nu)$ and $\mathfrak{I}^{\bullet}(\nu)$ are δ-rings, $\mathfrak{I}^0(\nu)$ is a σ-ring, and $\nu^{\bullet}$, as well as ν if auto-determined, are measures on $\mathfrak{I}^0(\nu)$.*

If moreover 1 is measurable, then $\mathfrak{I}^0(\nu)$ is a σ-algebra, and we have :

(ii) *For every integrable function $f \le 0$ and $\gamma < 0$, the sets $\{f < \gamma\}$ and $\{f \le \gamma\}$ are integrable.*

(iii) *A set is moderated iff it is covered by a sequence of integrable sets. In particular, ν is moderated iff ν is σ-finite, i.e. iff X is covered by a sequence of integrable sets.*

By Proposition 5.10.(i), countable unions of measurable sets are measurable, and by the Monotone Convergence Theorem, countable intersections of (essentially) integrable sets are (essentially) intergable, since $\nu^\bullet$ is also an upper integral by Corollary 5.2. Assertion (i) now follows from Propositions 11.4 and 13.1.

By Theorem 11.7 and Remark 10.3.2, f is $\mathfrak{I}^0(\nu)$–measurable. The above sets therefore are measurable and thus integrable since contained in an integrable set.

If A is moderated, i.e. contained in a set $\{f > 0\}$ for some integrable function $f \geq 0$, then $\{f \geq \frac{1}{n}\} = \{-nf \leq -1\}$ is integrable and

$$A \subset \bigcup \{f \geq \tfrac{1}{n}\} .$$

Conversely, if $A \subset \bigcup A_n$ for some integrable sets A_n, then $A \subset \{f > 0\}$, where

$$f := \sum \frac{1}{2^n \cdot (\nu(A_n) + 1)} \cdot 1_{A_n}$$

is integrable by the Monotone Convergence Theorem 5.4. □

NOTATION. For an integral μ, the systems of all μ–integrable, essentially μ–integrable and μ–measurable sets will be denoted respectively by

$$\mathfrak{L}^\times(\mu) \ , \ \mathfrak{L}^\bullet(\mu) \quad \text{and} \quad \mathfrak{L}^0(\mu) .$$

If μ is regular, then we use the notation $\mathfrak{L}^*(\mu) := \mathfrak{I}(\mu^{\sigma*})$.

THROUGHOUT THE REST OF THIS SECTION,
m IS A FINITE MEASURE ON $\mathfrak{K}$.

THEOREM.

(i) *The upper integral* $\int^\times \cdot\, dm$ *is auto–determined, the set of functions* $-n \cdot 1_K$ *with* $n \in \mathbb{N}$ *and* $K \in \mathfrak{K}$ *is almost coinitial in* $\overline{\mathbb{R}}^X$ *w.r.t.* $\int^\times \cdot\, dm$ *and* $\int^\bullet \cdot\, dm$, *and every* $\mathfrak{K}_\delta$*–measurable function is* m*–measurable.*

(ii) *The* m*–measurable sets form a* σ*–algebra* $\mathfrak{L}^0(m)$ *containing* $\mathfrak{B}(\mathfrak{K}_\delta)$, *and* *in particular* $\mathfrak{K}$, *on which* $\int^{\times} \cdot dm$ *and* $\int^{\bullet} \cdot dm$ *are measures extending* m . *Furthermore,* $\mathfrak{L}^{\times}(m)$ *and* $\mathfrak{L}^{\bullet}(m)$ *are the* δ*-rings of* m*-measurable sets* A *with respectively*

$$\int^{\times} A\, dm < \infty \quad \text{and} \quad \int^{\bullet} A\, dm < \infty .$$

(iii) *We have*

$$\mathfrak{L}^{0}(m) = \mathfrak{S}(\mathfrak{L}^{\times}(m)) .$$

More precisely, a set A *is measurable iff* $A \cap K$ *is integrable for every* $K \in \mathfrak{K}$. *In this case*

$$\int_{\bullet} A\, dm = \int^{\bullet} A\, dm = \sup_{K \in \mathfrak{K}} \int^{\times} A \cap K\, dm .$$

(iv) *A function is* m*-measurable iff it is* $\mathfrak{L}^{0}(m)$*-measurable.*

The first assertion follows from Proposition 9.6. In fact, every $\mathfrak{K}_{\delta}$-measurable function is $\mathscr{E}(\mathfrak{K}_{\delta})$-measurable by Theorem 10.10, and by the Monotone Convergence Theorem 5.4, we have $\mathscr{E}_{-}(\mathfrak{K}_{\delta}) \subset \mathscr{L}^{\times}_{-}(m)$, hence $\mathscr{E}_{-}(\mathfrak{K}_{\delta}) \supset \mathscr{E}_{-}(\mathfrak{K})$ is almost coinitial in $\mathscr{L}^{\bullet}_{-}(m)$.

Assertion (ii) follows immediately from (i) of the above proposition, having in mind Proposition and Corollary 11.4. To prove (iii), one also has to consider Proposition 4.3, Theorem 3.6.(i), and the formula

$$\min(1_{A}, n \cdot 1_{K}) = 1_{A \cap K} .$$

Finally, by (iii) and Corollary 10.3, a function f is $\mathfrak{L}^{0}(m)$-measurable iff f is $\mathfrak{L}^{\times}(m)$-semicontinuous, i.e. $\mathscr{E}(\mathfrak{L}^{\times}(m))$-measurable by Theorem 10.10. This implies that f is m-measurable, using Proposition 4.11. Conversely, we may assume $f \leq 0$. For every integrable set A and $\gamma < 0$ we have

$$\{f \leq \gamma\} \cap A = \{\max(f, \gamma \cdot 1_{A}) \leq \gamma\} \in \mathfrak{L}^{\times}(m)$$

by (ii) of the above proposition, since $\max(f, \gamma \cdot 1_{A})$ is m-integrable. This proves (iv). □

Using (iii), we may proceed as in Corollary 11.9 to prove Carathéodory's characterization of measurability :

COROLLARY. *For every set* A *, the following assertions are equivalent :*

(i) A *is* m*-measurable.*

(ii) *For every set* B *, equivalently every* $B \in \mathfrak{K}$ *, we have*

$$\int_{\times} B\,dm = \int_{\times} B \cap A\,dm + \int_{\times} B \setminus A\,dm .$$

(iii) *For every set* B *, equivalently every* $B \in \mathfrak{K}$ *, we have*

$$\int^{\times} B\,dm = \int^{\times} B \cap A\,dm + \int^{\times} B \setminus A\,dm .$$

13.7 Next we prove that m has a unique extension to a measure m_δ on $\mathfrak{K}_\delta$ with the same Daniell integration theory. If m is semiregular, then we shall see in Theorem 13.8 that moreover the essential upper integral for m coincides with the essential upper functional for m_δ .

PROPOSITION.

(i) *The restriction* m_δ *of* $m^\times$ *to* $\mathfrak{K}_\delta$ *is the only extension of* m *to a (finite) measure on* $\mathfrak{K}_\delta$ *, and we have*

$$\int^{\times} \cdot\,dm_\delta = \int^{\times} \cdot\,dm \ , \quad m_\delta{}^\sigma = m^\sigma \quad \text{and} \quad m_\delta{}^\times = m^{\sigma\times} .$$

Moreover

$$m_{\delta *} = m^\sigma{}_* \quad \text{on } \overline{\mathbb{R}}_+^X .$$

(ii) *For every set* A *we have*

$$m_*(A) \le m_{\delta *}(A) \le m_{\delta \bullet}(A) \le \int_{\times} A\,dm \le \int^{\bullet} A\,dm \le m_\delta{}^\bullet(A) \le m_\delta{}^*(A) .$$

By Proposition 9.8 we have $m_\delta = m_\times = \int^\bullet \cdot\,dm$ on $\mathcal{E}_-(\mathfrak{K}_\delta) \subset \mathcal{E}_-(\mathfrak{K},\mathfrak{G})_\sigma$. This proves that m_δ is a measure on $\mathfrak{K}_\delta$ extending m , and the only one by the Daniell property. Using Proposition 9.11.(ii), we have $m_\delta = \int^\bullet \cdot\,dm$ on $\mathcal{E}(\mathfrak{K}_\delta,\mathfrak{G}_\sigma)$ by the uniqueness statement, and

$$\int^{\times} \cdot\,dm_\delta = \int^{\times} \cdot\,dm .$$

Application of Proposition 9.8 to m_δ and m therefore yields

$$m_\delta{}^\sigma = \int^\bullet \cdot \, dm_\delta = \int^\bullet \cdot \, dm = m^\sigma \quad \text{on } \mathcal{E}(\mathfrak{K}_\delta, \mathfrak{G}_\sigma)_\sigma = \mathcal{E}(\mathfrak{K}, \mathfrak{G})_\sigma .$$

In view of Theorem 10.10, we have

$$\mathcal{E}_{-\sigma} := \mathcal{E}_-(\mathfrak{K})_\sigma = \mathcal{E}_-(\mathfrak{K}_\delta)_\sigma = \mathcal{E}_-^\wedge(\mathfrak{K}_\delta) \subset \mathfrak{R}_-^\times(m_\delta) .$$

Since $m^\sigma = m_\delta{}^\sigma$ coincides with $m_{\delta\times} = m_\delta{}^\times$ on $\mathcal{E}_{-\sigma}$, for every function f we have

$$m_\delta{}^\times(f) \geq m^{\sigma\times}(f) = \inf_{d,e \in \mathcal{E}_{-\sigma},\, d-e \geq f} m^\sigma(d) - m^\sigma(e) =$$

$$= \inf_{d,e \in \mathcal{E}_{-\sigma},\, d-e \geq f} m_\delta{}^\times(d - e) \geq m_\delta{}^\times(f) .$$

Moreover, since m_δ coincides with m^σ on $\mathcal{E}_-(\mathfrak{K}_\delta)$, we infer from Lemma 7.4 that $m_\delta{}^* = m^\sigma$ on $\mathcal{E}_{-\sigma}$. For any function $f \geq 0$ we thus have

$$m^\sigma{}_*(f) = \sup_{s \in \mathcal{E}_{-\sigma},\, -s \leq f} -m^\sigma(s) = \sup_{s \in \mathcal{E}_{-\sigma},\, -s \leq f} m_{\delta *}(-s) \leq m_{\delta *}(f) \leq m^\sigma{}_*(f) .$$

Assertion (ii) follows from Corollaries 11.4 and 9.6, having in mind that $\int^\times \cdot \, dm_\delta = \int^\times \cdot \, dm$. □

13.8 THEOREM.

(i) *The semiregularity of* m_δ *is equivalent to that of* m^σ .

(ii) *If* m *is semiregular, then* m_δ *is semiregular, and we have*

$$m_\delta{}^\bullet = \int^\bullet \cdot \, dm .$$

In particular, $m_\delta{}^\bullet$ *is an upper integral, and* $\mathfrak{L}^0(m) = \mathfrak{R}^0(m_\delta)$.

Furthermore, for $G \in \mathfrak{G}_\sigma$ *and every set* A *we have*

$$m_\delta(G) = m_*(G) \quad \text{and} \quad m_{\delta *}(A) = \int_\bullet A \, dm .$$

If A *is* m*-measurable, then*

$$m_{\delta *}(A) = \int^\bullet A \, dm ,$$

i.e. $\int^\bullet \cdot \, dm$ *is an inner* $\mathfrak{K}_\delta$*-regular measure on* $\mathfrak{L}^0(m)$.

Assertion (i) follows immediately from Propositions 6.7 and 13.7.(i). The first part of (ii) is a consequence of (i) and Proposition 9.9, whereas the second one follows from Theorem 9.10, since $m_\delta^{\times} = m^{\sigma\times}$ and therefore

$$m_\delta^{\bullet} = m^{\sigma\bullet} = \int^{\bullet} \cdot\, dm \,.$$

For $G \in \mathfrak{G}_\sigma$ we have $m_\delta(G) = m^\sigma(G) = m_\times(G) = m_*(G)$ by Proposition 6.7, and from Proposition 6.8 we infer that

$$m_{\delta *}(A) = m_{\delta \bullet}(A) = \int_{\bullet} A \, dm$$

holds for every set A. The last assertion in (ii) therefore follows from Proposition 9.6 and Corollary 12.1.(i). □

PROPOSITION. *If m is $\mathfrak{G}$-regular, then $\int^* \cdot\, dm$ is an auto-determined upper integral with*

$$\int^{\bullet} \cdot\, dm \le \int^{*} \cdot\, dm \le \int^{\times} \cdot\, dm \,.$$

The δ-ring $\mathfrak{L}^*(m)$ consists of all m-measurable sets A with $\int^* A\, dm < \infty$. Furthermore, m_δ is a $\mathfrak{G}_\sigma$-regular measure with

$$\int^{*} \cdot\, dm_\delta = \int^{*} \cdot\, dm \quad \textit{and} \quad m_\delta^{\bullet} = \int^{\bullet} \cdot\, dm \,,$$

and for every set A we have

$$m_{\delta *}(A) = \int_{*} A \, dm \quad \textit{and} \quad m_\delta^{*}(A) = \int^{*} A \, dm \,.$$

In particular, $\int^* \cdot\, dm$ is an outer $\mathfrak{G}_\sigma$-regular measure on $\mathfrak{L}^0(m) = \mathfrak{G}(\mathfrak{L}^*(m))$. Moreover,

$$\int^{*} A\, dm = \int^{\times} A\, dm \quad \textit{whenever} \quad \int^{\times} A\, dm < \infty \,,$$

and

$$\int^{\bullet} A\, dm = \int^{*} A\, dm \quad \textit{whenever} \quad \int^{*} A\, dm < \infty \,,$$

i.e. $\int^{\bullet} \cdot dm$ *is outer* $\mathfrak{G}_\sigma$*-regular on* $\{\int^{*} \cdot dm < \infty\}$.

The first assertions are consequences of Proposition 9.3 and Theorem 9.7. By Theorem 9.2, m^σ is regular and therefore $\mathcal{E}_\sigma$-bounded below. Thus, for $K \in \mathfrak{K}_\delta$ there exists a function $s \in \mathcal{E}_\sigma$ with $-s \le -1_K$ and $m^\sigma{}_*(s) < \infty$. By Remark 10.6.1 and Example 10.7.1, the set $G := \{s > \frac{1}{2}\}$ belongs to $\mathfrak{G}_\sigma$, and we have $G \supset K$ and $1_G \le 2s$, hence $m_{\delta *}(G) \le 2m_{\delta *}(s) = 2m^\sigma{}_*(s) < \infty$ by Proposition 13.7.(i). This proves that m_δ is $\mathfrak{G}_\sigma$-bounded and therefore $\mathfrak{G}_\sigma$-regular. The equality $\int^{*} \cdot dm_\delta = \int^{*} \cdot dm$ follows from $m_\delta{}^\sigma = m^\sigma$, whereas the equality $m_\delta{}^\bullet = \int^{\bullet} \cdot dm$ is a consequence of (ii) of the theorem and of Theorems 6.4 and 9.7.

The formula $m_{\delta *}(A) = \int_{*} A \, dm$ is a special case of Proposition 13.7.(i). The dual formula follows from Addendum 12.1, applied to m^σ. In fact,

$$\int^{*} A \, dm = \inf_{G \in \mathfrak{G}_\sigma,\, G \supset A} m^\sigma(G) = \inf_{G \in \mathfrak{G}_\sigma,\, G \supset A} m_\delta(G) = m_\delta{}^*(A) .$$

From Proposition 13.6.(i) we infer that $\int^{*} \cdot dm$ is an outer $\mathfrak{G}_\sigma$-regular measure on $\mathfrak{L}^0(m) = \mathfrak{M}^0(m_\delta)$.

The last formulas to be proved are a consequence of Theorem 9.7 and Proposition 9.3. □

COROLLARY. *If* m *is semiregular, a set* A *is* m*-measurable iff*

$$m_{\delta *}(B) = m_{\delta *}(B \cap A) + m_{\delta *}(B \setminus A)$$

for every set B *, or equivalently for every* $B \in \mathfrak{K}$.

If m *is* $\mathfrak{G}$*-regular, then* A *is* m*-measurable iff*

$$m_\delta{}^*(B) = m_\delta{}^*(B \cap A) + m_\delta{}^*(B \setminus A)$$

for every set B *, or equivalently for every* $B \in \mathfrak{G}$.

This follows immediately from $\mathfrak{L}^0(m) = \mathfrak{N}^0(m_\delta)$ and Corollary 12.4. Note that by the Daniell property of $\int^\bullet \cdot dm$, respectively $\int^* \cdot dm$, the formulas for $B \in \mathfrak{K}$, respectively $B \in \mathfrak{G}$, are sufficient since

$$m_{\delta *}(\bigcap A_n) = \lim m_{\delta *}(A_n)$$

for every decreasing sequence of sets A_n with $m_{\delta *}(A_n) < \infty$, respectively $m_\delta{}^*(\bigcup A_n) = \lim m_\delta{}^*(A_n)$ for every increasing sequence A_n of sets. □

REMARK. In the case of a ring $\mathfrak{K}$, the classical concept of Carathéodory measurability coincides with the one of m-measurability.

Note that m is $\mathfrak{K}$-regular and $m_\delta{}^*$ is the Carathéodory outer measure.

13.9 LEMMA. *Let l be a (not necessarily finite) measure on a lattice of sets $\mathfrak{L}$ and denote by l^∇ its finite part, defined on the lattice $\mathfrak{L}^\nabla$ of sets in $\mathfrak{L}$ with finite measure. Then we have*

$$\int_\times \cdot dl^\nabla = \int^\bullet \cdot dl^\nabla \le l^{\nabla\bullet} \le l = \int^\times \cdot dl^\nabla \quad \text{on } \mathfrak{L},$$

with equality holding on $\mathfrak{L}^\nabla$. Equality holds at $L \in \mathfrak{L}$ iff l is semifinite at L.

The inequalities

$$l^\nabla_\times \le \int_\times \cdot dl^\nabla = \int^\bullet \cdot dl^\nabla \le \int^\times \cdot dl^\nabla \le l^{\nabla\times} = l \quad \text{and} \quad \int^\bullet \cdot dl^\nabla \le l^{\nabla\bullet} \le l \quad \text{on } \mathfrak{L},$$

as well as the corresponding equalities on $\mathfrak{L}^\nabla$, follow from 9.6 and Lemma 12.6, since all sets in $\mathfrak{L}$ are l^∇-measurable. It therefore remains to prove that $l(L) = \int^\times L \, dl^\nabla$ whenever $L \in \mathfrak{L}$ and $\int^\times L \, dl^\nabla < \infty$. But for any such L there exists an increasing sequence (s_n) in $\mathcal{E}_-(\mathfrak{L}^\nabla) - \mathcal{E}_-(\mathfrak{L}^\nabla)$ with $1_L \le \sup s_n$. Since $s_n \le \alpha_n 1_{K_n}$ for suitable $\alpha_n > 0$ and $K_n \in \mathfrak{L}^\nabla$, there exists an increasing sequence (L_n) in $\mathfrak{L}^\nabla$ with $L \subset \bigcup_n L_n$. By Proposition 13.1 we have

$$l(L) = \sup l(L \cap L_n) = \sup \int^\times L \cap L_n \, dl^\nabla = \int^\times L \, dl^\nabla .$$

If l is semifinite at L, then equality holds at L since $l(L) = l_\times^\nabla(L)$. Conversely, from $l(L) = l^{\nabla\bullet}(L)$ we infer that l is semifinite at L. □

If confusion appears unlikely, instead of l^∇ we use the notation l for this functional, as we did in 12.6.

PROPOSITION. *Let $\mathfrak{L}$ be a lattice containing $\mathfrak{K}$.*

(i) *For every extension l of m to a measure on $\mathfrak{L}$, we have*

$$\int^\times \cdot\, dl \le \int^\times \cdot\, dm \;; \quad \text{in particular,} \quad \int_\times \cdot\, dm \le l \le \int^\times \cdot\, dm \quad \text{on } \mathfrak{L}.$$

If all sets in $\mathfrak{L}$ are m-measurable, e.g. $\mathfrak{L}$ is contained in the σ-algebra generated by $\mathfrak{S}(\mathfrak{K}_\delta)$, then the following assertions hold:

(ii) *Among all extensions l of m to a measure on $\mathfrak{L}$, the set function $q := \int_\times \cdot\, dm = \int^\bullet \cdot\, dm$ is the smallest, and the only one with*

$$l(L) = \sup_{K \in \mathfrak{K}} l(L \cap K) \quad \text{for all } L \in \mathfrak{L},$$

whereas $p := \int^\times \cdot\, dm$ is the largest.

Moreover, q is semifinite, and we have

$$\int^\bullet \cdot\, dq = \int^\bullet \cdot\, dm \quad \text{and} \quad \int^\times \cdot\, dp = \int^\times \cdot\, dm .$$

Since the finite part of l extends m to an integral on $\mathcal{E}_-(\mathfrak{L}^\nabla) \supset \mathcal{E}_-(\mathfrak{K})$, the inequality $\int^\times \cdot\, dl \le \int^\times \cdot\, dm$ follows from Proposition 9.11.(i). This implies (i). In fact, by the lemma we have

$$\int_\times \cdot\, dm \le \int_\times \cdot\, dl \le \int^\times \cdot\, dl = l \le \int^\times \cdot\, dm \quad \text{on } \mathfrak{L}.$$

If all sets in $\mathfrak{L}$ are m-measurable, (i) and Theorem 13.6.(ii) show that q is the smallest and p is the largest extension of m to a measure on $\mathfrak{L}$. Moreover, q satisfies the formula on $\mathfrak{L}$ by Theorem 13.6.(iii) and thus is semifinite.

Conversely, since $L \cap K \in \mathfrak{L}^\times(m)$ for all $L \in \mathfrak{L}$ and $K \in \mathfrak{K}$, we have

$l(L \cap K) = \int_X L \cap K \, dm$, hence $l = q$. Since $\mathcal{E}_-(\{q < \infty\}) \subset \mathcal{L}^\bullet_-(m)$, the equality $\int^\bullet \cdot dq = \int^\bullet \cdot dm$ is a consequence of Corollary 9.11. The last assertion in (ii) results from Proposition 9.11.(ii), since $\mathcal{E}_-(\{r < \infty\}) \subset \mathcal{L}^\times_-(m)$. □

THEOREM. *Let $\mathfrak{L}$ be a lattice of m-measurable sets containing $\mathfrak{K}$, e.g. $\mathfrak{L}$ is contained in the σ-algebra generated by $\mathfrak{G}(\mathfrak{K}_\delta)$.*

If m is semiregular and $\mathfrak{K}_\delta \subset \mathfrak{L}$, then $q = m_\delta{}^\bullet = \int^\bullet \cdot dm$ is the only extension of m to a measure on $\mathfrak{L}$ which is inner $\mathfrak{K}_\delta$-regular, and its finite part is semiregular.

If m is $\mathfrak{G}$-regular and $\mathfrak{K}_\delta \subset \mathfrak{L}$, then $r := m_\delta{}^* = \int^* \cdot dm$ is an extension of m to a measure on $\mathfrak{L}$ with

$$r^\bullet = \int^\bullet \cdot dr = m_\delta{}^\bullet .$$

Its finite part is semiregular, and regular w.r.t. every lattice compatible with $\mathfrak{L}$ and containing $\mathfrak{G}_\sigma$.

If moreover $\mathfrak{G}_\sigma \subset \mathfrak{L}$, then among all extensions of m to a measure on $\mathfrak{L}$, the set function r is the largest one which is inner $\mathfrak{K}_\delta$-regular on $\mathfrak{G}_\sigma$, and the smallest one which is outer $\mathfrak{G}_\sigma$-regular. In particular it is the only one which is inner $\mathfrak{K}_\delta$-regular on $\mathfrak{G}_\sigma$ and outer $\mathfrak{G}_\sigma$-regular.

We can apply Theorem 12.7 to the unique measure m_δ extending m , since

$$\int^\bullet \cdot dm = m_\delta{}^\bullet \quad \text{respectively} \quad \int^* \cdot dm = m_\delta{}^* \quad \text{on } \mathfrak{L}$$

by Theorem and Proposition 13.8. Only the formula

$$\int^\bullet \cdot dr = \int^\bullet \cdot dm ,$$

remains to be proved. This follows from Corollary 9.11, since $r = \int^\bullet \cdot dm$ on $\mathcal{E}_-(\{r < \infty\}) \subset \mathcal{L}^*_-(m)$. □

REMARK. The formula for l characterizing the measure q is equivalent to the semifiniteness and the $\mathfrak{E}(\mathfrak{K})$-tightness of l on $\mathfrak{E}_{-}(\mathfrak{L}^{\nabla})$.

This is a weakening of Remark 12.7.1.

13.10 Corresponding to Corollary 12.8.2, we infer from (iii) the

COROLLARY 1. *Let $\mathfrak{K}$ be a δ-lattice and let $\mathfrak{R}$ be a ring which contains $\mathfrak{K}$ and which is contained in the σ-algebra generated by $\mathfrak{G}(\mathfrak{K})$. Then restriction to $\mathfrak{K}$ defines a bijection between the inner $\mathfrak{K}$-regular measures on $\mathfrak{R}$ which are finite on $\mathfrak{K}$, and the semiregular measures on $\mathfrak{K}$. The essential upper integrals of each measure and of its restriction coincide with the essential upper functional of the restriction.*

A modification of this result without the hypothesis $\mathfrak{K} = \mathfrak{K}_{\delta}$ will be given in Theorem 16.5.

COROLLARY 2. *Every semifinite measure l admits a smallest extension, which is semifinite, and a largest extension to a measure on any lattice whose sets are l-measurable.*

This follows from Proposition 13.9.(i) and Lemma 13.9 since every measure extension of the finite part is infinite on sets of infinite measure, and therefore an extension. □

COROLLARY 3. *Every measure can be additively decomposed into a semifinite measure and a measure taking only the values 0 and ∞, such that semiregularity is inherited by each summand.*

Let l be a measure on a lattice $\mathfrak{L}$ and define q and r as in Corollary 12.8.4. It only remains to prove that r is a measure. Let $K_n, L_n \in \mathfrak{L}$ with $L_n \subset K_n$ and $r(L_n) < \infty$ and suppose that $(K_n \setminus L_n)$ is a sequence of pairwise disjoint sets with $\bigcup_{n=1}^{\infty} (K_n \setminus L_n) \in \mathfrak{L}$. We may assume that

$$\sum_{n=1}^{\infty} [r(K_n) - r(L_n)] = 0 \,, \quad \text{i.e.} \quad r(K_n) = r(L_n) = 0 \text{ for all } n \,,$$

and have to prove that

$$r(\bigcup_{n=1}^{\infty} [K_n \setminus L_n]) = 0 \,,$$

i.e. $q(H) = l(H)$ for all $H \in \mathfrak{L}$ with $H \subset \bigcup_{n=1}^{\infty} (K_n \setminus L_n)$. But

$$q(H) = q(\bigcup_{n=1}^{\infty} [H \cap K_n \setminus L_n]) = \sum_{n=1}^{\infty} [q(H \cap K_n) - q(H \cap L_n)] =$$

$$= \sum_{n=1}^{\infty} [l(H \cap K_n) - l(H \cap L_n)] = l(\bigcup_{n=1}^{\infty} [H \cap K_n \setminus L_n]) = l(H) \,. \quad \square$$

EXAMPLE. Let m be a Radon measure on a Hausdorff space X. Then all Borel sets, i.e. all sets which belong to the *Borel σ-algebra* $\mathfrak{B}(X)$ generated by $\mathfrak{F}(X)$, are m-measurable. Therefore, $m^{\bullet} = m_*$ on $\mathfrak{B}(X)$ is the only inner $\mathfrak{K}(X)$-regular measure which extends m to $\mathfrak{B}(X)$. Every Borel set A with $m^*(A) < \infty$ is m-integrable with

$$m^{\bullet}(A) = \inf_{G \in \mathfrak{G}(X),\, G \supset A} m(G) \,.$$

The measure $m^{\bullet}$ on $\mathfrak{B}(X)$ is called the *regular Borel measure* associated with m.

The measure m^* on $\mathfrak{B}(X)$ is the only measure extending m which is outer $\mathfrak{G}(X)$-regular on $\mathfrak{B}(X)$ and inner $\mathfrak{K}(X)$-regular on $\mathfrak{G}(X)$. This measure is sometimes called the *Riesz measure* associated with m.

Since $\mathfrak{K}(X)$ is a δ-lattice, the essential upper functional for m and the essential upper integrals for the corresponding regular Borel and Riesz measures all coincide.

13.11 We next discuss uniqueness problems.

PROPOSITION. *Let $\mathfrak{A}$ be a ring, $\mathfrak{E} \subset \mathfrak{A}$ a system which is stable with respect to* *finite unions or finite intersections,* $\mathfrak{R}$ *the ring generated by* $\mathfrak{E}$ *, and suppose that* $\mathfrak{A}$ *is contained in the* σ*-algebra* $\mathfrak{S}(\mathfrak{R}_\delta)$ *, e.g.* $\mathfrak{A}$ *is the* σ*-algebra generated by* $\mathfrak{E}$ *. If* m_1 *and* m_2 *are measures on* $\mathfrak{A}$ *which coincide and are finite on* $\mathfrak{E}$ *, and if for* $i = 1,2$ *and for all* $A \in \mathfrak{A}$

$$m_i(A) = \sup_{R\in\mathfrak{R}} m_i(A \cap R) ,$$

holds, then $m_1 = m_2$.

Denote by m the restriction of m_1 to $\mathfrak{R}$. Then m is a finite measure which by Proposition 12.9 coincides with m_2 on $\mathfrak{R}$. By hypothesis, for every $R \in \mathfrak{R}$, the ring $\mathfrak{A} \cap R$ of all sets $A \cap R$ with $A \in \mathfrak{A}$ consists of m-measurable and even m-integrable sets, since

$$\int^{\times} A \cap R \, dm \leq m(R) < \infty .$$

By Proposition 13.9.(i), the measures m_1 and m_2 coincide on $\mathfrak{A} \cap R$ and therefore by hypothesis on $\mathfrak{A}$. □

COROLLARY. *Let* m_1 *and* m_2 *be measures on a* σ*-algebra* $\mathfrak{A}$ *, generated by a system* $\mathfrak{E}$ *which is stable with respect to finite unions or finite intersections. If* X *is covered by a sequence* (E_n) *in* $\mathfrak{E}$ *and if* m_1 *and* m_2 *coincide and are finite on* $\mathfrak{E}$ *, then they are equal.*

If $\mathfrak{E}$ *is stable with respect to finite intersections, the finiteness condition has only to be imposed on the sets* E_n .

By the proposition, m_1 and m_2 coincide on the ring $\mathfrak{R}$ generated by all sets in $\mathfrak{E}$ on which m_1 and m_2 are finite. We may therefore suppose that $\mathfrak{E}$ is stable with respect to finite intersections.

For $E \in \mathfrak{E}$, we have $E = \bigcup E_n \cap E \in \mathfrak{R}_\sigma$. Therefore $\mathfrak{A}$ is contained in the σ-algebra $\mathfrak{S}(\mathfrak{R}_\delta)$. Moreover, for $i = 1,2$ and $A \in \mathfrak{A}$ we have

$$m_i(A) = \sup_n m_i(A \cap [E_1 \cup ... \cup E_n]) = \sup_{R\in\mathfrak{R}} m_i(A \cap R) .$$

In view of the proposition, this proves the corollary. □

In general, it is not sufficient to impose the finiteness condition only on the sets E_n , even if this sequence is increasing. This is shown in the following

EXAMPLE. For $k,l \in \mathbb{N}$, denote by E_k and F_l respectively the subsets

$$\{n \in \mathbb{N} : n \leq 2k + 1\} \quad \text{and} \quad \{n \in \mathbb{N} : n \geq l\}$$

of $\mathbb{N}$, and by $\mathfrak{E}$ the system of all unions $E_k \cup F_l$.

On the σ-algebra $\mathfrak{P}(\mathbb{N})$ generated by $\mathfrak{E}$, the measures

$$m_1 : A \longmapsto \mathfrak{card}(A)$$

and

$$m_2 : A \longmapsto 2 \cdot \mathfrak{card}(A \cap 2 \cdot \mathbb{N})$$

are different. However, they coincide on $\mathfrak{E}$, are finite on each E_n, and $\mathbb{N}$ is the union of this increasing sequence.

13.12 DEFINITION. An increasing set function m on $\mathfrak{K}$ with $m(\emptyset) = 0$ is said to have the *Bourbaki property*, respectively the *Bourbaki property at the empty set*, if for every downward directed family (K_i) in $\mathfrak{K}$ with $m(K_i) < \infty$ and $\bigcap K_i \in \mathfrak{K}$, respectively $\bigcap K_i = \emptyset$, one has

$$m(\bigcap K_i) = \inf m(K_i) .$$

PROPOSITION. *The Bourbaki property of* m *is equivalent to the Bourbaki property of the functional* m *on* $\mathcal{E}_-(\mathfrak{K})$.

The proof of this is analogous to the one given in 13.3. □

REMARKS.

(1) *The equality* $\sup m(e_i) = 0$ *holds for every upward directed family* (e_i) *in* $\mathcal{E}_-(\mathfrak{K})$ *with* 0 *as upper envelope iff* m *has the Bourbaki property at the empty set.*

This equivalence is implicit in the proof of 13.3 in which under the present circumstances $e = 0$. □

(2) *If* m *is semiregular, then the Bourbaki property at the empty set is equivalent to the Bourbaki property.*

This can be proved easily in the set-theoretic set-up, whereas in view of Theorem 8.1 the functional analytic proof needs the additional requirement that m be $\mathfrak{S}(\mathfrak{K})$-regular (cf. Corollary 12.3).

Example. *Every Radon measure has the Bourbaki property.*

This follows from the preceding Remark 2, since a downward directed family of compact sets with empty intersection contains the empty set. □

§ 14 REPRESENTATION BY CONTENTS

THROUGHOUT THIS SECTION,
τ IS AN INCREASING LINEAR FUNCTIONAL
ON A LATTICE CONE $\mathcal{T}$ OF FUNCTIONS,
$\mathfrak{K}$ AND $\mathfrak{G}$ ARE COMPATIBLE LATTICES OF SETS.

14.1 This section is devoted to the investigation of the representability of τ by a finite content on $\mathfrak{K}$, i.e. by an increasing linear functional on $\mathcal{E}(\mathfrak{K},\mathfrak{G})$. Accordingly, we first give a description of $\mathcal{E}(\mathfrak{K},\mathfrak{G})$-tightness (cf. § 7) which shows that in this special case our general definition coincides with the usual concept.

DEFINITION. τ is said to be $\mathfrak{K}$*-tight* if

$$\sup_{K\in\mathfrak{K}} \tau^{\times}(1_{CK}\cdot t) = 0$$

holds for all $t \in \mathcal{T}_-$.

We say that τ satisfies the *truncation formula at* $-\infty$, respectively *at* 0, if this holds for $\tau^{\times}$ on $\mathcal{T}_-$, i.e. if

$$\tau(t) = \inf_{n\in\mathbb{N}} \tau^{\times}(\max(t,-n)),$$

respectively

$$\sup_{n\geq 1} \tau^{\times}(\max(t, -\tfrac{1}{n})) = 0$$

for all $t \in \mathcal{T}_-$. Note that in these formulas $\tau^{\times}$ can be replaced by τ, if $\mathcal{T}_-$ is Stonian.

REMARK. *If τ is semiregular, then $\mathfrak{K}$-tightness is equivalent to*

$$\sup_{K\in\mathfrak{K}} \tau^{*}(1_{CK}\cdot t) = 0$$

for every $t \in \mathcal{T}_-$, i.e. for $\varepsilon > 0$ there exists a set $K \in \mathfrak{K}$ such that $\tau(u) \geq -\varepsilon$ for all $u \in \mathcal{T}_-$ with $u \geq t$ and $u = 0$ on K.

This follows from Proposition 6.7. □

EXAMPLES.

(1) We say that $\mathcal{T}$ is *$\mathfrak{K}$-adapted*, if for every $t \in \mathcal{T}_-$ there exists an $u \in \mathcal{T}_-$ such that the following condition holds :

For every $\varepsilon > 0$ there exists $K \in \mathfrak{K}$ with $\varepsilon u \leq t$ on $\complement K$.

Note that $\mathcal{T}$ is $\mathfrak{K}$-adapted if $\mathcal{T}$ is $\mathcal{E}_-(\mathfrak{K})$-adapted. The converse is true if every function in $\mathcal{T}_-$ is bounded on every $K \in \mathfrak{K}$.

If $\mathcal{T}$ is $\mathfrak{K}$-adapted, e.g. if every function in $\mathcal{T}$ is positive outside some set from $\mathfrak{K}$, then every increasing linear functional on $\mathcal{T}$ is $\mathfrak{K}$-tight.

This follows immediately from the inequality $1_{\complement K} \cdot t \geq \varepsilon u$. □

(2) *Let X be a Hausdorff space and $\mathcal{T}$ a lattice cone of lower semicontinuous functions on X. Then τ is $\mathfrak{K}(X)$-tight iff τ is $\mathcal{S}(X)$- or $\mathcal{E}(\mathfrak{K}(X),\mathfrak{G}(X))$-tight, and $\mathcal{T}$ is $\mathfrak{K}(X)$-adapted iff $\mathcal{T}$ is $\mathcal{S}(X)$-adapted.*

Note that $\mathcal{E}_-(\mathfrak{K}(X))$ is coinitial in $\mathcal{S}_-(X)$. The first part therefore is a reformulation of Example 7.1.2, whereas the second part follows from the above example. □

PROPOSITION. *If every set from $\mathfrak{K}(\mathcal{T})$ is contained in some set from $\mathfrak{K}$, and if τ satisfies the truncation formula at 0, then τ is $\mathfrak{K}$-tight.*

Conversely, if every set from $\mathfrak{K}$ is contained in some set from $\mathfrak{K}(\mathcal{T})$, and if τ is $\mathfrak{K}$-tight, then τ satisfies the truncation formula at 0.

The first assertion follows since for $t \in \mathcal{T}_-$ we have

$$\max(t, -\frac{1}{n}) \leq 1_{\complement\{t \leq -1/n\}} \cdot t .$$

In the second case, the truncation formula at 0 necessarily holds since for $K \in \mathfrak{K}$ there exists $s \in \mathcal{T}_-$ with $K \subset \{s \leq -1\}$, hence

$$\max(t, -\frac{1}{n}) \geq \frac{1}{n} \cdot s + 1_{\complement K} \cdot t$$

and therefore

$$\tau^{\times}(\max(t, -\frac{1}{n})) \geq \frac{1}{n} \cdot \tau(s) + \tau^{\times}(1_{\complement K} \cdot t)$$

by Proposition 2.7. □

EXAMPLE.

(3) *Let $\mathcal{T}$ be a lattice cone of lower semicontinuous functions on a Hausdorff space X, such that $\mathcal{T}_-$ is Stonian and every function in $\mathcal{T}_-$ vanishes at infinity.*

If $\tau^\bullet$ is finite on $\mathfrak{K}(X)$, e.g. if for every $K \in \mathfrak{K}(X)$ there exists $t \in \mathcal{T}$ with $t \leq -1_K$, and if τ is semiregular, respectively regular, then the following assertions are equivalent:

(i) *τ is $\mathfrak{K}(X)$-tight.*

(ii) *τ is an integral, respectively a Bourbaki integral.*

(iii) *τ satisfies the truncation formula at 0.*

By the preceding Example 2, assertion (i) is equivalent to the $\mathcal{S}(X)$-tightness of τ, which implies (ii) by Example 9.5, respectively Proposition 8.2. Obviously, (ii) yields (iii), and since $\mathfrak{K}(\mathcal{T}) \subset \mathfrak{K}(X)$, the remaining implication follows from the first part of the proposition. □

14.2 PROPOSITION. *Let $\mathcal{T}_-$ be Stonian. Then τ is $\mathcal{E}(\mathfrak{K},\mathfrak{G})$-tight iff τ is $\mathfrak{K}$-tight and satisfies the truncation formula at $-\infty$.*

By the hypothesis, we have $\max(t,-n) \in \mathcal{T}_-$ for any $t \in \mathcal{T}_-$ and $n \in \mathbb{N}$. Since for every $e \in \mathcal{E}_-$ there exist $n \in \mathbb{N}$ and $K \in \mathfrak{K}$ with $-n\cdot 1_K \leq e$, the $\mathcal{E}$-tightness is equivalent to

$$\tau(t) = \inf_{n,K} \tau_\times(\max(t, -n\cdot 1_K))$$

or, by Remark 7.1.1, to

$$\sup_{n,K} \tau^\times([t + n\cdot 1_K]_-) = 0$$

for all $t \in \mathcal{T}_-$.

Suppose first that τ is $\mathcal{E}$-tight. Since

$$\max(t,-n) \leq \max(t, -n\cdot 1_K) \quad \text{and} \quad 1_{\complement K}\cdot t \geq (t + n\cdot 1_K)_- ,$$

we get

$$\tau(t) = \inf_{n\in\mathbb{N}} \tau(\max(t,-n))$$

and

$$\sup_{K\in\mathfrak{K}} \tau^\times(1_{\complement K}\cdot t) = 0 .$$

Conversely, from

$$\max(t, -n\cdot 1_K) + 1_{\complement K}\cdot t \leq \max(t,-n)$$

and Theorem 2.2.(v) we infer that

$$\tau_{\times}(\max(t, -n\cdot 1_K)) + \tau^{\times}(1_{\complement K}\cdot t) \le \tau^{\times}(\max[t, -n\cdot 1_K] + 1_{\complement K}\cdot t) \le \tau(\max(t,-n)),$$

hence

$$\tau(t) \le \inf_{n,K} \tau_{\times}(\max(t, -n\cdot 1_K)) \le$$

$$\le \inf_{n\in\mathbb{N}} \tau(\max(t,-n)) - \sup_{K\in\mathfrak{K}} \tau^{\times}(1_{\complement K}\cdot t) = \tau(t)\,.$$

This proves that τ is $\mathcal{E}$-tight. □

Remark. If every function in $\mathcal{T}$ is bounded below, then we may omit the Stonian hypothesis on $\mathcal{T}$, and τ is $\mathfrak{K}$-tight iff τ is $\mathcal{E}(\mathfrak{K},\mathfrak{G})$-tight. If moreover $-1 \in \mathcal{T}$, then tightness is equivalent to the condition

$$\inf_{K\in\mathfrak{K}} \tau_{\times}(\complement K) = 0$$

or, if τ is semiregular, to

$$\inf_{K\in\mathfrak{K}} \tau_{*}(\complement K) = 0\,.$$

This follows from Remark 14.1. □

Corollary. *Let $\mathcal{T}_-$ be Stonian.*

If every set from $\mathfrak{K}(\mathcal{T})$ is contained in some set from $\mathfrak{K}$, and if τ satisfies both truncation formulas, then τ is $\mathcal{E}(\mathfrak{K},\mathfrak{G})$-tight.

Conversely, if every set from $\mathfrak{K}$ is contained in some set from $\mathfrak{K}(\mathcal{T})$, and if τ is $\mathcal{E}(\mathfrak{K},\mathfrak{G})$-tight, then τ satisfies both truncation formulas.

This is a consequence of the last two propositions. □

14.3 Theorem. *Suppose that $\mathcal{T}_-$ is Stonian and consists of $\mathfrak{K}$-measurable functions.*

If τ is difference-regular, $\mathfrak{K}$-tight, satisfies the truncation formula at $-\infty$, and if $\tau^{\bullet} < \infty$ on $\mathfrak{K}$, then there exists a finite content m on $\mathfrak{K}$ representing τ.

If moreover $\mathcal{T}$ consists of $\mathfrak{K}$-measurable functions and τ is regular, then m can be chosen to be semiregular.

This follows immediately from Proposition 14.2, Theorem 7.7 and Corollary 7.8, since all functions in $\mathcal{T}_-$, respectively $\mathcal{T}$, are $\mathcal{E}(\mathfrak{K})$-measurable by Theorem 10.10. □

REMARKS.

(1) In the second part of the theorem, m is $\mathfrak{G}$-regular if for every $K \in \mathfrak{K}$ there exists $G \in \mathfrak{G}$ with $K \subset G$ and $\tau^*(G) < \infty$.

This is a consequence of Remark 7.8.1. □

(2) The finiteness assumption on $\tau^\bullet$ can be replaced by the truncation formula at 0 if one admits non-finite contents.

It suffices to apply the theorem to the lattice $\mathfrak{K}^{\mathcal{T}}$ of all $K \in \mathfrak{K}$ contained in some set from $\mathfrak{K}(\mathcal{T})$ and then, by Proposition 12.7, (ii) and (iii), to extend m from $\mathfrak{K}^{\mathcal{T}}$ to a possibly non-finite content on $\mathfrak{K}$, preserving semiregularity. Indeed, $\mathfrak{K}$-measurable functions are always $\mathfrak{K}^{\mathcal{T}}$-measurable, and we have :

If τ is $\mathfrak{K}$-tight and satisfies the truncation formula at 0, then τ is $\mathfrak{K}^{\mathcal{T}}$-tight.

To prove this assertion, let $t \in \mathcal{T}_-$ and $\varepsilon > 0$ be given. Choose $K \in \mathfrak{K}$ and $n \geq 1$ such that

$$\tau^\times(1_{\complement K} \cdot t) \geq -\varepsilon \quad \text{and} \quad \tau(\max(t, -\tfrac{1}{n})) \geq -\varepsilon .$$

By the $\mathfrak{K}$-measurability of t, there exists $L \in \mathfrak{K}$ with

$$\{t \leq -\tfrac{1}{n}\} \cap K \subset L \subset \{t \leq -\tfrac{1}{2n}\} \cap K .$$

We then have $L \in \mathfrak{K}^{\mathcal{T}}$, and from

$$1_{\complement L} \cdot t \geq 1_{\complement K} \cdot t + \max(t, -\tfrac{1}{n})$$

the result follows. □

If moreover every set from $\mathfrak{K}(\mathcal{T})$ is contained in some set from $\mathfrak{K}$, then $\mathfrak{K}$-tightness is superflous. This follows from Proposition 14.1.

COROLLARY. *Let* $\mathfrak{L}$ *be a sublattice of* $\mathfrak{G}(\mathfrak{K})^c$. *If* l *is a finite content on* $\mathfrak{L}$ *with*

$$\inf_{K \in \mathfrak{K}} l_{\times}(L \setminus K) = 0 ,$$

then there exists a possibly non-finite content m *on* $\mathfrak{K}$ *representing* l.

If moreover l *is* $\mathfrak{G}(\mathfrak{L})$*-regular, then* m *can be chosen to be semiregular.*

From the assumptions and Remark 10.3.3 we infer that all functions in $\mathcal{E}(\mathfrak{L})$ are $\mathfrak{K}^{\mathcal{E}\text{-}(\mathfrak{L})}$–measurable. The formula is by Proposition 14.2 and Remark 7.1.1 equivalent to the $\mathfrak{K}$–tightness of l. We then proceed as in Remark 2. □

EXAMPLE. *Let* $\mathfrak{L}$ *be a sublattice of* $\mathfrak{K}$. *Then every bounded semiregular content on* $\mathfrak{L}$ *can be extended to a bounded semiregular content on* $\mathfrak{K}$.

By Proposition 12.7.(iii), we may replace $\mathfrak{L}$ and $\mathfrak{K}$ by $\mathfrak{L} \cup \{X\}$ and $\mathfrak{K} \cup \{X\}$ respectively, since the restriction to $\mathfrak{K}$ of a semiregular content on $\mathfrak{K} \cup \{X\}$ is semiregular and bounded. The result now follows because bounded semiregular contents on $\mathfrak{L}$ are $\mathfrak{G}(\mathfrak{L})$–regular by Theorem 12.3. □

Note that it is not difficult to weaken boundedness to semifiniteness (cf. *Adamski* [1984], Theorem 2.5).

14.4 To obtain representation theorems without use of the Hahn–Banach theorem, i.e. without recourse to Theorem 7.7, we have to impose stronger conditions. The simplest one is essential integrability of all sets from $\mathfrak{K}$. This will be replaced later on by separation properties.

PROPOSITION. *Suppose that all sets in* $\mathfrak{K}$ *are essentially integrable w.r.t.* τ, *and that every function in* $\mathcal{T}_-$ *is* $\mathfrak{K}$*-measurable.*

The restriction m *of* $\tau^\bullet$ *to* $\mathfrak{K}$ *is a* $\mathcal{T}$*-tight finite content. If* τ *is semiregular, then* m *is semiregular.*

The content m *represents* τ *iff* τ *is difference-regular and* $\mathcal{E}(\mathfrak{K},\mathfrak{G})$*-tight. In this case, we have* $\tau^\bullet = m^\bullet$. *If* τ *is semiregular, then* m *is the only* $\mathcal{T}$*-tight semiregular content on* $\mathfrak{K}$ *representing* τ. *If moreover* τ^* *is finite on* $\mathfrak{K}$, *then* m *coincides with* τ^* *on* $\mathfrak{K}$ *and with* τ_* *on* $\mathcal{E}(\mathfrak{K},\mathfrak{S})$ *, and is the only semi-regular content on* $\mathfrak{K}$ *representing* τ *.*

All functions from $\mathcal{E}_-$ are essentially integrable w.r.t. τ , and by the hypothesis and Theorem 10.10, all functions from $\mathcal{T}_-$ are $\mathcal{E}$–measurable. We can now apply Corollary 7.5. □

Sufficient conditions for measurability are given in the following

LEMMA. *Let* $\mathcal{T}_-$ *be Stonian, and suppose that*

$$\mathfrak{K}(\mathcal{T}) \cap \mathfrak{N}^{\times}(\tau) \subset \mathfrak{K} .$$

Then all functions in $\mathcal{T}_-$ *are* $\mathfrak{K}$*–measurable, every set from* $\mathfrak{K}(\mathcal{T})$ *is contained in some set from* $\mathfrak{K}$ *, and all functions in* $\mathcal{T}$ *are* $\mathfrak{K}(\mathcal{T})\cap\mathfrak{N}^{\times}(\tau)$*–measurable.*

If moreover $\mathfrak{G}(\mathcal{T}) \cap \mathfrak{N}^{\bullet}(\tau) \subset \mathfrak{G}(\mathfrak{K})$ *and* τ^* *is finite on* $\mathfrak{K}$ *, then all functions in* $\mathcal{T}$ *are* $\mathfrak{K}$*–measurable.*

Indeed, 1 is measurable since -1 is $\mathcal{T}$–measurable and, by Theorem 11.7 applied to $\tau^{\times}$, for $t \in \mathcal{T}_-$ we have

$$\{t \le \gamma\} \in \mathfrak{K}(\mathcal{T}) \cap \mathfrak{N}^{\times}(\tau) \subset \mathfrak{K}$$

for all but at most countably many $\gamma < 0$. Thus, $\mathfrak{K}$–measurability follows from Remark 10.3.2, and every set from $\mathfrak{K}(\mathcal{T})$ is contained in some set from $\mathfrak{K}$ choosing a suitable $\gamma > -1$. Since every $t \in \mathcal{T}$ is measurable, Corollary 11.7 shows that for every $K \in \mathfrak{K}(\mathcal{T}) \cap \mathfrak{N}^{\times}(\tau)$, and all but at most countably many $\gamma \in \mathbb{R}$, we have

$$\{t \le \gamma\} \cap K \in \mathfrak{N}^{\times}(\tau) .$$

Due to the $\mathfrak{K}(\mathcal{T})$–semicontinuity of t (cf. Proposition 10.8), these sets also belong to $\mathfrak{K}(\mathcal{T})$ and therefore to $\mathfrak{K}(\mathcal{T}) \cap \mathfrak{N}^{\times}(\tau)$. This in turn implies that t is $\mathfrak{K}(\mathcal{T})\cap\mathfrak{N}^{\times}(\tau)$–measurable by Remark 10.3.2.

To prove the $\mathfrak{K}$–measurability of $t \in \mathcal{T}$ in the last case, we use again Remark 10.3.2. Since $t_- \in \mathcal{R}^{\times}_-(\tau)$, we get

$$\{t \le \gamma\} = \{t_- \le \gamma\} \in \mathfrak{K}$$

for all but at most countably many $\gamma < 0$ by the above. On the other hand, let

$K \in \mathfrak{K}$ be given and choose $s \in \mathcal{T}$ with $s \geq 1_K$ and $\tau(s) < \infty$. Since for $n \in \mathbb{N}$ we have

$$\min(t,ns) \in \mathcal{T} \cap \mathcal{R}^\bullet(\tau),$$

from Theorem 11.7, applied to $-\min(t^+,ns)$, we infer that for all but at most countably many $\gamma \in]0,n[$ we have

$$\{\min(t,ns) > \gamma\} \in \mathfrak{G}(\mathcal{T}) \cap \mathfrak{R}^\bullet(\tau) \subset \mathfrak{G}(\mathfrak{K})$$

and therefore

$$\{t \leq \gamma\} \cap K = K \setminus \{\min(t,ns) > \gamma\} \in \mathfrak{K}.$$

Finally, one has only to take the union of the exceptional sets. □

COROLLARY. *If* $\mathcal{T}_-$ *is Stonian and*

$$\mathfrak{K}(\mathcal{T}) \cap \mathfrak{R}^\times(\tau) \subset \mathfrak{K} \subset \mathfrak{R}^\bullet(\tau),$$

then the content $m := \tau^\bullet_{|\mathfrak{K}}$ *represents* τ *iff* τ *is difference-regular and satisfies both truncation formulas.*

The truncation formulas necessarily hold by Example 11.6.2. Their validity is sufficient by the proposition, in view of the lemma and Corollary 14.2. □

REMARK. If τ is *regular*, then the corollary can be applied to $\mathfrak{R}^*(\tau)$, since

$$\mathfrak{R}^\times(\tau) \subset \mathfrak{R}^*(\tau) \subset \mathfrak{R}^\bullet(\tau).$$

As $\mathfrak{R}^*(\tau)$ is a ring, we then have :

The restriction m *of* τ^* *to the ring* $\mathfrak{R}^*(\tau)$ *is a finite content representing* τ *iff* τ *satisfies both truncation formulas.*

In this case, m *is the only finite content on* $\mathfrak{R}^*(\tau)$ *representing* τ, *and we have* $\tau^\bullet = m^\bullet$.

Applied to a Stonian vector lattice of functions, this improves results of *L.H. Loomis* (cf. 16.2). A similar result can be obtained from the next proposition.

14.5 Since regularity of contents is, by Theorem 12.3, equivalent to semiregularity and boundedness, the following will be useful.

LEMMA. *Let $\mathcal{T}_-$ be Stonian and suppose that*

$$\mathfrak{G}(\mathcal{T}) \cap \mathfrak{R}^{\bullet}(\tau) \subset \mathfrak{G}.$$

If the restriction m of τ^* to $\mathfrak{K}$ is a finite content, then m is $\mathfrak{G}$-bounded.

For $K \in \mathfrak{K}$ there exists $t \in \mathcal{T}_+$ with $t \geq 1_K$ and $\tau(t) < \infty$. Since t is essentially integrable and 1 is measurable w.r.t. τ, by Theorem 11.7 there exists $\gamma > 1$ with

$$G := \{t > \tfrac{1}{\gamma}\} \in \mathfrak{G}(\mathcal{T}) \cap \mathfrak{R}^{\bullet}(\tau) \subset \mathfrak{G}$$

and $K \subset G$. Since $1_G \leq \gamma \cdot t$, for $L \in \mathfrak{K}$ with $L \subset G$ we have

$$m(L) = \tau^*(L) \leq \tau^*(G) \leq \gamma \cdot \tau(t),$$

hence

$$m_*(G) \leq \gamma \cdot \tau(t) < \infty. \quad \square$$

REMARKS.

(1)   If $\mathfrak{G}(\mathcal{T}) \subset \mathfrak{G}$, then we can omit the Stonian condition.

(2)   *If τ is regular, then we may replace $\mathfrak{R}^{\bullet}(\tau)$ in the above condition by $\mathfrak{R}^*(\tau)$, and if moreover every set from $\mathfrak{K}$ is contained in some set from $\mathfrak{K}(\mathcal{T})$, then τ^* is finite on $\mathfrak{K}$.*

Indeed, in the proof of the lemma, t is even integrable w.r.t. τ^*. Moreover, for $K \in \mathfrak{K}$ and $u \in \mathcal{T}_-$ with $K \subset \{u \leq -1\}$ we have

$$\tau^*(K) = -\tau_*(-1_K) \leq -\tau_*(u) = -\tau(u) < \infty. \quad \square$$

We now investigate the relationship between τ and m in the case where representation is not ensured, i.e. the truncation formulas are not necessarily satisfied.

PROPOSITION. *Let $\mathcal{T}$ be Stonian. If τ is regular, then the restriction m of τ^* to $\mathfrak{K}(\mathcal{T}) \cap \mathfrak{R}^*(\tau)$ is a $\mathfrak{G}(\mathcal{T}) \cap \mathfrak{R}^*(\tau)$-regular content. We have*

$$m^{\bullet}(f) \leq \tau^{\bullet}(f)$$

for every function f with a minorant in $\mathcal{E}_-(\mathfrak{K}(\mathcal{T}) \cap \mathfrak{R}^*(\tau))$, and every function *which is measurable w.r.t.* τ *is measurable w.r.t.* m. *Furthermore,*

$$\tau^* \le m^* ,$$

in particular $\mathcal{R}^*(m) \subset \mathcal{R}^*(\tau)$ *and* $\mathfrak{N}^*(m) = \mathfrak{N}^*(\tau)$.

In view of the lemma and the preceding Remark 2, the first assertion follows, having in mind that both lattices

$$\mathfrak{K} := \mathfrak{K}(\mathcal{T}) \cap \mathfrak{N}^*(\tau) \quad \text{and} \quad \mathfrak{G} := \mathfrak{G}(\mathcal{T}) \cap \mathfrak{N}^*(\tau)$$

are compatible by Theorem 10.6.

Suppose first that $\pm f \in \mathcal{E}_-^{\uparrow}$. We prove that $m^*(f) \le \tau^*(f)$. Let $t \in \mathcal{T}$ with $t \ge f$ and $\tau(t) < \infty$ be given. We have to show that for any $\varepsilon > 0$ there exists $e \in \mathcal{E}$ with $e \ge f$ and $\tau(t) \ge m(e) - \varepsilon$. Since $t \in \mathcal{R}^*(\tau)$ and f is bounded, by Lemma 10.9 and Theorem 11.7, for sufficiently large $n \in \mathbb{N}$ there exists $e_n \in \mathcal{E}$ with

$$e_n + \max(t_-, -\frac{1}{2^{n-1}}) \le \max(t,-n) = t$$

and

$$f \le \min(t,n) \le e_n + \min(t^+, \frac{1}{2^{n-1}}) .$$

Choose $K \in \mathfrak{K}$ and $G \in \mathfrak{G}$ with $G \supset K \supset \{f \ne 0\}$. Then

$$e_n - \frac{1}{2^{n-1}} \cdot 1_K \le t \quad \text{and} \quad f \le e_n + \frac{1}{2^{n-1}} \cdot 1_G =: e \in \mathcal{E} ,$$

since $t \ge f = 0$ on $\complement K$ and $e_n \ge 0$ on $\{t \ge 0\}$ by construction of e_n. This gives

$$\tau(t) \ge \tau^*(e_n) - \frac{1}{2^{n-1}} \tau^*(K) = \tau^*(e) - \frac{1}{2^{n-1}} [\tau^*(K) + \tau^*(G)] \ge$$

$$\ge \tau^*(e) - \varepsilon = m(e) - \varepsilon$$

for sufficiently large n, since $\mathcal{E} \subset \mathcal{R}^*(\tau)$ and $m = \tau_* = \tau^*$ on $\mathcal{E}$. In fact, by definition of m we only have to prove that $m(G) = \tau_*(G)$ for $G \in \mathfrak{G}$. By Corollary 12.3, we have

$$m(G) = \sup_{K \in \mathfrak{K},\ K \subset G} \tau_*(K) \le \tau_*(G) .$$

Conversely, for $t \in \mathcal{T}_-$ with $-t \le 1_G$ and $\varepsilon > 0$, by Theorem 11.7 there exists $\gamma \in]0,1[$ with $K := \{t \le -\gamma\} \in \mathfrak{K}$ and $\gamma \cdot \tau^*(G) \le \varepsilon$. But $K \subset G$ and

$$-t \le 1_K + \gamma \cdot 1_G ,$$

hence

$$-\tau(t) \leq \tau^*(K) + \gamma \cdot \tau^*(G) \leq m(G) + \varepsilon ,$$

which yields $\tau_*(G) \leq m(G)$.

Suppose now that $f \in \mathcal{E}_-^{\uparrow}$. For any $v \in \mathcal{E}_-$, we have $\pm\min(f,-v) \in \mathcal{E}_-^{\uparrow}$ and therefore by the above

$$m^*(\min(f,-v)) \leq \tau^*(\min(f,-v)) .$$

This yields

$$m^\bullet(f) = \sup_{v \in \mathcal{E}_-} m^*(\min(f,-v)) \leq \sup_{v \in \mathcal{R}^*_-(\tau)} \tau^*(\min(f,-v)) = \tau^\bullet(f) .$$

If $f \in \mathcal{R}^0(\tau)$, then $g := \mathrm{med}(f,-v,u) \in \mathcal{R}^*(\tau)$ for $u,v \in \mathcal{E}_-$ and

$$\tau_*(g) \leq m_*(g) \leq m^*(g) \leq \tau^*(g) = \tau_*(g)$$

by the first part, hence $g \in \mathcal{R}^*(m)$ and thus $f \in \mathcal{R}^0(m)$.

For arbitrary functions f, from $m = \tau^*$ on $\mathcal{E}$ we infer that

$$m^*(f) = \inf_{e \in \mathcal{E},\, e \geq f} \tau^*(e) \geq \tau^*(f) .$$

This proves that $\mathcal{R}^*(m) \subset \mathcal{R}^*(\tau)$. The reverse inclusion for integrable sets follows from the Integrability Criterion 4.5, since $\mathfrak{R}^0(\tau) \subset \mathfrak{R}^0(m)$ and $m^*(A) < \infty$ for any $A \in \mathfrak{R}^*(\tau)$. Indeed, there exists $t \in \mathcal{T}$ with $t \geq 1_A$ and $\tau(t) < \infty$, hence

$$A \subset \{t > \gamma\} =: G \in \mathfrak{G}$$

for some $\gamma \in]0,1[$ by Theorem 11.7, which implies

$$m^*(A) \leq m(G) = \tau^*(G) < \infty . \quad \square$$

If X is a locally compact space and $\mathcal{T}$ is a Stonian vector lattice of functions in $\mathscr{C}^0(X)$, the regularity of the content can be used to prove some results of *H. Bauer* emphasizing the relationship between the content τ^* on $\mathfrak{R}^*(\tau)$ and the topology (cf. 16.3).

14.6 As an application, we can now describe the upper functionals which are represented by contents.

PROPOSITION. *Let $\mathcal{T}$ be Stonian and ν an upper functional determined by $\mathcal{T}$ such that $\mathcal{T}_- \subset \mathcal{I}_-(\nu)$.*

Then the restriction of ν *to* $\mathfrak{K}(\mathcal{T}) \cap \mathfrak{I}(\nu)$ *is a* $\mathfrak{S}(\mathcal{T}) \cap \mathfrak{I}^\bullet(\nu)$*-regular content. It is the only semiregular content* m *with* $\nu^\bullet = m^\bullet$ *iff* ν *satisfies both truncation formulas on* $\mathcal{T}_-$ *.*

Let τ denote the restriction of $\nu^\bullet$ to $\mathcal{T}$. Then τ is regular by Proposition 6.2, and $\nu^\bullet = \tau^\bullet \leq \tau^* \leq \nu$.

Next we prove that

$$\mathfrak{K}(\mathcal{T}) \cap \mathfrak{I}(\nu) = \mathfrak{K}(\mathcal{T}) \cap \mathfrak{R}^*(\tau) .$$

Indeed, for $K \in \mathfrak{K}(\mathcal{T}) \cap \mathfrak{R}^*(\tau)$ there exists $t \in \mathcal{T}_-$ with $t \leq -1_K$. Since -1_K is measurable and t is integrable w.r.t. ν, the set K is integrable w.r.t. ν. The reverse inclusion follows from $\mathfrak{I}(\nu) \subset \mathfrak{R}^*(\tau)$.

Thus, by Proposition 14.5, the restriction of τ^* to $\mathfrak{K}(\mathcal{T}) \cap \mathfrak{R}^*(\tau)$, which obviously coincides with ν, is a $\mathfrak{S}(\mathcal{T}) \cap \mathfrak{I}^\bullet(\nu)$-regular content, since $\mathfrak{R}^*(\tau)$ is contained in $\mathfrak{I}^\bullet(\nu)$. From Corollary and Proposition 14.4 we infer that m is the only semiregular content with $m^\bullet = \tau^\bullet = \nu^\bullet$. □

COROLLARY. *Let* ν *be an auto-determined upper functional. Then the restriction of* ν *to the ring* $\mathfrak{I}(\nu)$ *is the only finite content* m *with* $\nu^\bullet = m^\bullet$ *iff* ν *satisfies both truncation formulas on* $\mathcal{I}_-(\nu)$ *and* 1 *is measurable.*

The necessity follows from Example 11.6.2, and the sufficiency by applying the proposition to the Stonian function cone $\tilde{\mathcal{I}}(\nu)$, noting that $\mathfrak{I}(\nu) \subset \mathfrak{K}(\tilde{\mathcal{I}}(\nu))$. □

14.7 **THEOREM.** *Let* $\mathcal{T}$ *be strongly Stonian with* $\mathcal{T} = \mathcal{T}_\sigma$. *Then there exists a bijection between the difference-regular linear functionals* τ *on* $\mathcal{T}$ *satisfying both truncation formulas, and the finite contents* m *on* $\mathfrak{K}(\mathcal{T})$ *with* $\mathcal{T}_- \subset \mathfrak{R}_-^\bullet(m)$, *given by*

$$\tau(t) = m^\bullet(t) \quad \textit{for all } t \in \mathcal{T}$$

and

$$m(K) = -\tau(-1_K) \quad \textit{for all } K \in \mathfrak{K}(\mathcal{T}) .$$

Furthermore, we have $\tau^\bullet = m^\bullet$, *and* τ *is semiregular iff* m *is.*

Note first that $\mathcal{E}_- \subset \mathcal{T}_-$ by Proposition 10.11 and thus

$$m(K) = -\tau(-1_K) = \tau^{\bullet}(K)$$

for all $K \in \mathfrak{K}(\mathcal{T})$. Secondly, all functions in $\mathcal{T}$ are $\mathfrak{K}(\mathcal{T})$-semicontinuous by Proposition 10.8, and therefore $\mathfrak{K}(\mathcal{T})$-measurable by Proposition 10.3.(iii). Thirdly, Corollary 14.2 shows that τ is $\mathcal{E}$-tight iff τ satisfies both truncation formulas.

Proposition 14.4 now implies that the mapping $\tau \longmapsto m$ is injective , and that m is semiregular if τ is. By Corollary 7.3, this mapping is surjective, and τ is semiregular if m is. □

Remark. This theorem gives an abstract version of the correspondence between Radon integrals and Radon measures proved in Theorem 12.5, more generally between the semiregular linear functionals μ on $\mathcal{S}(X)$ and the semiregular contents m on $\mathfrak{K}(X)$.

Indeed, both truncation formulas are trivial since the functions $-n \cdot 1_K$ with $n \in \mathbb{N}$ and $K \in \mathfrak{K}(X)$ are coinitial in $\mathcal{S}_-(X)$. Furthermore, $\mathcal{S}_-(X)$ is contained in $\mathcal{R}^{\bullet}_-(m)$ by the Integrability Criterion 4.5. Finally, μ is $\mathcal{S}(X)$-bounded iff m is $\mathfrak{G}(X)$-bounded. □

14.8 We next discuss the problem whether τ is representable by the restriction of τ^* to more general lattices $\mathfrak{K}$ of sets than those encountered so far. To be able to apply Addendum 7.6, we must prove that τ_* is linear on $\mathcal{E}_-(\mathfrak{K})$. The separation properties to be defined below will prove to be decisive for this linearity. The only way we know for securing these properties is via max-stability of $\mathcal{T}$ (cf. the first lines of 7.1).

Definition 1. The lattice of sets $\mathfrak{K}$ is said to be *τ-separated by $\mathcal{T}$* if for any two disjoint sets $K_1, K_2 \in \mathfrak{K}$ and $\varepsilon > 0$ there exist functions $t_1, t_2 \in \mathcal{T}$ such that

$$t_i \geq 1_{K_i} \quad \text{and} \quad \tau(\min(t_1, t_2)) \leq \varepsilon .$$

This concept only depends on the positive functions in $\mathcal{T}$.

Remarks.

(1) *If* τ_* *is finite and linear on* $\mathcal{E}_-(\mathfrak{K})$ *, then the restriction* m *of* τ^* *to* $\mathfrak{K}$ *is a finite content.*

(2) *If the restriction* m *of* τ^* *to* $\mathfrak{K}$ *is a finite content, then* $\mathfrak{K}$ *is* τ*-separated by* $\mathcal{T}$ *.*

Indeed, for $i = 1,2$, let functions $t_i \in \mathcal{T}$ be given such that

$$t_i \geq 1_{K_i} \quad \text{and} \quad \tau(t_i) \leq m(K_i) + \varepsilon \, .$$

Then

$$\tau(\min(t_1,t_2)) = \tau(t_1) + \tau(t_2) - \tau(\max(t_1,t_2)) \leq$$

$$\leq m(K_1) + m(K_2) + 2\varepsilon - \tau^*(\max(1_{K_1},1_{K_2})) =$$

$$= m(K_1) + m(K_2) - m(K_1 \cup K_2) + 2\varepsilon = 2\varepsilon \, . \quad \square$$

The following concept is more suitable for applications :

Definition 2. The lattice $\mathfrak{K}$ of sets is said to be *Urysohn-separated by* $\mathcal{T}$ if for any two disjoint sets $K,L \in \mathfrak{K}$ there exists $t \in \mathcal{T}$ with

$$K \subset \{t \geq 1\} \quad \text{and} \quad L \subset \{t \leq 0\} \, .$$

Proposition. *Let* τ *be semiregular and suppose that* τ^* *is finite on* $\mathfrak{K}$ *. If* $\mathfrak{K}$ *is Urysohn-separated by* $\mathcal{T}$ *, then* $\mathfrak{K}$ *is also* τ*-separated by* $\mathcal{T}$ *.*

By hypothesis, for any two disjoint sets $K_1,K_2 \in \mathfrak{K}$ there exist functions

$$t_i \in \{\tau < \infty\} \quad \text{with} \quad t_i \geq 1_{K_i} \, , \quad t_1 = 0 \text{ on } K_2 \quad \text{and} \quad t_2 = 0 \text{ on } K_1 \, .$$

By Proposition 6.8 we have $\tau(t_i) = \tau_*(t_i)$, hence for $\varepsilon > 0$ there are functions $s_i \in \mathcal{T}$ such that $-s_i \leq t_i$ and $\tau(s_i + t_i) \leq \varepsilon$. This implies that

$$(t_1 + s_2)^+ \geq 1_{K_1} \, , \; (t_2 + s_1)^+ \geq 1_{K_2}$$

and

$$\tau(\min\,[(t_1 + s_2)^+,(t_2 + s_1)^+\,]) \leq \tau(t_1 + s_1 + t_2 + s_2) \leq 2\varepsilon \, . \quad \square$$

14.9 EXAMPLES. Let X be a Hausdorff space.

(1) $\mathfrak{Z}(X)$ *is Urysohn-separated by* $\mathscr{C}^b(X)$.

Indeed, if $K = \{s = 0\}$ and $L = \{t = 0\}$ are disjoint with $s,t \in \mathscr{C}_+^b(X)$, then $\frac{t}{s+t}$ has the asserted property (cf. Example 10.7.3). □

(2) *If* X *is normal, then* $\mathfrak{F}(X)$ *is Urysohn-separated by* $\mathscr{C}^b(X)$.

This is just Urysohn's Lemma. □

(3) $\mathfrak{K}(X)$ *is Urysohn-separated by* $\mathscr{S}(X)$.

In fact, for $L \in \mathfrak{K}(X)$, the indicator function $1_{\complement L}$ belongs to $\mathscr{S}(X)$. □

(4) *Let* $\mathcal{T}$ *be a lattice cone of continuous functions. If* X *is not a one-point space, and if for any two different points* $x,y \in X$ *there exists* $t \in \mathcal{T}$ *with* $t(x) > 0$ *and* $t(y) < 0$, *then* $\mathfrak{K}(X)$ *is Urysohn-separated by* $\mathcal{T}$.

Indeed, if $K,L \in \mathfrak{K}(X)$ are disjoint sets, then by the compactness of K and the max-stability of $\mathcal{T}$, for $y \in L$ there exists $t_y \in \mathcal{T}$ with $K \subset \{t_y > 1\}$ and $t_y(y) < 0$. By the compactness of L and the min-stability of $\mathcal{T}$, this proves the existence of an Urysohn-function for K and L. □

(5) *Let* $\mathcal{T}$ *be a vector lattice of continuous functions. Then* $\mathfrak{K}(X)$ *is Urysohn-separated by* $\mathcal{T}$ *iff* $\mathcal{T}$ *is linearly separating.*

In particular, if X *is locally compact, respectively completely regular, then* $\mathfrak{K}(X)$ *is Urysohn-separated by* $\mathcal{K}(X)$ *and* $\mathscr{C}^0(X)$, *respectively by* $\mathscr{C}^b(X)$.

This follows from Example 4, Remark 8.7 and Example 8.7.1. □

14.10 LEMMA. *If the restriction* m *of* τ^* *to* $\mathfrak{K}$ *is finite and semiregular, then* m *is a content iff* $\mathfrak{K}$ *is* τ*-separated by* $\mathcal{T}$.

In this case, m *coincides with* τ_* *on* $\mathscr{E}_-(\mathfrak{K},\mathfrak{G})$.

Obviously, m is increasing and subadditive since τ^* is sublinear on $\bar{\mathbb{R}}_+^X$ by Lemma 1.5.

By Remark 14.8.2, the separation condition is necessary for m to be a content. Conversely, from the separation property it follows that m is additive. In fact, if $K_1,K_2 \in \mathfrak{K}$ are disjoint, and if $\varepsilon > 0$, then the separation property furnishes $t_i \in \mathcal{T}$ with $t_i \geq 1_{K_i}$ and $\tau(\min(t_1,t_2)) \leq \varepsilon$. Therefore, for all $t \in \mathcal{T}$ with $t \geq 1_{K_1 \cup K_2}$ we have

$$m(K_1) + m(K_2) \leq \tau(\min(t,t_1)) + \tau(\min(t,t_2)) \leq \tau(t + \min(t_1,t_2)) \leq \tau(t) + \varepsilon ,$$

hence

$$m(K_1) + m(K_2) \leq m(K_1 \cup K_2) .$$

Consequently, m is a semiregular content by Remark 12.2.3.

We next prove that

$$m(e) = \inf_{d \in \mathcal{D},\, d \geq e} m(d)$$

for all $e \in \mathscr{E}_-$, where $\mathcal{D} := \bigcup_k \mathcal{D}_k$ and $\mathcal{D}_k$ denotes the subset of all functions in $\mathscr{E}_-$ having a representation of the form $-\sum_{i=1}^{k} \alpha_i \cdot 1_{L_i}$ with $\alpha_i \in \mathbb{R}_+$ and k pairwise disjoint sets $L_i \in \mathfrak{K}$.

To this end, let $e \in \mathscr{E}_-$ and $\omega_{-n} < \ldots < \omega_{-1} < \omega_0 := 0$ denote the finitely many values of e together with the possible value 0 . With $K_j := \{e \leq \omega_j\}$ and $K_{-n-1} := \emptyset$, we have

$$e = -\sum_{j=-n}^{-1} (\omega_{j+1} - \omega_j)\cdot 1_{K_j} ,$$

hence

$$m(e) = -\sum_{j=-n}^{-1} (\omega_{j+1} - \omega_j)\cdot m(K_j) = \sum_{j=-n}^{-1} \omega_j \cdot [m(K_j) - m(K_{j-1})] =$$

$$= \sum_{j=-n}^{-1} \omega_j \cdot m_*(K_j \setminus K_{j-1}) = \sum_{j=-n}^{-1} \omega_j \cdot \sup_{L_j \in \mathfrak{K},\, L_j \subset K_j \setminus K_{j-1}} m(L_j) .$$

Since τ_* is superlinear on $\mathscr{E}_-$, it is now sufficient to prove that

$$\tau_*(e) \le m(e)$$

for all $e \in \mathscr{D}_k$ and $k \in \mathbb{N}$. For $k = 1$, this is the definition of m. Suppose that the inequality holds on $\mathscr{D}_k$, and let $e \in \mathscr{D}_{k+1}$ be given. Then $e = d - \alpha \cdot 1_K$ for some $d \in \mathscr{D}_k$, $\alpha := \|e\|_\infty$ and $K := \{e \le -\alpha\}$; the set $L := \{d < 0\} \in \mathfrak{K}$ is disjoint from K. For $\varepsilon > 0$ and $\beta := \|d\|_\infty$, the τ-separation of $\mathfrak{K}$ by $\mathscr{T}$ furnishes functions $t,t' \in \mathscr{T}$ with

$$t \ge \alpha \cdot 1_K \quad , \quad t' \ge \beta \cdot 1_L \quad \text{and} \quad \tau(\min(t,t')) \le \varepsilon .$$

For $u \in \mathscr{T}$ with $-u \le e$ and $\tau(u) < \infty$, we have

$$-u \le e \le \min(-\alpha \cdot 1_K, d) \quad \text{and} \quad -t \le -\alpha \cdot 1_K \ , \ -t' \le d .$$

Since $\mathscr{T}$ is min-stable, without loss of generality we may suppose that $t,t' \le u$. But then

$$-\tau(u) \le -\tau(\max(t,t')) - \tau(\min(t,t')) + \varepsilon =$$

$$= -\tau(t) - \tau(t') + \varepsilon \le \tau_*(-\alpha \cdot 1_K) + \tau_*(d) + \varepsilon ,$$

hence

$$\tau_*(e) \le \tau_*(-\alpha \cdot 1_K) + \tau_*(d) + \varepsilon = m(-\alpha \cdot 1_K) + m(d) + \varepsilon = m(e) + \varepsilon . \quad \square$$

PROPOSITION. *Suppose that all functions in $\mathscr{T}_+$ are $\mathfrak{K}$-measurable and that the restriction m of τ^* to $\mathfrak{K}$ is finite. Then m is a content iff $\mathfrak{K}$ is τ-separated by $\mathscr{T}$.*

In this case, m is semiregular and coincides with τ_* on $\mathscr{E}_-(\mathfrak{K},\mathfrak{S})$. In particular, τ_* is linear on $\mathscr{E}_-(\mathfrak{K},\mathfrak{S})$.

By Theorem 10.10, all functions in $\mathscr{T}_+$ are $\mathscr{E}(\mathfrak{K},\mathfrak{S})$-measurable. Lemma 7.6 shows that the restriction μ of τ_* to $\mathscr{E}_-(\mathfrak{K},\mathfrak{S})$ is superlinear and semiregular. For $K \in \mathfrak{K}$ we have

$$m(K) = \tau^*(1_K) = -\tau_*(-1_K) = -\mu(-1_K) = \mu_*(K) ,$$

which in view of Lemma 12.1 gives $\mu_*(A) = m_*(A)$ for any set A. This implies the semiregularity of m, since for $K,L \in \mathfrak{K}$ with $K \subset L$ we have

$$m(L) = -\mu(-1_L) \le -\mu(-1_K) + \mu_*(-1_K - (-1_L)) = m(K) + m_*(L \setminus K) .$$

The result now follows from the lemma. $\square$

14.11 Our next goal is to generalize Proposition 14.4.

THEOREM. *Suppose that all functions in* $\mathcal{T}$ *are* $\mathfrak{K}$*-measurable and that the restriction* m *of* τ^* *to* $\mathfrak{K}$ *is finite.*

If τ *is semiregular and* $\mathfrak{K}$ *is* τ*-separated by* $\mathcal{T}$ *, then* m *is a* $\mathcal{T}$*-tight semiregular content on* $\mathfrak{K}$ *, which represents* τ *iff* τ *is* $\mathfrak{E}(\mathfrak{K},\mathfrak{G})$*-tight. In this case,* m *is the only semiregular content on* $\mathfrak{K}$ *representing* τ *.*

By Theorem 10.10, all functions in $\mathcal{T}$ are $\mathfrak{E}(\mathfrak{K},\mathfrak{G})$-measurable. The assertion therefore follows from Proposition 14.10 and Addendum 7.6. □

COROLLARY. *Let* $\mathcal{T}$ *be a lattice cone of lower semicontinuous functions on a Hausdorff space* X *, and* τ *a semiregular linear functional on* $\mathcal{T}$ *such that the restriction* m *of* τ^* *to* $\mathfrak{K}(X)$ *is finite.*

Then m *is a Radon measure representing* τ *iff* $\mathfrak{K}(X)$ *is* τ*-separated by* $\mathcal{T}$ *and* τ *is* $\mathfrak{K}(X)$*-tight. In this case,* m *is unique.*

Obviously, all functions in $\mathcal{T}$ are $\mathfrak{K}(X)$-semicontinuous, hence $\mathfrak{K}(X)$-measurable. Furthermore, by Example 14.1.2, the functional τ is $\mathfrak{E}(\mathfrak{K}(X),\mathfrak{G}(X))$-tight iff it is $\mathfrak{K}(X)$-tight. Finally, m is $\mathfrak{G}(X)$-bounded and hence a Radon measure by Remark 14.5.1. □

REMARK. The corollary shows that the conditions in Theorem 8.4 or Remark 8.4.2 imply τ-separation. The proof of linearity of τ_* using these conditions is much simpler than the proof given in Lemma 14.10.

PROPOSITION. *Let* $\mathcal{T}_-$ *be Stonian,*

$$\mathfrak{K}(\mathcal{T}) \cap \mathfrak{N}^\times(\tau) \subset \mathfrak{K} \quad \text{and} \quad \mathfrak{G}(\mathcal{T}) \cap \mathfrak{N}^\bullet(\tau) \subset \mathfrak{G} \,,$$

and suppose that the restriction m *of* τ^* *to* $\mathfrak{K}$ *is finite.*

If τ *is semiregular and* $\mathfrak{K}$ *is* τ*-separated by* $\mathcal{T}$ *, then* m *is a* $\mathcal{T}$*-tight* $\mathfrak{G}$*-regular content which represents* τ *iff* τ *satisfies both truncation formulas.*

In this case, m *is the only semiregular content on* $\mathfrak{K}$ *representing* τ *.*

We can apply the theorem, since by Lemma 14.4 all functions in $\mathcal{T}$ are

$\mathfrak{K}$–measurable. The $\mathfrak{G}$–boundedness of m is obtained from Lemma 14.5. The truncation formulas are necessary for representation (cf. Example 11.6.2). Conversely, τ is $\mathscr{E}(\mathfrak{K},\mathfrak{G})$– tight by Lemma 14.4 and Corollary 14.2. □

14.12 We intend to prove, via the preceding theorem, that there exists a bijection between the set

> $\mathcal{M}_{\mathfrak{K}}(\mathcal{T})$ *of all semiregular* $\mathscr{E}(\mathfrak{K},\mathfrak{G})$*–tight linear functionals* τ *on* $\mathcal{T}$ *such that for every* $K \in \mathfrak{K}$ *there exists a function* $t \in \mathcal{T}$ *with* $t \geq 1_K$ *and* $\tau(t) < \infty$,

and a subset of the set

> $\mathcal{M}_{\mathcal{T}}(\mathfrak{K})$ *of all semiregular contents* m *on* $\mathfrak{K}$ *with* $\mathcal{T}_- \subset \mathfrak{R}^{\bullet}_-(m)$ *and such that for every* $K \in \mathfrak{K}$ *there exists a function* $t \in \mathcal{T}$ *with* $t \geq 1_K$ *and* $m^{\bullet}(t) < \infty$.

This is formulated in the following

PROPOSITION. *Suppose that* $\mathcal{T}$ *consists of* $\mathfrak{K}$*–measurable functions and Urysohn–separates* $\mathfrak{K}$. *Then the mapping* $\tau \longmapsto \tau^*{}_{|\mathfrak{K}}$ *is a bijection of* $\mathcal{M}_{\mathfrak{K}}(\mathcal{T})$ *onto the set of all* $\mathcal{T}$*–tight contents* $m \in \mathcal{M}_{\mathcal{T}}(\mathfrak{K})$, *for which the restriction of* $m^{\bullet}$ *to* $\mathcal{T}_-$ *is semiregular. The inverse mapping is* $m \longmapsto m^{\bullet}{}_{|\mathcal{T}}$.

For $\tau \in \mathcal{M}_{\mathfrak{K}}(\mathcal{T})$, Proposition 14.8 shows that $\mathfrak{K}$ is τ–separated by $\mathcal{T}$. Therefore, $m := \tau^*{}_{|\mathfrak{K}}$ is a $\mathcal{T}$–tight semiregular content on $\mathfrak{K}$ representing τ by Theorem 14.11. This proves that the mapping $\tau \longmapsto m$ is injective into the above subset of $\mathcal{M}_{\mathcal{T}}(\mathfrak{K})$.

To prove surjectivity, let m be in this subset. Since all functions in $\mathcal{T}$ are measurable w.r.t. m by Corollary 11.4, the functional $\tau := m^{\bullet}{}_{|\mathcal{T}}$ is represented by m (cf. Proposition 6.3) . By Theorem 7.2, the two additional hypotheses imposed on m imply that τ is difference–regular and even semiregular (cf. Remark 6.7). Proposition 7.2 then shows that τ is $\mathscr{E}(\mathfrak{K},\mathfrak{G})$–tight and hence

$\tau \in \mathcal{M}_{\mathfrak{K}}(\mathcal{T})$. Finally, the uniqueness statement in Theorem 14.11 proves that τ^* coincides with m on $\mathfrak{K}$. □

REMARKS.

(1) *If* $\mathfrak{G}(\mathcal{T}) \subset \mathfrak{G}$ *, then every content in* $\mathcal{M}_{\mathcal{T}}(\mathfrak{K})$ *is* $\mathfrak{G}$*-regular.*

This follows from Remark 14.5.1. □

(2) Note that in the proposition the semiregular contents on $\mathfrak{K}$ may be replaced by inner $\mathfrak{K}$-regular contents on the algebra generated by $\mathfrak{K}$, or more generally on any ring which contains $\mathfrak{K}$ and which is contained in the algebra generated by $\mathfrak{G}(\mathfrak{K})$ (cf. Corollary 12.8.2).

Although, since measure theory for lattices was not available, historically representation theorems have been formulated in this context, we shall not follow this line.

(3) *If any set from* $\mathfrak{K}$ *is contained in some set from* $\mathfrak{K}(\mathcal{T})$ *, then every* $m \in \mathcal{M}_{\mathcal{T}}(\mathfrak{K})$ *is* $\mathcal{T}$*-tight.*

This follows from Example 7.1.1. In fact, if $e \in \mathcal{E}_-$ vanishes outside $K \in \mathfrak{K}$ and if $t \in \mathcal{T}_-$ is such that $K \subset \{t \leq -1\}$, then $nt \leq e$ for some $n \in \mathbb{N}$. □

(4) *If*

$$(\mathcal{T}_- - \mathcal{T}_-)_- = \mathcal{T}_- ,$$

then $m^\bullet$ *is semiregular on* $\mathcal{T}_-$ *for every* $m \in \mathcal{M}_{\mathcal{T}}(\mathfrak{K})$.

This is shown in Corollary 6.7. □

We do not know of any application of this remark other than to the trivial case of a vector lattice.

14.13 EXAMPLES.

(1) Let $\mathfrak{L}$ and $\mathfrak{H}$ be the compatible lattices of all finite unions of compact respectively bounded open rectangles in $\mathbb{R}^n$.

There is a bijection between the semiregular or $\mathfrak{H}$-regular contents l on $\mathfrak{L}$ and the Radon measures m on $\mathbb{R}^n$, given by

$$l(L) = m(L) \quad \textit{for all } L \in \mathfrak{L}$$

and

$$m(K) = l^*(K) \quad \textit{for all } K \in \mathfrak{K}(\mathbb{R}^n)\,.$$

Note first that any content on $\mathfrak{L}$ is $\mathfrak{H}$-bounded since for any $L \in \mathfrak{L}$ there exists $H \in \mathfrak{H}$ with $L \subset H$, and we have $\overline{H} \in \mathfrak{L}$.

Only the Urysohn separation property and the semiregularity of the restriction l of m to $\mathfrak{L}$ remain to be checked. To this end, note that for any two disjoint compact sets K and L, by compactness of K, there exists $H \in \mathfrak{H}$ with

$$K \subset H \quad \text{and} \quad \overline{H} \cap L = \emptyset\,.$$

Thus, 1_H Urysohn-separates K and L. Furthermore, for any $L,M \in \mathfrak{L}$ with $L \subset M$ and $\varepsilon > 0$, by the inner $\mathfrak{K}(\mathbb{R}^n)$-regularity of m there exists $K \in \mathfrak{K}(\mathbb{R}^n)$ with $K \subset M \smallsetminus L$ and

$$m(M) \le m(L) + m(K) + \varepsilon\,.$$

Choosing now $H \in \mathfrak{H}$ as above, we have $K \subset \overline{H} \cap M \subset M \smallsetminus L$, which gives

$$l(M) = m(M) \le m(L) + m(\overline{H} \cap M) + \varepsilon \le l(L) + l_*(M \smallsetminus L) + \varepsilon\,. \quad \square$$

This gives another way to construct the Lebesgue measure on $\mathbb{R}^n$. In fact we have

$$\mathscr{E}(\mathfrak{L},\mathfrak{H}) \subset \mathscr{T}(\mathbb{R}^n)\,,$$

and the volume functional v on $\mathscr{T}(\mathbb{R}^n)$ induces an $\mathfrak{H}$-regular content on $\mathfrak{L}$. This can be seen as in Example 1.10.2. However, we would prefer to define the Lebesgue measure on $\mathbb{R}^n$ as a product measure.

The description of general Radon measures on $\mathbb{R}^n$ as rectangle functions is complicated and not worth-while, in contrast to the one-dimensional case :

(2) *There exists a bijection between the Radon measures m on $\mathbb{R}$ and the increasing and right-continuous functions $w : \mathbb{R} \longrightarrow \mathbb{R}$ with $w(0) = 0$, uniquely determined by*

$$m^*(]a,b]) = w(b) - w(a) \quad \textit{for all } a,b \in \mathbb{R}\,.$$

For any such w, we define

$$l([a,b]) := w(b) - w(a-)$$

for all $a,b \in \mathbb{R}$ with $a \le b$. Since any $L \in \mathfrak{L}$ is a union of finitely many disjoint compact intervals, l extends uniquely to an increasing additive set function on $\mathfrak{L}$, again denoted by l. To prove that l is a semiregular content, we use Remark 12.2.3. Indeed, due to the above simple representation of elements in $\mathfrak{L}$, we can almost fill the gaps of $L \setminus K$ by compact intervals since

$$w(x-) = \lim_{y \to x-} w(y)$$

at the right-hand side of the gap and

$$w(x) = \lim_{y \to x+} w(y-)$$

at the left-hand side of the gap, w being right-continuous. This proves the required inequality.

By Example 1, there is a Radon measure m with

$$m^*(]a,b]) = l([a,b]) - l(\{a\}) = w(b) - w(a) .$$

As $w(0) = 0$, this shows that the mapping $w \longmapsto m$ is injective since

$$w(x) = \begin{cases} m^*(]0,x]) & \text{for } x \ge 0 \\ - m_*(]x,0]) & \text{for } x < 0 \end{cases} .$$

Conversely, let m be a Radon measure on $\mathbb{R}$, and define w as above. Then w is increasing with $w(0) = 0$. Furthermore, by the outer $\mathfrak{G}(\mathbb{R})$-regularity of m^* at $\emptyset$ and $\{x\}$, we have

$$w(x+) - w(x) = \inf_{y>x} [w(y) - w(x)] = \inf_{y>x} m^*(]x,y]) = 0$$

and

$$w(x) - w(x-) = \inf_{y<x} [w(x) - w(y)] = \inf_{y<x} m^*(]y,x]) = m(\{x\})$$

respectively. The first formula shows that w is right continuous. If l is defined as in the first part, then

$$l([a,b]) = w(b) - w(a-) = w(b) - w(a) + w(a) - w(a-) =$$

$$= m^*(]a,b]) + m(\{a\}) = m([a,b])$$

by the second formula. Example 1 then shows the surjectivity of the mapping $w \longmapsto m$. □

14.14 COROLLARY. *Let $\mathcal{T}$ be a vector lattice of real-valued $\mathfrak{K}$-measurable functions which Urysohn-separates $\mathfrak{K}$. Then there exists a bijection between the set $\mathcal{M}_{\mathfrak{K}}(\mathcal{T})$ of all $\mathcal{E}(\mathfrak{K},\mathfrak{G})$-tight positive linear forms τ on $\mathcal{T}$ and the set $\mathcal{M}_{\mathcal{T}}(\mathfrak{K})$ of all semiregular contents m on $\mathfrak{K}$ with $\mathcal{T} \subset \mathcal{R}^{\bullet}(m)$, given by*

$$\tau(t) = m^{\bullet}(t) \quad \text{for all } t \in \mathcal{T}$$

and

$$m(K) = \tau^*(K) \quad \text{for all } K \in \mathfrak{K}.$$

By the Urysohn separation property, for every $K \in \mathfrak{K}$ there exists $t \in \mathcal{T}$ such that $t \geq 1_K$. Since every increasing linear functional τ on $\mathcal{T}$ is a (regular) positive linear form, we have $\tau(t) < \infty$. On the other hand, if $\mathcal{T} \subset \mathcal{R}^{\bullet}(m)$ then $m^{\bullet}(t) < \infty$, and since $-t \in \mathcal{T}$ and $-t \leq -1_K$, Example 7.1.1 proves that m is $\mathcal{T}$-tight. □

REMARK. Note that essential integration has to be used, since functions in $\mathcal{T}$ may not vanish outside some set from $\mathfrak{K}$ or may be unbounded.

EXAMPLES. Let X be a Hausdorff space.

(1) **ALEXANDROFF'S REPRESENTATION THEOREM.** *There exists a bijection between the positive linear forms τ on $\mathcal{C}^b(X)$ and the $\mathfrak{Z}(X)^{\complement}$-regular contents m on $\mathfrak{Z}(X)$, given by*

$$\tau(t) = m^*(t) \quad \text{for all } t \in \mathcal{C}^b(X)$$

and

$$m(K) = \tau^*(K) \quad \text{for all } K \in \mathfrak{Z}(X).$$

By Proposition 10.8 and Example 10.7.3, a functions in $\mathcal{C}^b(X)$ is $\mathfrak{Z}(X)$-semicontinuous and hence $\mathfrak{Z}(X)$-measurable. Furthermore, $\mathcal{C}^b(X)$ Urysohn-separates $\mathfrak{Z}(X)$ by Example 14.9.1, and $\mathfrak{Z}(X)^{\complement} = \mathfrak{G}(\mathcal{C}^b(X))$. By Proposition and Remark 14.2, every positive linear form on $\mathcal{C}^b(X)$ is $\mathcal{E}(\mathfrak{Z}(X),\mathfrak{Z}(X)^{\complement})$-tight, since $X \in \mathfrak{Z}(X)$.

The result now follows from the above corollary and Remark 14.12.1, because m^* is finite and therefore coincides with $m^{\bullet}$ on $\mathcal{C}^b(X)$ for any content m on $\mathfrak{Z}(X)$. □

By the same kind of arguments, using Urysohn's Lemma, we obtain

(2) MARKOFF'S REPRESENTATION THEOREM. *If* X *is normal, then there exists a bijection between the positive linear forms* τ *on* $\mathscr{C}^b(X)$ *and the* $\mathfrak{G}(X)$*-regular contents* m *on* $\mathfrak{F}(X)$ *, given by*

$$\tau(t) = m^*(t) \quad \text{for all } t \in \mathscr{C}^b(X)$$

and

$$m(K) = \tau^*(K) \quad \text{for all } K \in \mathfrak{F}(X) .$$

(3) RIESZ REPRESENTATION THEOREM. *If* X *is locally compact, then there exists a bijection between the positive linear forms* τ *on* $\mathscr{K}(X)$ *and the Radon measures* m *on* X *, given by*

$$\tau(t) = m^*(t) \quad \text{for all } t \in \mathscr{K}(X)$$

and

$$m(K) = \tau^*(K) \quad \text{for all } K \in \mathfrak{K}(X) .$$

Since $\mathfrak{K}(X)$-measurability by Example 10.3 means lower semicontinuity w.r.t. the toplogy $\mathfrak{G}(\mathfrak{K}(X)) \supset \mathfrak{G}(X)$, every function in $\mathscr{K}(X)$ has this property. Furthermore, Example 14.9.5 shows that $\mathscr{K}(X)$ Urysohn-separates $\mathfrak{K}(X)$. Every positive linear form on $\mathscr{K}(X)$ being obviously $\mathfrak{K}(X)$-tight, the result now follows immediately from the corollary and Remark 14.12.1. □

We already have proved this result, even more : In Theorem 8.2 we have shown that $\mu \longmapsto \mu^\phi$ is a bijection between the positive linear forms on $\mathscr{K}(X)$ and the Radon integrals on $\mathscr{I}(X)$. Using Theorem 12.5, we infer that for a positive linear form on $\mathscr{K}(X)$, essential Bourbaki integration coincides with essential integration for the corresponding Radon measure.

A generalization is the following which also could be deduced from Theorem 8.8.

(4) *Let* $\mathscr{T} \subset \mathscr{C}(X)$ *be a linearly separating Stonian vector lattice. Then there exists a bijection between the positive linear forms* τ *on* $\mathscr{T}$ *which are* $\mathfrak{K}(X)$*-tight and the Radon measures* m *on* X *with* $\mathscr{T} \subset \mathfrak{R}^\bullet(m)$ *, given by*

$$\tau(t) = m^\bullet(t) \quad \text{for all } t \in \mathscr{T}$$

and

$$m(K) = \tau^*(K) \quad \textit{for all } K \in \mathfrak{K}(X) .$$

This follows by the same arguments as before having in mind the Examples 14.9.5 and 14.1.2. □

In particular, if X is completely regular, then there exists a bijection between the $\mathfrak{K}(X)$–tight positive linear forms on $\mathscr{C}^b(X)$ and the bounded Radon measures on X, i.e. those for which X is essentially integrable.

Using Example 14.1.3, we get a representation which will be studied in more detail in 16.3.

(5) BAUER'S REPRESENTATION THEOREM. *Let X be locally compact and $\mathcal{T} \subset \mathscr{C}^0(X)$ be a linearly separating Stonian vector lattice. Then there exists a bijection between the positive linear forms τ on $\mathcal{T}$ satisfying the truncation formula at 0, and the Radon measures m on X with $\mathcal{T} \subset \mathcal{R}^\bullet(m)$, given by*

$$\tau(t) = m^\bullet(t) \quad \textit{for all } t \in \mathcal{T}$$

and

$$m(K) = \tau^*(K) \quad \textit{for all } K \in \mathfrak{K}(X) .$$

§ 15 REPRESENTATION BY MEASURES

THROUGHOUT THIS SECTION,
$\mathcal{T}$ IS A LATTICE CONE OF FUNCTIONS,
τ IS AN INTEGRAL ON $\mathcal{T}$,
$\mathfrak{K}$ AND $\mathfrak{G}$ ARE COMPATIBLE LATTICES OF SETS.

15.1 We intend to apply Theorem 9.13 to get *τ-representing (finite) measures* on $\mathfrak{K}$, i.e. τ-representing integrals on $\mathcal{E}(\mathfrak{K},\mathfrak{G})$. Note that $\mathcal{E}(\mathfrak{K},\mathfrak{G})$-measurability means $\mathfrak{K}$-measurability by Theorem 10.10 and that, if $\mathcal{T}_-$ is Stonian, $\mathcal{E}(\mathfrak{K},\mathfrak{G})$-tightness of τ^σ is equivalent to $\mathfrak{K}$-tightness by Proposition 14.2 and the Monotone Convergence Theorem (Addendum 5.4).

PROPOSITION. *Let $\mathcal{T}_-$ be Stonian and τ a semiregular integral. Suppose that all sets in $\mathfrak{K}$ are essentially τ-integrable and that every function in $\mathcal{T}_{\sigma-}$ is $\mathfrak{K}$-measurable.*

The restriction m of $\int^\bullet \cdot d\tau$ to $\mathfrak{K}$ is a measure. It is τ-representing iff τ^σ is $\mathfrak{K}$-tight. In this case, m is the only semiregular τ-representing measure on $\mathfrak{K}$ for which m_δ is $\mathcal{T}$-tight, and we have

$$\int^\bullet \cdot d\tau = m^\bullet = \int^\bullet \cdot dm \ .$$

If furthermore $\tau^{\sigma *}$ is finite on $\mathfrak{K}$, then m coincides with $\tau^{\sigma *}$ on $\mathfrak{K}$ and is the only semiregular τ-representing measure on $\mathfrak{K}$. It is $\mathfrak{G}$-regular if $\mathfrak{G} \supset \mathfrak{G}(\mathcal{T})_\sigma \cap \mathcal{L}^\bullet(\tau)$.

We only have to note that by Remark 7.1.2 and Proposition 13.7.(i) the $\mathcal{T}$-tightness of m^σ is equivalent to that of m_δ. The regularity assertion follows from Lemma 14.5, applied to τ^σ, since $\mathfrak{G}(\mathcal{T}_\sigma) = \mathfrak{G}(\mathcal{T})_\sigma$ and $\mathcal{L}^\bullet(\tau) = \mathfrak{M}^\bullet(\tau^\sigma)$ by Theorem 9.10. □

Instead of Theorem 9.13, we could also have applied Proposition 14.4 to τ^{σ}.

Remark. *If $\mathfrak{K}$ is a δ-lattice, it is sufficient to suppose that all functions in $\mathcal{T}_{-}$ are $\mathfrak{K}$-measurable.*

This follows from Remark 9.13.1, since $\mathcal{E}\hat{}_{-}(\mathfrak{K}) = \mathcal{E}_{-}(\mathfrak{K})_{\sigma}$ by Theorem 10.10. $\square$

15.2 Proposition. *Suppose that $\mathcal{T}_{-}$ is Stonian and that all functions in $\mathcal{T}_{\sigma}$ are $\mathfrak{K}$-measurable. Let τ be a semiregular integral such that the restriction m of $\tau^{\sigma *}$ to $\mathfrak{K}$ is finite and has the Daniell property at the empty set.*

Then m is a τ-representing measure on $\mathfrak{K}$ iff $\mathfrak{K}$ is τ^{σ}-separated by $\mathcal{T}_{\sigma}$ and τ^{σ} is $\mathfrak{K}$-tight. In this case, m is $\mathcal{T}$-tight and the only semiregular τ-representing measure on $\mathfrak{K}$.

We can apply Theorem 14.11 to the integral τ^{σ}, which is semiregular by Proposition 9.9. One has just to note that separation is necessary by Remark 14.8.2, that m is $\mathcal{T}$-tight by coinitiality, and a measure by Corollary 13.4. $\square$

15.3 Proposition. *Let $\mathcal{T}_{-}$ be Stonian with $\mathfrak{K}(\mathcal{T})_{\delta} \subset \mathfrak{K}$ and $\mathfrak{G}(\mathcal{T})_{\sigma} \subset \mathfrak{G}$. Suppose that τ is a semiregular integral such that the restriction m of $\tau^{\sigma *}$ to $\mathfrak{K}$ is finite and has the Daniell property at the empty set.*

Then m is a τ-representing measure on $\mathfrak{K}$ iff $\mathfrak{K}$ is τ^{σ}-separated by $\mathcal{T}_{\sigma}$. In this case, m is the only semiregular τ-representing measure on $\mathfrak{K}$, and m is $\mathfrak{G}$-regular.

This follows using Proposition 14.11. $\square$

15.4 Proposition. *Let $\mathcal{T}$ be Stonian and ν an upper integral determined by*

$\mathcal{T}$ for which $\mathcal{T}_- \subset \mathcal{I}_-(\nu)$. Then the restriction of ν to $\mathfrak{K}(\mathcal{T})$ is the only $\mathfrak{G}(\mathcal{T})$-regular measure m with $\nu^\bullet = m^\bullet = \int^\bullet \cdot dm$.

We can apply Proposition 14.6. Indeed, by Propositions 10.6 and 3.5.(ii), all constant functions are ν-measurable, hence $\mathfrak{K}(\mathcal{T}) \subset \mathfrak{I}(\nu)$ by Proposition 13.6.(ii). Since the truncation formulas are a consequence of the Monotone Convergence Theorem, we infer that the restriction m of ν to $\mathfrak{K}(\mathcal{T})$ is the only $\mathfrak{G}(\mathcal{T})$-regular content with $\nu^\bullet = m^\bullet$, which by Corollary 13.4 is a measure.

Moreover, the restriction of ν to $\mathfrak{K}(\mathcal{T}_\sigma) = \mathfrak{K}(\mathcal{T})_\delta$ coincides with m_δ, which, by the same arguments applied to $\mathcal{T}_\sigma$, proves that

$$\nu^\bullet = m_\delta^\bullet .$$

The last equality follows from Theorem 13.8.(ii). □

COROLLARY. *Let ν be an auto-determined upper integral for which 1 is measurable. Then the restriction of ν to the δ-ring $\mathfrak{I}(\nu)$ is the only finite measure m on $\mathfrak{I}(\nu)$ with $\nu^\bullet = m^\bullet$, and we have*

$$\nu = \int^* \cdot dm .$$

In particular, a set is ν-measurable iff it is locally ν-integrable, and a function is ν-measurable iff it is $\mathfrak{J}^0(\nu)$-measurable.

Noting that $\mathfrak{K}(\tilde{\mathfrak{J}}(\nu)) = \mathfrak{I}(\nu)$ by Proposition 13.6.(ii) and having in mind Theorems 13.6 and 13.8, this follows upon applying the proposition to the Stonian function cone $\tilde{\mathfrak{J}}(\nu)$, except for the second formula. Setting $\mathcal{E} := \mathcal{E}(\mathfrak{I}(\nu),\mathfrak{I}(\nu))$, it obviously holds on $\mathcal{E}_\sigma$, and since for $g \in \tilde{\mathfrak{J}}(\nu)$ we have $g^+ \in \mathcal{E}_\sigma$ by Proposition 10.8 and Corollary 10.9, hence

$$\int^* g \, dm \leq \int^* g^+ \, dm = \nu(g^+) < \infty ,$$

we get

$$\nu(g) = \nu^\bullet(g) = \int^\bullet g \, dm = \int^* g \, dm$$

by Corollary 3.9. Finally, for any function f we have

$$\nu(f) = \inf_{g\in\tilde{\mathfrak{I}},\, g\geq f} \nu(g) = \inf_{g\in\tilde{\mathfrak{I}},\, g\geq f} \int^* g\, dm \geq \int^* f\, dm =$$

$$= \inf_{g\in\mathcal{E}_\sigma,\, g\geq f} \int^* g\, dm = \inf_{g\in\mathcal{E}_\sigma,\, g\geq f} \nu(g) \geq \nu(f)\,. \quad \square$$

REMARK. The second formula had to be proved directly, since representation theory uses essential integration. Though reduction to integration via the formula $\nu^\bullet = \nu$ on $\{\nu < \infty\}$ seems possible at first glance, in general uncontrollable finiteness problems arise.

Note that the coincidence of ν and $\int^* \cdot dm$ on $\tilde{\mathfrak{I}}(\nu)$ could also have been proved using the Monotone Convergence Theorem, which leads to an independent proof of $\nu = \int^* \cdot dm$, however not of $\nu^\bullet = m^\bullet$.

15.5 THEOREM OF DANIELL–STONE. *Let $\mathcal{T}$ be Stonian and suppose that τ is a regular integral.*

Then the restriction m of $\int^* \cdot d\tau$ to $\mathfrak{K}(\mathcal{T})_\delta$ is the only finite τ-representing measure on $\mathfrak{K}(\mathcal{T})_\delta$. It is $\mathfrak{G}(\mathcal{T})_\sigma$-regular, and we have

$$\int^\bullet \cdot d\tau = m^\bullet\,.$$

This follows from Proposition 15.1 applied to $\mathfrak{K}(\mathcal{T})_\delta$, which by Proposition 13.6.(ii) is contained in $\mathfrak{L}^\times(\tau)$. Note that $\mathcal{T}_\sigma$ consists of $\mathfrak{K}(\mathcal{T})_\delta$-measurable functions by Proposition 10.8, and that τ^σ is $\mathfrak{K}(\mathcal{T})_\delta$-tight by Proposition 14.1. In particular, we have $\mathfrak{L}^0(\tau) = \mathfrak{L}^0(m)$, since $m^\bullet = \int^\bullet \cdot dm$ by Theorem 13.8.(ii). The uniqueness assertion is a consequence of the subsequent lemma, applied to $\mathfrak{K}(\mathcal{T})_\delta$, and Lemma 13.9. $\square$

We could also have applied Corollary 14.4.

LEMMA. *For any τ-representing measure l on a lattice $\mathfrak{L}$ which contains $\mathfrak{K}(\mathcal{T})_\delta$, and $K \in \mathfrak{K}(\mathcal{T})_\delta$ we have*

$$\int^\bullet K\,dl = \int^\bullet K\,d\tau .$$

By Remark 9.12, l is τ^σ-representing. Choosing $t \in \mathcal{T}_{\sigma-}$ such that $t \geq -1$ and $K = \{t = -1\}$, the function

$$-1_K = \sup_{n\geq 1}\,[n\cdot(t+1)-1]_- = \sup_{n\geq 1}\, n\cdot[t - \max(t,-1+\tfrac{1}{n})]$$

is essentially l-integrable by the Monotone Convergence Theorem 5.4, and we have

$$\int^\bullet -1_K\,dl = \sup_{n\geq 1}\, n\cdot[\tau^\sigma(t) - \tau^\sigma(\max(t,-1+\tfrac{1}{n}))] = \int^\bullet -1_K\,d\tau . \quad \square$$

COROLLARY. *Let $\mathfrak{L}$ be a lattice of τ-measurable sets which contains $\mathfrak{K}(\mathcal{T})_\delta$, e.g. the σ-algebra $\mathfrak{A}(\mathcal{T})$ generated by $\mathfrak{K}(\mathcal{T}) \cup \mathfrak{G}(\mathcal{T})$, which is the smallest one such that all functions in $\mathcal{T}$ are measurable. Then every τ-representing measure on $\mathfrak{L}$ which is finite on $\mathfrak{K}(\mathcal{T})$ extends m.*

Among all τ-representing measures on $\mathfrak{L}$, the set function $q := \int^\bullet \cdot\,d\tau$ is the smallest, and the only one which is finite on $\mathfrak{K}(\mathcal{T})$ and inner $\mathfrak{K}(\mathcal{T})_\delta$-regular. Moreover, we have

$$\int^\bullet \cdot\,d\tau = q^\bullet = \int^\bullet \cdot\,dq .$$

Note that by Corollary 15.4 every function in $\mathcal{T}$ is $\mathfrak{L}^0(\tau)$-measurable. Therefore, $\mathfrak{A}(\mathcal{T})$ is contained in $\mathfrak{L}^0(\tau)$.

Let l be a τ-representing measure on $\mathfrak{L}$. Since q coincides with $m^\bullet$ on $\mathfrak{L}$, hence is inner $\mathfrak{K}(\mathcal{T})_\delta$-regular, and

$$q(K) = m(K) = \int^\bullet K\,d\tau = \int^\bullet K\,dl \leq l(K)$$

by the lemma and Lemma 13.9, we have $q \leq l$. If moreover l is finite on $\mathfrak{K}(\mathcal{T})$, hence on $\mathfrak{K}(\mathcal{T})_\delta$, then

$$\int_{\bullet} K\, dl = l(K)$$

by Lemma 13.9. This proves the first part and the uniqueness assertion.

The formula, and hence the fact that q is a τ-representing measure, follows from Propositions 13.8, 12.7.(ii) and 13.9.(ii). Indeed,

$$q^{\bullet} = m^{\bullet} = \int^{\bullet} \cdot\, dm \quad \text{and} \quad \int^{\bullet} \cdot\, dq = \int^{\bullet} \cdot\, dm\,. \quad \square$$

15.6 PROPOSITION. *Let $\mathcal{T}$ be strongly Stonian, suppose that τ is a regular integral and $\mathfrak{L}$ is a lattice of τ-measurable sets which contains $\mathfrak{K}(\mathcal{T})_{\delta}$, e.g. the σ-algebra $\mathfrak{A}(\mathcal{T})$.*

A τ-representing measure l on $\mathfrak{L}$ is finite on $\mathfrak{K}(\mathcal{T})$ iff $\mathcal{T}_{-} \subset \mathcal{L}^{\times}_{-}(l)$.

Among all theses measures, the set function $p := \int^{\times} \cdot\, d\tau$ is the largest, and we have

$$\int^{\times} \cdot\, d\tau = \int^{\times} \cdot\, dp\,.$$

Note first that p and τ are measures on $\mathfrak{L}$ by Proposition 13.6.(i), both extending the measure m of Theorem 15.5 by definition. Since $\mathcal{T}_{\sigma}$ is strongly Stonian, Proposition 10.11 shows that, setting $\mathcal{E} := \mathcal{E}(\mathfrak{K}(\mathcal{T})_{\delta}, \mathfrak{G}(\mathcal{T})_{\sigma})$, we have $\mathcal{E}_{\sigma} = \mathcal{T}_{\sigma} \cap \mathcal{E}_{-}^{\uparrow}$. For $t \in \mathcal{T}_{-}$, the sequence of functions

$$t_n := \max\,(t, -n\cdot 1_{\{t \leq -1/n\}}) \in \mathcal{T}_{\sigma} \cap \mathcal{E}_{-\sigma}$$

decreases to t. Since $\int^{\times} \cdot\, dm$ and $\int^{\times} \cdot\, d\tau$ coincide on $\mathcal{E}_{-}$ and hence on $\mathcal{E}_{-\sigma}$, the Monotone Convergence Theorem (Addendum 5.4) yields

$$\int^{\times} t\, d\tau = \inf_n \int^{\times} t_n\, d\tau = \inf_n \int^{\times} t_n\, dm = \int^{\times} t\, dm\,.$$

Therefore, $\mathcal{T}_{-} \subset \mathcal{L}^{\times}_{-}(m)$, and $\int^{\times} t\, d\tau = \int^{\times} t\, dm$ on $(\mathcal{T}_{-} - \mathcal{T}_{-})_{\sigma}$, which proves that $\int^{\times} \cdot\, d\tau \geq \int^{\times} \cdot\, dm$. The converse inequality follows from Proposition

9.11.(i), since τ^σ is an integral on $\mathcal{T}_\sigma$ with $\int^\times \cdot d\tau^\sigma = \int^\times \cdot d\tau$ by Remark 9.8, which coincides with m on $\mathcal{E}_-$. This shows that $p = \int^\times \cdot dm$ on $\mathfrak{L}$. From Proposition 13.9.(ii) we infer that p is the largest extension of m to a measure on $\mathfrak{L}$ and

$$\int^\times \cdot dp = \int^\times \cdot dm = \int^\times \cdot d\tau\ ;$$

in particular, the measure p is τ–representing, finite on $\mathfrak{K}(\mathcal{G})$ and therefore the largest such measure by Corollary 15.5.

Let now l be a τ–representing measure on $\mathfrak{L}$. Since for $t \in \mathcal{T}_-$ and $K := \{t \leq -1\}$, we have $t \leq -1_K$ and hence

$$l(K) = \int^\times K\, dl = -\int_\times -1_K\, dl \leq -\int_\times t\, dl$$

by Lemma 13.9, we infer from $\mathcal{T}_- \subset \mathcal{L}^\times_-(l)$ that l is finite on $\mathfrak{K}(\mathcal{T})$. Conversely, by Corollary 15.5, l extends m, which by Proposition 13.9.(i) gives

$$\int^\times \cdot dl \leq \int^\times \cdot dm = \int^\times \cdot d\tau\ .$$

For $t \in \mathcal{T}_-$, this yields

$$-\infty < \tau(t) = \int_\times t\, d\tau \leq \int_\times t\, dl \leq 0$$

and proves that $t \in \mathcal{L}^\times_-(l)$ since $t \in \mathcal{L}^\bullet_-(l)$. □

Corollary 1. *If $\mathfrak{G}(\mathcal{T})_\sigma \subset \mathfrak{L}$, then among all τ-representing measures on $\mathfrak{L}$, the set function $r := \int^* \cdot d\tau$ is the largest one which is semifinite on $\mathfrak{G}(\mathcal{T})$, and the smallest one which is outer $\mathfrak{G}(\mathcal{T})_\sigma$-regular. In particular, it is the only one which is semifinite on $\mathfrak{G}(\mathcal{T})$ and outer $\mathfrak{G}(\mathcal{T})_\sigma$-regular.*

Moreover, r is finite on $\mathfrak{K}(\mathcal{T})$, inner $\mathfrak{K}(\mathcal{T})_\delta$-regular on $\mathfrak{G}(\mathcal{T})_\sigma$, and we have

$$\int^\bullet \cdot d\tau = r^\bullet = \int^\bullet \cdot dr\ .$$

By the regularity of τ^σ and the second part of Addendum 12.1, for $G \in \mathfrak{G}(\mathcal{T})_\sigma$ we have

$$r(G) = \int^* G\,d\tau = \tau^\sigma{}_*(G) = \sup_{K \in \mathfrak{K}(\mathcal{T}_\sigma),\, K \subset G} \tau^\sigma{}_*(K) = m_*(G) = m(G)\,,$$

since by τ-integrability

$$\tau^\sigma{}_*(K) = \int^* K\,d\tau = m(K)\,.$$

This proves that r is inner $\mathfrak{K}(\mathcal{T})_\delta$-regular on $\mathfrak{G}(\mathcal{T})_\sigma$. Furthermore, by the first part of Addendum 12.1 and Proposition 13.8, for $L \in \mathfrak{L}$ we get

$$r(L) = \int^* L\,d\tau = \inf_G \tau^\sigma(G) = \inf_G m(G) = \int^* L\,dm\ ,$$

where G runs through all sets in $\mathfrak{G}(\mathcal{T}_\sigma) = \mathfrak{G}(\mathcal{T})_\sigma$ with $G \supset L$. Thus, from Theorem 13.9 we infer that

$$r^\bullet = \int^\bullet \cdot\,dr = m^\bullet\,,$$

in particular r represents τ.

Let now l be a τ-representing measure on $\mathfrak{L}$. For any $L \in \mathfrak{L}$, we then have

$$r(L) = \inf_G \tau^\sigma(G) = \inf_G \int^\bullet G\,dl \le \inf_G l(G) \ge l(L)\,,$$

since $1_G \in \mathcal{T}_\sigma$. By Lemma 13.9, equality holds at the first place iff l is semifinite on $\mathfrak{G}(\mathcal{T})_\sigma$, i.e. on $\mathfrak{G}(\mathcal{T})$, respectively at the second place iff l is outer $\mathfrak{G}(\mathcal{T})_\sigma$-regular. This finally proves the characterizations of r. □

COROLLARY 2. *If there exists a sequence* (t_n) *in* $\mathcal{T}_-$ *with*

$$\inf t_n(x) < 0 \quad \textit{for all } x \in X\,,$$

then a τ*-representing measure on* $\mathfrak{L}$ *which is finite on* $\mathfrak{K}(\mathcal{T})$ *is uniquely determined. It is inner* $\mathfrak{K}(\mathcal{T})_\delta$*-regular and, if* $\mathfrak{G}(\mathcal{T})_\sigma \subset \mathfrak{L}$*, outer* $\mathfrak{G}(\mathcal{T})_\sigma$*-regular.*

Let l be a τ-representing measure on $\mathfrak{L}$ which is finite on $\mathfrak{K}(\mathcal{T})$. Since X is covered by the sequence of sets $\{t_n \le -\frac{1}{k}\} \in \mathfrak{L}^\times(\tau)$ with $k,n \in \mathbb{N}$, the integral $\int^\times \cdot\,d\tau$ is moderated. Thus

$$\int^{\bullet} \cdot d\tau = q \leq l \leq p = \int^{\times} \cdot d\tau = \int^{\bullet} \cdot d\tau \quad \text{on } \mathfrak{L}$$

by Corollary 15.5, the proposition and Proposition 5.9. The last assertion follows from Corollary 1. □

COROLLARY 3. *Let $\mathcal{T}$ be a Stonian vector lattice and τ an integral on $\mathcal{T}$.*

Among all τ-representing measures on $\mathfrak{A}(\mathcal{T})$ for which all functions in $\mathcal{T}$ are integrable, the set function $\int^{\bullet} \cdot d\tau$ is the smallest and the only inner $\mathfrak{K}(\mathcal{T})_\delta$-regular one, whereas the set function $\int^{*} \cdot d\tau$ is the largest and the only outer $\mathfrak{G}(\mathcal{T})_\sigma$-regular one.

If there exists a strictly positive function in $\mathcal{T}_\sigma$, then there is only one τ-representing measure on $\mathfrak{A}(\mathcal{T})$ for which all functions in $\mathcal{T}$ are integrable.

This follows from Corollary 15.5 and Corollaries 1 and 2. Note that $\mathcal{T}$ is strongly Stonian by Remark 10.11.1, and that every τ-representing measure l on $\mathfrak{A}(\mathcal{T})$, integrating all functions in $\mathcal{T}$, is finite on $\mathfrak{K}(\mathcal{T})$ and $\mathfrak{G}(\mathcal{T})$. □

COROLLARY 4. *Let $\mathcal{T}$ be a Stonian vector lattice, such that there exists a strictly positive function in $\mathcal{T}_\sigma$. Then there is a bijection between the integrals τ on $\mathcal{T}$ and the measures m on $\mathfrak{A}(\mathcal{T})$ with $\mathcal{T} \subset \mathcal{L}^*(m)$.*

Every measure m on $\mathfrak{A}(\mathcal{T})$ with $\mathcal{T} \subset \mathcal{L}^*(m)$ is inner $\mathfrak{K}(\mathcal{T})_\delta$-regular and outer $\mathfrak{G}(\mathcal{T})_\sigma$-regular.

One has just to note that every measure m on $\mathfrak{A}(\mathcal{T})$ with $\mathcal{T} \subset \mathcal{L}^*(m)$ induces an integral on $\mathcal{T}$ by Theorem 9.7. □

EXAMPLE. Let X be a Hausdorff space. The *Baire* σ-algebra $\mathfrak{A}(X)$ is the σ-algebra generated by $\mathscr{C}^b(X)$, i.e. by $\mathfrak{Z}(X)$. From the last corollary we infer that there is a bijection between the integrals on $\mathscr{C}^b(X)$ and the finite measures on $\mathfrak{A}(X)$, and that any such measure is inner $\mathfrak{Z}(X)$- and outer $\mathfrak{Z}(X)^{\complement}$-regular. This last assertion is also a consequence of Remark 13.1.

Note that if X is metrisable, then $\mathfrak{A}(X)$ is the σ-algebra of all Borel sets, since $\mathfrak{Z}(X) = \mathfrak{F}(X)$ by Example 10.7.3.

APPENDIX

§ 16 NOTES AND SPECIAL APPLICATIONS

In this first part of the appendix, we comment on some of the notions encountered in the text, discuss relations of our abstract Riemann type approach to integration and Riesz representation to previous work, and give some special applications.

16.1 Historical notes and comments.

Starting point of our approach is the idea of *Darboux* [1875] , p. 65, to use upper and lower Darboux sums instead of arbitrary Riemann sums, originally considered by *Riemann* [1854], p. 225, and to introduce the upper and lower Darboux functional as in 1.1 in order to define Riemann integrability.

The function cone $\mathcal{S}(X)$ has been introduced by *Young* [1911] for a compact interval $X \subset \mathbb{R}$ to treat the Lebesgue integral. In contrast to the original definition of *Lebesgue* [1902], p. 454, Young considers the upper and the lower integral and defines integrability via these functionals as in 2.5.

Daniell [1917/18] generalizes Young's construction. His starting point is an integral μ defined on a vector lattice $\mathcal{S}$ and its canonical extension μ^{σ} to $\mathcal{S}_{\sigma}$ (in the form of Remark 9.2.1). The upper and lower integrals $\mu^{\sigma *}$ and μ^{σ}_{*} are introduced by *Daniell* [1917/18], p. 287, and abstract Daniell integration with respect to $\mu^{\sigma *}$ is performed.

Loomis [1954] starts with a positive linear form μ on $\mathcal{S}$, possibly without the Daniell property, introduces μ^{*} and μ_{*} , and develops the abstract Riemann integration theory. We comment on Loomis' approach in 16.2.

The fact that upper functionals have to be considered as the corner stones of integration theory is alluded to in the early work of *Stone* [1948], p. 338. In *Alfsen* [1958], *Ionescu Tulcea* [1969], and *Bourbaki* [1969], the special case of upper integrals is considered. Alfsen uses upper and lower integrals without imposing a linear structure, whereas Bourbaki and Ionescu Tulcea work with upper integrals for positive functions only. Their approach therefore ammounts to a certain lack of symmetry. The relation between Bourbaki and Ionescu Tulcea integration theory is studied in *Bassan* [1985].

Essential integration originates in *Riemann's* "Habilitationsschrift" [1854] (improper integral, p. 226) and *Young* [1911] (absolutely convergent improper integral, p. 33 et seq.). *Loomis* [1954] and *Ionescu Tulcea* [1969] discuss essential integration, as does *Bourbaki* [1956], § 2, [1967], § 1, for locally compact spaces, and [1969], § 1, for arbitrary Hausdorff spaces, in the form of Example 3.8.2.

Auto-determined upper functionals play a role in *Ionescu Tulcea* [1969], p. 7, under the name of "regular" upper integrals.

Measurability in functional analytic terms is introduced in *Daniell* [1919/20], § 2. Our definition 4.1 is that of *Stone* [1948], p.448. The **integrability criterion** is to be found in *Stone* [1948], p. 452.

In *Daniell* [1917/18], p. 280, the property bearing this author's name is introduced in the dual form of Theorem 9.2.(ii). The **fundamental lemma** 5.1 (in the form of Remark 5.1) can be found in that paper (p. 283) as well as the **convergence theorems** (p. 289–290). Without imposing linear structure, analogues are proved by *Alfsen* [1958], p. 69–71, and [1963], Proposition 3.1, p. 427.

The introduction of **integrals** in § 9 reflects the fact that the starting function cone $\mathcal{S}$ is neither assumed to be max-stable nor a vector lattice (cf. Remark 9.2.1). The condition in Remark 9.5.3 is property (I_3) of *Alfsen* [1958], p. 68 (cf. also *Alfsen* [1963], Theorem 5, p. 429). Its inadequacy for defining integrals is the reason of the somewhat technical approach in the non-regular case.

Lattice measurability as introduced in Definition 10.3 corresponds to the property (M_0) of *Günzler* [1975], p. 117. For the special case of an algebra, there is an extensive study of measurability and integrability by *Dunford and Schwartz* [1958] and by *Bhaskara Rao* [1983]. We comment on their results in 16.4. For the study of **compatible lattices**, the reader is referred to *Topsoe* [1970 a,b].

The **Stonian condition** is to be found in the article of *Stone* [1948], p. 452. The **truncation formulas** appear in *Loomis* [1954], p. 176, and in *Bauer* [1956], p. 468 and p. 476, and later on e.g. in *Günzler* [1973], p. 173. The history of Theorems 11.7 and 11.8 goes at least back to the early work of *Stone* [1948], p. 454, and *Loomis* [1954], p. 170.

The importance of **inner regularity** has already been stressed by *Young*

[1911], p. 15. *Srinivasan* [1955] and *Topsoe* [1974] lay emphasis on inner instead on outer regularity. Our approach supports this point of view.

Semiregular contents are studied by *Choksi* [1958], p. 390, under the term of "regular contents" (cf. Remark 12.2.3). The misleading term "tightness" for what we prefer to call semiregularity seems to originate in the works of *Kisynski* [1968 b] and *Topsoe* [1970 a,b] (cf. Remark 12.7.2). These authors use the term "semiregular" for what we call $\mathfrak{G}(X)$– respectively $\mathfrak{G}(\mathfrak{K})$–regular. *Kelley and Srinivasan* [1984], p. 120, use the term "exact" instead of semiregular. The functional analytic Definition 6.6 is to be found in the paper of *Topsoe* [1976], p. 240. Theorems 6.9 and 12.3 show that semiregularity and **regularity** are closely related. They indicate why we propose these terms for the corresponding set–theoretic notions.

Definition 12.5 of **Radon measures** is by Proposition 12.5 equivalent to that given by *Choquet* [1955], p. 207.

The **extension theorem for contents** Corollary 12.8.1 is to be found in a paper of *Smiley* [1944], p. 441. It was rediscovered by *Pettis* [1951], Corollary 1.2.1, p. 189, and *Kisynski* [1968 a], p. 327. The topological version of Corollary 12.8.2 is due to *Kisynski* [1968 b], Theorem 1.2, p. 142, its general form to *Topsoe* [1970 a], Theorem 4.1, p. 15.

Other extension theorems may be viewed as representation theorems, e.g. Example 14.3, which is due to *Bachman and Sultan* [1980], Theorem 2.1, p. 390, and generalized to the unbounded semifinite case by *Adamski* [1984], Theorem 2.5, p. 358.

The **measure extension theorem** Corollary 13.10.1 is to be found in *Topsoe* [1970 b], Theorem 1, p. 198. It was rediscovered by *Kelley and Srinivasan* [1971], Proposition 9, p. 238 (cf. *Kelley, Nayak and Srinivasan* [1973], Theorem 1, p. 158). In a further article by *Kelley and Srinivasan* [1984], Proposition 3, p. 124, the method of proof is related to ours. The general form of the extension theorem given by *König* [1985], Theorem 3.4, p. 92, will be proved in 16.5. The definition 13.1 of a measure is equivalent, but formally weaker, than the one given by *Kelley and Srinivasan* [1971] (cf. Remark 13.2 and their Theorem 4, p. 235).

The decomposition theorems Corollary 12.8.4 and Corollary 13.10.3 extend a result of *Luther* [1968].

The functional analytic representation theorems in § 7 and § 8 are perhaps the most interesting results (cf. the remarks in *Topsoe* [1982], p. 304–305). The set–theoretical counterpart (cf. Example 7.1.2) of our basic definition 7.1 of **tightness**, used for the definition of bounded Radon integrals **as tight positive linear forms** on the function space $\mathscr{C}^b(X)$ for X completely regular, goes back to *LeCam* [1957], Lemma 3, p. 213, *Schwartz* [1964/65], Définition, p. 8, and *Varadarajan* [1965], Theorem 29, p. 179. The explicit representation as regular Borel measures is also to be found in Schwartz's lectures (Théorème, p. 138) and in *Bourbaki* [1969], Proposition 5, p. 58 and Théorème 2, p. 46. Bourbaki denotes the tightness condition by (M).

In the general context of a rich space of continuous sections, the Representation Theorem 8.8 is due to *Portenier* [1970/71 b], Théorème 16, p. 4. Note that our functional analytic approach to integration theory, which makes no use of indicator functions in $\mathscr{S}$, may be formulated in the context of line bundles. In the topological situation (cf. *Portenier* [1970/71 a]), one can define the analogue of $\mathscr{S}(X)$ as some cone of lower semicontinuous sections, as well as the concept of Radon integrals as regular linear functionals on this cone, without using projective systems. Then one can prove the Representation Theorems 8.4 and 8.8 in exactly the same way. Note that Choquet's conical measures, defined on a convex cone C and having properties like Radon measures, may then be introduced without the notion of localization as Radon integrals on the canonical line bundle associated with $C \setminus \{0\}$.

The set–theoretical versions of the representation theorems in § 14 owe much to the work of *Pollard and Topsoe* [1975] and *Topsoe* [1976], which is commented in 16.6 and 16.8. The **separation condition** in Definition 14.8.1 is condition (A 6) of *Topsoe* [1976], p. 240. The Representation Theorem 14.3 contains in particular a result of *Heiden* [1978], Theorem 1, p. 211, as Remark 14.3.2 shows.

The **classical representation theorems** are respectively found in *Riesz* [1909], p. 976 and [1911], p. 43, *Markoff* [1938], Theorem 22, p. 184 , and *Alexandroff* [1941], Theorem 1, p. 577 together with Theorem 1, p. 563. *Radon* [1913], p. 1333, has generalized Riesz' theorem, originally stated for compact intervals, to compact subsets of higher dimensional Euclidean spaces. We comment on *Bauer's* Representation Theorem in 16.3.

Adapted function spaces (in the set–theoretical meaning of Example 14.1.1) are studied in *Choquet* [1962], p. 4, and in an article by *Arens* [1963]. The condition is also to be found in the early work of *Loomis* [1954], Lemma 5(b), p. 176.

In the formulation of the **Daniell–Stone Theorem** 15.2, whose classical predecessors are in *Daniell* [1919/20], p. 210, *Goldstine* [1941], p. 618–619, and *Stone* [1948], p. 454, we have been guided by *Bauer* [1990], 29.3 and 29.4, p. 207–208.

The method to prove **semiregularity via minimality** in the representation theorem Corollary 7.7 is adapted from *Lembcke* [1970] and [1977]. It is based on a generalization of *Andenaes'* [1970] version of the Hahn–Banach theorem. In the setting of function cones, this general form is proved in § 17 of the appendix. It is also used to prove the Theorem of *Henry* [1969] in 16.9.

Proposing the quick approach to the **ordinary Lebesgue integral** in Example 8.2, we are however aware of the fact that the function cone $\mathcal{T}(\mathbb{R}^n)$ is unfortunately not suitable for iterated integration.

Theorem 8.18 on **inverse images** of Radon measures is to be found in *Lembcke* [1970], Satz 6.9, p. 87 and Bemerkung 6.7.1, p. 85, and in the bounded case under more general conditions in *Bourbaki* [1969], Proposition 8, p. 32, whereas Corollary 8.18 is in *Schwartz* [1973], Theorem 12, p. 39. However, our approach to this theorem via a representation theorem seems to be new.

16.2 L.H. Loomis' abstract Riemann integration theory.

Loomis [1954] considers *a positive linear form* τ *on a Stonian vector lattice* $\mathcal{T}$ of real-valued functions and calls the set $\mathcal{R}^*(\tau)$ of all functions integrable w.r.t. τ the *two-sided completion* of $\mathcal{T}$ w.r.t. τ. A set is said to have *content* if it is integrable w.r.t. τ, i.e. belongs to the ring $\mathfrak{R}^*(\tau)$. A function f is called *Riemann summable* if it is integrable with respect to the *content* m *induced by* τ^* *on* $\mathfrak{R}^*(\tau)$, and its abstract Riemann integral $m^*(f)$ is denoted by $\int f\,dm$.

Loomis proves in sections 1 and 2 (cf. in particular *Loomis* [1954], Lemma 2, p. 172 and Theorem 1, p. 173) the following

PROPOSITION. *$\mathcal{R}^*(m)$ consists of all bounded functions $g \in \mathcal{R}^*(\tau)$ vanishing outside some set integrable w.r.t. τ, and $m^* = \tau^*$ on $\mathcal{R}^*(m)$. In particular,*

$$\mathfrak{R}^*(\tau) = \mathfrak{R}^*(m).$$

Indeed, if κ denotes the restriction of τ^* to the above vector lattice $\mathcal{K} \subset \mathcal{R}^*(\tau)$ of functions, then κ is a positive linear form with $\mathcal{R}^*(\kappa) = \mathcal{K}$, hence $\mathfrak{R}^*(\kappa) = \mathfrak{R}^*(\tau)$. Since κ trivially satisfies the truncation formulas, and $\mathcal{T}_-$ consists of $\mathfrak{R}^*(\tau)$-measurable functions by Remark 10.3.2 and Corollary 11.7, Remark and Proposition 14.4 prove that $\kappa^\bullet = m^\bullet$. The result now follows from the Integrability Criterion 4.5, since $\kappa^*(|g|) < \infty$ is equivalent to $m^*(|g|) < \infty$ and therefore $\kappa^* = m^*$, in particular $\mathfrak{R}^*(m) = \mathfrak{R}^*(\kappa) = \mathfrak{R}^*(\tau)$. □

To prove this result, Loomis uses the following concept. A function $f \geq 0$ is said to be *measurable* w.r.t. $\mathfrak{R}^*(\tau)$ if $\{f > \gamma\} \in \mathfrak{R}^*(\tau)$ for all but at most countably many $\gamma > 0$. This condition is very restrictive, e.g. for the linear form $\tau = 0$ on $\mathcal{T} = \{0\}$, we have $\mathcal{R}^*(\tau) = \{0\}$, hence $\mathfrak{R}^*(\tau) = \emptyset$, and therefore 0 is the only function which is measurable w.r.t. $\mathfrak{R}^*(\tau)$. However, we have

$$\mathcal{R}^0(\tau) = \mathcal{R}^\bullet(\tau) = \overline{\mathbb{R}}^X .$$

In section 3, Loomis' concept of measurability leads to the definition of f being *improperly Riemann summable* w.r.t. $\mathfrak{R}^*(\tau)$, which means that $f^\pm$ are measurable and

$$\int f^\pm \, dm := \sup_{g \in \mathcal{R}^*(m),\ g \leq f^\pm} m^*(g) < \infty .$$

Loomis [1954], Corollary 1, p. 175, proves that for every improperly Riemann summable function $f \geq 0$, one has

$$\int f \, dm = \sup_{0<a<b} m^*[\min(f,b) - \min(f,a)] =$$
$$= \lim_{n \to \infty} \tau^*[\min(f,n) - \min(f,1/n)] .$$

Moreover, every $t \in \mathcal{T}$ is improperly Riemann summable, and one has $\int t \, dm = \tau(t)$ for all $t \in \mathcal{T}$ iff τ satisfies both truncation formulas. In this case, the *one-sided completion* of $\mathcal{T}$ w.r.t. τ, defined implicitely as $\mathcal{R}^\bullet_{\mathbb{R}}(\tau)$ through the Essential Integrability Criterion 4.4, coincides with the set of all functions which are improperly Riemann summable w.r.t. $\mathfrak{R}^\bullet(\tau)$.

This result (cf. *Loomis* [1954], Theorem 5, p. 180) is actually generalized in Remark 14.4 :

THEOREM. *If τ satisfies both truncation formulas, then $\tau^\bullet = m^\bullet$. In particular,*

$$\mathcal{R}^0(\tau) = \mathcal{R}^0(m) \quad \textit{and} \quad \mathcal{R}^\bullet(\tau) = \mathcal{R}^\bullet(m) .$$

We use the more general concept of measurability w.r.t. τ or m (cf. Theorem 11.8), which coincides with Loomis' concept if $1 \in \mathcal{R}^*(\tau)$, e.g. $1 \in \mathcal{T}$.

Finally, it is interesting to note that Loomis' condition

$$\text{for all } t \in \mathcal{T}_+ \text{ there exists } s \in \mathcal{T}_+ \text{ with } t/s \to 0 \text{ if } t \to 0$$

implies that $\mathcal{T}$ is $\mathfrak{K}(\mathcal{T})$-adapted, which in turn implies the truncation formula at 0 by Example 14.1.1 and Proposition 14.1.

16.3 Representation theorem of H. Bauer.

We now study Example 14.14.5 (cf. *Bauer* [1956], Korollar, p. 469) for a locally compact space X in more detail, relating the representing content to the topology in the sense of *Haupt and Pauc* [1950].

In the following, $\mathcal{T} \subset \mathscr{C}^0(X)$ is a *Stonian vector lattice* such that for any $x \in X$ there is some $t \in \mathcal{T}$ with $t(x) \neq 0$, and $\mathcal{T}$ is *point-separating*, i.e. for any two different points $x,y \in X$ there exists $t \in \mathcal{T}$ with $t(x) \neq t(y)$.

LEMMA. *$\mathcal{T}$ is linearly separating.*

Let $x,y \in X$ be different points and choose $s \in \mathcal{T}_+$ with $s(x) > 0$. If $s(x) = s(y)$, one only has to choose a function in $\mathcal{T}$ which separates x from y.

If $s(x) \neq s(y)$, we may assume that $s(x) < s(y)$. Since there exists $t \in \mathcal{T}_+$ with $t(x), t(y) > 0$, defining $u := \min(t,t(x)) \in \mathcal{T}_+$, we get $u(x) = t(x)$ and therefore

$$s(x)u(y) \leq s(x)t(x) < s(y)t(x) = s(y)u(x) . \quad \square$$

COROLLARY. *$\mathcal{T}$ is uniformly dense in $\mathscr{C}^0(X)$, and $\mathcal{K}_-(X) \subset \hat{\mathcal{T}}_-$.*

This follows from the Stone–Weierstrass theorem in $\mathscr{C}^0(X)$ and Example 4.8.2. $\square$

Let now τ be a *positive linear form on* $\mathcal{T}$, and denote by m the finite *content induced by* τ^* on $\mathfrak{K} := \mathfrak{K}(\mathcal{T}) \cap \mathfrak{R}^*(\tau)$. For $\mathfrak{G} := \mathfrak{G}(\mathcal{T}) \cap \mathfrak{R}^*(\tau)$,

Proposition 14.5 implies the following

PROPOSITION. *The content* m *is* $\mathfrak{G}$*-regular,* $\tau^* \le m^*$ *and* $m^\bullet(f) \le \tau^\bullet(f)$ *for every function* f *which is bounded below and positive outside some compact set. Every measurable function w.r.t.* τ *is measurable w.r.t.* m *, and* $\mathfrak{R}^*(m) = \mathfrak{R}^*(\tau)$ *.*

One only has to note that $\mathfrak{K} \subset \mathfrak{K}(X)$, and that every compact set K is contained in a set $\{t \le \gamma\} \in \mathfrak{K}$ for some $t \in \mathcal{T}_-$ and $\gamma \in]-1,0[$ by Theorem 11.7, since for every $x \in X$ there exists $u \in \mathcal{T}$ with $u(x) < -1$. □

REMARKS.

(1) Every function in $\mathcal{K}(X)$ is integrable w.r.t. τ , hence integrable w.r.t. m , and both integrals coincide.

This follows from the above corollary, since $\hat{\mathcal{T}_-} \subset \mathcal{R}^*_-(\tau)$. □

(2) These results show that the finite content m induced by τ^* on $\mathfrak{R}^*(\tau)$ is *adapted to the topology* , as stated by *Bauer* [1956], Satz 7 or Korollar, p. 463. This mainly means that m is inner $\mathfrak{K}$-regular as well as outer $\mathfrak{G}$-regular on $\mathfrak{R}^*(\tau) = \mathfrak{R}^*(m)$, and that $\mathfrak{G}$ is a base for the topology of X .

The first property follows from Proposition and Theorem 12.4, and the second one from the above remark. In fact, for any $x \in X$ and any open neighbourhood U of x , there exists $k \in \mathcal{K}_+(X) \subset \mathcal{R}^*(\tau)$ with $k(x) = 1$ and $\mathrm{supp}(k) \subset U$, hence $x \in \{k > \gamma\} \subset U$ for some $\gamma > 0$ with $\{k > \gamma\} \in \mathfrak{G}$ by Theorem 11.7.

Since every set in $\mathfrak{R}^*(\tau)$ is relatively compact, the last adaptation property to be proved (cf. *Bauer* [1956], 3.2(3), p. 462) follows immediately from integrability (cf. Proposition 2.5) :

For $A \in \mathfrak{R}^*(\tau)$ and $\varepsilon > 0$, there exist $s,t \in \mathcal{T}$ with $s \le 1_A \le t$ and $\tau(t) - \tau(s) \le \varepsilon$, which by

$$s \le 1_{A^\circ} \le 1_A \le 1_{\overline{A}} \le t$$

shows that $A^\circ, \overline{A} \in \mathfrak{R}^*(\tau)$ with $\tau^*(A^\circ) = \tau^*(\overline{A}) = \tau^*(A)$. □

Let now μ denote the ***positive linear form induced on*** $\mathcal{K}(X)$ ***by*** m^* or τ^* (cf. Remark 1). By Theorem 8.2, μ can be identified with a Radon integral on X.

THEOREM. *The upper functionals* μ^* *and* m^* *coincide. Furthermore,* τ *is represented by* m *(or* μ*) iff* τ *satisfies the truncation formula at* 0. *In this case, we have*

$$\tau^\bullet = m^\bullet = \mu^\bullet .$$

For every function f, obviously

$$\mu^*(f) = \inf_{k \in \mathcal{K}(X),\, k \geq f} m^*(k) \geq m^*(f) .$$

Conversely, for $e \in \mathcal{E}$ with $e \geq f$ we have

$$m(e) = \tau^*(e) = \inf_{t \in \mathcal{T},\, t \geq e} \tau(e) ,$$

having in mind that $\mathfrak{K}, \mathfrak{G} \subset \mathfrak{N}^*(\tau) = \mathfrak{N}^*(m)$ by the proposition. Since every set in $\mathfrak{N}^*(\tau)$ is relatively compact, $\mathrm{supp}(e)$ is compact. Choose $g \in \mathcal{K}$ with

$$1_{\mathrm{supp}(e)} \leq g \leq 1 .$$

Every $t \in \mathcal{T}$ with $t \geq e$ is positive outside $\mathrm{supp}(e)$, hence

$$e \leq g \cdot t \leq t$$

and $g \cdot t \in \mathcal{K}(X)$. Therefore,

$$\inf_{t \in \mathcal{T},\, t \geq e} \tau(e) = \inf_{k \in \mathcal{K}(X),\, k \geq e} \tau^*(k) = \mu^*(e) \geq \mu^*(f) ,$$

which proves that $m^*(f) \geq \mu^*(f)$.

The last assertion follows from Corollary 14.4, all functions in $\mathcal{T}$ being bounded. □

The theorem improves results of *Bauer* [1956], sections 3.3 and 3.5.

16.4 Measurability and integrability in the sense of N. Dunford and J.T. Schwartz, as developed by K.P.S. and M. Bhaskara Rao.

The above authors treat a finitely additive integration theory for the special case of a *finite content* m *defined on an algebra* $\mathfrak{A}$ of subsets of X. Obviously, m is $\mathfrak{A}$–regular, and so m is regular on

$$\mathscr{E} := \mathscr{E}(\mathfrak{A}) = \mathscr{E}_-(\mathfrak{A}) - \mathscr{E}_-(\mathfrak{A}) .$$

In *Bhaskara Rao* [1983], Theorem 4.4.7, p. 101, for a real–valued function f, the following measurability concepts are proved to be equivalent :

(1) There exists a sequence (e_n) in $\mathscr{E}(\mathfrak{A})$ *converging hazily* to f, i.e.

$$\lim_{n\to\infty} m^*(\{|e_n - f| > \varepsilon\} = 0$$

for any $\varepsilon > 0$.

(2) For any $\varepsilon > 0$, there exists a partition $(A_i)_{i=0,\ldots,p}$ of X in $\mathfrak{A}$ such that $m(A_0) \le \varepsilon$ and

$$|f(x) - f(y)| \le \varepsilon$$

for all $x,y \in A_i$ and all $i = 1,\ldots,p$.

We will refer to these concepts, termed T_1– respectively T_2–measurability in *Bhaskara Rao* [1983], Definitions 4.4.5 and 4.4.6, p. 101, as *measurability in the sense of Bhaskara Rao*. Note that the first concept is that of total measurability in *Dunford and Schwartz* [1958], Definition III.2.10, p. 106.

PROPOSITION. *A real–valued function* f *is measurable in the sense of Bhaskara Rao iff* f *is measurable w.r.t.* m *and satisfies*

$$\lim_{\alpha\to\infty} m^*(\{|f| > \alpha\}) = 0 .$$

The above condition, called smoothness, is necessary by (2). It remains to prove that, for $n \in \mathbb{N}$, the function

$$f_n := \text{med}(f,n,-n)$$

is integrable w.r.t. m, since $\mathbb{N}\cdot(-1)$ is coinitial in $\mathscr{R}^*_-(m)$. Let $\delta > 0$ be given and choose $\varepsilon > 0$ such that

$$2n\varepsilon + \varepsilon\cdot m(X) \le \delta .$$

Let $(A_i)_{i=0,\ldots,p}$ be a partition of X in $\mathfrak{A}$ corresponding to ε as in (2). Then

$$e := n\cdot 1_{A_0} + \sum_{i=1}^{p} \sup f_n(A_i)\cdot 1_{A_i} \quad \text{and} \quad d := -n\cdot 1_{A_0} + \sum_{i=1}^{p} \inf f_n(A_i)\cdot 1_{A_i}$$

belong to $\mathscr{E}$, and we have

$$d \leq f_n \leq e \quad \text{and} \quad e - d \leq 2n \cdot 1_{A_0} + \varepsilon \cdot \sum_{i=1}^{p} 1_{A_i} \cdot$$

This yields

$$m(e-d) \leq 2n \cdot m(A_0) + \varepsilon \cdot \sum_{i=1}^{p} m(A_i) \leq 2n\varepsilon + \varepsilon \cdot m(X) \leq \delta ,$$

hence $f_n \in \mathcal{R}^*(m)$.

Conversely, let $f \in \mathcal{R}^0(m)$ be smooth. For $\varepsilon > 0$, there exists $\alpha > 0$ such that

$$m^*(\{|f| > \alpha\}) \leq \frac{\varepsilon}{2} .$$

Choose $\gamma_i \in \mathbb{R}$ according to Theorem 11.8.(iv) such that

$$\gamma_1 < -\alpha < \gamma_2 < \ldots < \gamma_p < \alpha < \gamma_{p+1} , \quad \gamma_{i+1} - \gamma_i \leq \varepsilon \quad \text{and} \quad \{f > \gamma_i\} \in \mathfrak{R}^*(m) .$$

By Corollary 12.1.(i), there exist decreasing finite sequences (K_i), (G_i) in $\mathfrak{A}$ with

$$K_i \subset \{f > \gamma_i\} \subset G_i \quad \text{and} \quad m(G_i) - m(K_i) \leq \frac{\varepsilon}{2(p+1)} .$$

If we define $A_i := K_i \setminus G_{i+1}$ and $A_0 := \complement G_1 \cup (\bigcup_{i=1}^{p+1} G_i \setminus K_i) \cup K_{p+1}$, then $(A_i)_{i=0,\ldots,p}$ is a partition of X in $\mathfrak{A}$ which satisfies (2), since

$$\complement G_1 \cup K_{p+1} \subset \{|f| > \alpha\} . \quad \square$$

Note that the set of real-valued measurable functions in the sense of Bhaskara Rao is stable with respect to addition (cf. *Bhaskara Rao* [1983], Corollary 4.4.9, p. 102), in contrast to the general situation encountered in Example 4.7.2. The smoothness condition on f is a little bit more general than the condition $m^\bullet(|f|) < \infty$ corresponding to the one encountered in Theorem 4.10, since

$$\alpha \cdot 1_{\{|f|>\alpha\}} \leq |f|$$

implies

$$m^*(\{|f| > \alpha\}) \leq \frac{1}{\alpha} \cdot m^\bullet(|f|) \to 0 \quad \text{for } \alpha \to \infty .$$

It is easy to see that the upper and lower integrals of a (bounded) function f introduced in ib., Definition 4.5.3, p. 116, coincide with $m^*(f)$ and $m_*(f)$. Thus, a function is *S-integrable* (ib., Definition 4.5.5, p. 116) iff $f \in \mathcal{R}^*(m)$. In this case, the S-integral of f coincides with $m_*(f) = m^*(f)$.

We now discuss the integral introduced by *Dunford and Schwartz* [1958], Definition III.2.17, p. 112, and termed D–integral by *Bhaskara Rao* :

Definition. A real-valued function f is *D-integrable* if there exists a sequence (e_n) in $\mathcal{E}(\mathfrak{A})$ converging to f hazily, such that

$$\lim_{p,q\to\infty} m(|e_p - e_q|) = 0 .$$

Then $\lim_{n\to\infty} m(e_n)$ exists in $\mathbb{R}$, is independent of the chosen sequence (cf. *Bhaskara Rao* [1983], Proposition 4.4.10, p. 102) and is called the *D-integral* of f. In fact, this is the essential integral of f w.r.t. m :

Theorem. *A real-valued function f is D-integrable iff f is essentially integrable w.r.t. m. In this case, both integrals coincide.*

Suppose that $f \in \mathcal{R}^{\bullet}(m)$. Given $n \in \mathbb{N}$, there exist $e_n, d_n \in \mathcal{E}$ such that

$$-n \le d_n \le f_n := \mathrm{med}(f,n,-n) \le e_n \le n ,$$

and

$$m(e_n - d_n) \le \frac{1}{n} .$$

We first prove that (e_n) converges hazily to f. Indeed, for $\varepsilon > 0$, with

$$A_n := \{e_n - d_n > \varepsilon\} \quad \text{and} \quad B_n := \{|f| > n\}$$

we have

$$\{|e_n - f| > \varepsilon\} \subset (\{-n \le f \le n\} \cap \{e_n - f_n > \varepsilon\}) \cup \{|f| > n\} \subset A_n \cup B_n ,$$

hence

$$m^*\{|e_n - f| > \varepsilon\}) \le m^*(A_n) + m^*(B_n) .$$

Since

$$\varepsilon \cdot 1_{A_n} \le e_n - d_n \quad \text{and} \quad 1_{B_n} \le |f| - \min(|f|, n-1) ,$$

we get

$$\varepsilon \cdot m^*(A_n) \le m(d_n - e_n) \le \frac{1}{n}$$

and

$$m^*(B_n) \le m^{\bullet}(|f| - \min(|f|, n-1)) = m^{\bullet}(|f|) - m^*(\min(|f|, n-1)) .$$

The assertion follows, since the last term also converges to 0 by the essential integrability of $|f|$.

Furthermore,

$$\lim_{p,q\to\infty} m(|e_p - e_q|) = 0$$

since

$$|e_p - e_q| \le |e_p - f_p| + |f_p - f| + |f - f_q| + |f_q - e_q| \le$$

$$\le e_p - d_p + |f| - \min(|f|,p) + |f| - \min(|f|,q) + e_q - d_q .$$

Finally, we have

$$|m^\bullet(f) - m(e_n)| \le |m^\bullet(f) - m^*(f_n)| + |m^*(f_n) - m(e_n)| \le$$

$$\le |m^\bullet(f) - m^*(f_n)| + m(e_n - d_n) .$$

Since $m^\bullet(f) = \lim_{n\to\infty} m^*(f_n)$ by Corollary 3.7, this proves that

$$m^\bullet(f) = \lim_{n\to\infty} m(e_n)$$

is the D-integral of f.

Conversely, let (e_n) in $\mathcal{E}$ converge hazily to f such that

$$\lim_{p,q\to\infty} m(|e_p - e_q|) = 0 .$$

By the proposition, f is measurable w.r.t. m. It thus remains to prove that $m^\bullet(|f|) < \infty$. Since

$$\big||e_p| - |e_q|\big| \le |e_p - e_q| ,$$

$(m(|e_n|))$ is a Cauchy sequence and therefore bounded by some $\beta < \infty$. On the other hand, with $A_k := \{|e_k - f| > 1\}$, we have

$$\min(|f|,n) \le n \cdot 1_{A_k} + 1 + |e_k| ,$$

hence

$$m^*(\min(|f|,n)) \le n \cdot m^*(A_k) + m(1) + m(|e_k|) \le 1 + m(1) + \beta$$

by hazy convergence, choosing k large enough. This shows that

$$m^\bullet(|f|) = \sup_n m^*(\min(|f|,n)) \le 1 + m(1) + \beta < \infty . \quad \square$$

The Integrability Criterion 4.5 shows that a function is integrable iff it is bounded and measurable. Since bounded functions are smooth, the proposition and the theorem yield Theorem 4.5.7 of *Bhaskara Rao* [1983], p. 117.

16.5 H. König's version of a basic measure extension theorem.

We now prove König's refinement of the fundamental measure extension theorem Corollary 13.10.1 (cf. *König* [1985], Theorem 3.4, p. 92). This amounts to the following

Theorem. *A finite content* m *on a lattice* $\mathfrak{K}$ *has a (unique) extension to a semiregular measure on* $\mathfrak{K}_\delta$ *iff* m *has the Daniell property at the empty set and satisfies*

$$m(L) \le m(K) + \sup_{A \in \mathfrak{K}_\delta,\, A \subset L \setminus K} \inf_{B \in \mathfrak{K},\, B \supset A} m(B)$$

for all $K, L \in \mathfrak{K}$ *with* $K \subset L$.

The conditions are necessary. In fact, there is at most one measure extension m_δ of m to $\mathfrak{K}_\delta$, and then

$$m_\delta(A) = \inf_{B \in \mathfrak{K},\, B \supset A} m(B)$$

holds for $A \in \mathfrak{K}_\delta$.

Conversely, we first note that m has the Daniell property (cf. *König* [1985], Proposition 2.13, p. 91), i.e.

$$\inf m(K_n) \le m(K)$$

for any decreasing sequence (K_n) in $\mathfrak{K}$ with intersection $K \in \mathfrak{K}$. Indeed, by the assumption applied to $K \subset K_1$, for $\varepsilon > 0$ there exist $A \in \mathfrak{K}_\delta$ with $A \subset K_1 \setminus K$ and a decreasing sequence (B_n) in $\mathfrak{K}$ with intersection A , such that $B_n \subset K_1$ and

$$m(K_1) \le m(K) + \inf m(B_n) + \varepsilon .$$

Since

$$K_n \cup B_n \subset K_1 \quad \text{and} \quad \bigcap K_n \cap B_n = K \cap A = \emptyset ,$$

we have

$$\inf m(K_n) + \inf m(B_n) = \inf m(K_n \cup B_n) + \inf m(K_n \cap B_n) \le$$
$$\le m(K_1) \le m(K) + \inf m(B_n) + \varepsilon ,$$

which proves our assertion.

Next, by Proposition 13.3, the functional m on $\mathscr{E}_-(\mathfrak{K})$ has the Daniell property, as does m^σ on $\mathscr{E}_-(\mathfrak{K})_\sigma$ by Remark 9.2.1. Through

$$m_\delta(A) := -\, m^\sigma(-\, 1_A)$$

we define a finite content m_δ on $\mathfrak{K}_\delta$ which extends m and has the Daniell property, hence satisfies the formula of the first part. Addendum 12.1 shows that $m_{\delta *}(A) = m^\sigma{}_*(A)$ for every set A, since $\mathfrak{K}(\mathfrak{E}\_(\mathfrak{K})_\sigma) = \mathfrak{K}_\delta$ by Remark 10.6.1.

Finally, by Corollary 13.4, m_δ is a measure if we can prove that m_δ is semiregular, i.e. that

$$m_\delta(B) \leq m_\delta(A) + m_{\delta *}(B \setminus A)$$

holds for all $A,B \in \mathfrak{K}_\delta$ with $A \subset B$. By the assumption, this formula holds for $A,B \in \mathfrak{K}$. But from Theorem 5.2 we know that $m^{\sigma *}$ is an upper integral, hence the set function $m_{\delta *}$ has the Daniell property. Therefore, the proof of the above formula can be reduced to the case $B \in \mathfrak{K}$. Let (A_n) be a sequence in $\mathfrak{K}$ decreasing to A. By the assumption, we have

$$m_\delta(B) \leq m_\delta(B \cup A_n) \leq m(A_n) + m_{\delta *}(B \cup A_n \setminus A_n) \leq m(A_n) + m_{\delta *}(B \setminus A)$$

for every $n \in \mathbb{N}$, which finally proves the semiregularity of m_δ. □

REMARKS.

(1) A purely set-theoretic proof can be given as follows. By the Daniell property, m extends uniquely to a finite content m_δ with the Daniell property on $\mathfrak{K}_\delta$, defining

$$m_\delta(\bigcap K_n) := \inf_n m(K_n)$$

for every decreasing sequence (K_n) in $\mathfrak{K}$.

Instead of using the fact that $m^{\sigma *}$ is an upper integral, the Daniell property of $m_{\delta *}$ can be proved by the set-theoretical version of Remark 5.1.

(2) The semiregularity condition is satisfied if $L \setminus K \in \mathfrak{K}_\delta$ for $K,L \in \mathfrak{K}$ with $K \subset L$, in particular if $\mathfrak{K}_\delta$ is a ring.

In fact, for $B \in \mathfrak{K}$ with $B \supset L \setminus K$, we have $L \subset K \cup B$ and hence

$$m(L) \leq m(K \cup B) \leq m(K) + m(B) . \;\square$$

(3) Let l be a semifinite content defined on a lattice $\mathfrak{L}$, and let l^∇ be its finite part, defined on the lattice $\mathfrak{L}^\nabla$ of the sets in $\mathfrak{L}$ having finite content. Then l extends uniquely to an inner $\mathfrak{L}_\delta$-regular measure on any ring $\mathfrak{R}$ containing $\mathfrak{L}_\delta$ which is contained in the σ-algebra generated by $\mathfrak{S}(\mathfrak{L}_\delta)$ iff l has the Daniell

property at the empty set and

$$l^{\nabla}(L) \leq l^{\nabla}(K) + l^{\nabla}_{\delta *}(L \setminus K)$$

holds for all $K, L \in \mathfrak{L}^{\nabla}$ with $K \subset L$.

This follows as in Corollary 13.10.2 from the theorem and Corollary 13.10.1, applied to l^{∇}. Note that $\mathfrak{G}(\mathfrak{L}_{\delta}) \subset \mathfrak{G}(\mathfrak{L}^{\nabla}{}_{\delta})$. □

16.6 Representation theorem of D. Pollard and F. Topsoe for cones of positive functions.

In *Pollard and Topsoe* [1975], the authors consider representation of an increasing linear functional $\rho : \mathcal{S} \longrightarrow \mathbb{R}_{+}$ defined on a Stonian lattice cone $\mathcal{S} \subset \mathbb{R}^{X}_{+}$ of positive real-valued functions by a content m on a lattice $\mathfrak{K}$ of sets. Their restrictive hypothesis is that $\mathcal{S}$ be *stable with respect to positive differences*, i.e.

$$(s - t)^{+} \in \mathcal{S}$$

for $s, t \in \mathcal{S}$. This condition is satisfied iff $\mathcal{S}$ is the positive cone of a vector lattice of functions. In fact,

$$\mathcal{T} := \mathcal{S} - \mathcal{S}$$

is a Stonian vector lattice with

$$\mathcal{T}_{+} = \mathcal{S} \quad \text{and} \quad \mathcal{T}_{-} = -\mathcal{S}.$$

In addition, they assume that $\mathfrak{G}(\mathcal{S}) \subset \mathfrak{G}(\mathfrak{K})$, and that $\mathfrak{K}$ is ρ-separated by $\mathcal{S}$.

To incorporate their result into our framework, extend ρ to a positive linear form τ on $\mathcal{T}$. Then $\mathfrak{K}$ is τ-separated, and $\mathcal{T}$ consists of $\mathfrak{K}$-semicontinuous functions. In fact, for any $t \in \mathcal{T}$ and $\gamma \in \mathbb{R}$ we have

$$\{t > \gamma\} \in \mathfrak{G}(\mathfrak{K}).$$

For $\gamma > 0$, this follows from the above assumption, since

$$\{t > \gamma\} = \{t^{+} > \gamma\}$$

and $t^{+} \in \mathcal{S}$. For $\gamma \leq 0$ and $K \in \mathfrak{K}$, we have to prove that

$$\{t \leq \gamma\} \cap K \in \mathfrak{K}.$$

Since $\mathcal{S}$ is Stonian, by the separation assumption there exists $r \in \mathcal{S}$ with

$$(-2\gamma+1)\cdot 1_K \leq r \leq -2\gamma+1 .$$

Now we have

$$\{t \leq \gamma\} \cap K = \{t_- \leq \gamma\} \cap K = \{(r + t_-)^+ \leq -\gamma+1\} \cap K \in \mathfrak{K} ,$$

since $(r + t_-)^+ \in \mathcal{S}$.

The separation assumption yields that the restriction m of τ^* to $\mathfrak{K}$ is finite. We can therefore apply Theorem 14.11 to infer that m is a semiregular content and that m represents τ iff τ is $\mathfrak{F}(\mathfrak{K})$–tight. By Proposition 14.2, this means that τ is $\mathfrak{K}$–tight (i.e. "$\mathfrak{K}$ exhausts ρ" in the terminology of Pollard and Topsoe), and that τ satisfies the truncation formula at $-\infty$, which is equivalent to the assumption

$$\rho(s) = \sup_n \rho(\min(s,n))$$

for all $s \in \mathcal{S}$, imposed by *Pollard and Topsoe* [1975], Theorem 2, p. 179. Note that in this case $\rho(s) = m_*(s)$ for all $s \in \mathcal{S}$, since m is semiregular and s is positive, hence $m_*(s) = m_\bullet(s)$, and that m is the only semiregular representing content. We also infer from Lemma 14.5 that m is $\mathfrak{G}(\mathfrak{K})$–regular, or $\mathfrak{G}$–regular if $\mathfrak{G}$ is compatible with $\mathfrak{K}$ and $\mathfrak{G}(\mathcal{S}) \subset \mathfrak{G}$.

Conditions equivalent to the Daniell property of m at the empty set are formulated in *Pollard and Topsoe* [1975], Theorem 3, p. 181, and yield representation by a measure.

16.7 Representation theorem of C. Berg, J.P.R. Christensen and P. Ressel for cones of positive functions.

In the following, we generalize 16.6 (cf. *Pollard and Topsoe* [1975], bottom of p. 176) and a Riesz representation theorem given by the above authors for cones of positive continuous functions on a Hausdorff space, which will be discussed in the final remark.

Let $\mathcal{S} \subset \bar{\mathbb{R}}_+^X$ be a Stonian lattice cone of positive functions such that $(s - t)^+ \in \mathcal{S}$ for all $s \in \mathcal{S}$ and $t \in \mathcal{S}^b$. We consider an increasing linear functional $\rho : \mathcal{S} \longrightarrow \bar{\mathbb{R}}_+$, denote by $\mathcal{S}^\nabla$ the lattice cone of bounded functions $t \in \mathcal{S}$ with $\rho(t) < \infty$, and suppose that

$$\rho(s) = \sup_{t \in \mathcal{S}^\nabla, t \leq s} \rho(t) \quad \text{for all } s \in \mathcal{S} .$$

This means that

$$\rho(s) = \sup_{t \in \{\rho<\infty\},\, t \leq s} \rho(t) \quad \text{for all } s \in \mathcal{S}$$

and

$$\rho(s) = \sup_n \rho(\min(s,n)) \quad \text{for all } s \in \{\rho < \infty\} .$$

Let $\mathfrak{K}$ and $\mathfrak{G}$ be compatible lattices of sets such that $\mathfrak{K}$ is ρ-separated by $\mathcal{S}$, and suppose that $\mathfrak{G}(\mathcal{S}) \subset \mathfrak{G}$.

PROPOSITION. *If the restriction m of ρ^* to $\mathfrak{K}$ is finite, and if for $t \in \mathcal{S}^\nabla$ and $\varepsilon > 0$ there exists $K \in \mathfrak{K}$ such that $\rho(u) \leq \varepsilon$ for all $u \in \mathcal{S}$ with $u \leq t$ and $u = 0$ on K, then m is the only $\mathfrak{G}$-regular content representing ρ, i.e.*

$$\rho(s) = m_*(s) = m^\bullet(s) \ \text{for all } s \in \mathcal{S} .$$

In fact, $\mathcal{T} := \mathcal{S} - \mathcal{S}^\nabla$ is a lattice cone, as the formula

$$\min(s_1 - t_1, s_2 - t_2) = \min(s_1 + t_2 , s_2 + t_1) - (t_1 + t_2)$$

and the corresponding formula for the maximum shows. Moreover,

$$\mathcal{T}_+ = \mathcal{S} \quad \text{and} \quad \mathcal{T}_- = -\mathcal{S}^\nabla ,$$

and $\mathcal{T}$ consists of $\mathfrak{K}$-semicontinuous functions by the argument in 16.6. Furthermore,

$$\tau : s - t \longmapsto \rho(s) - \rho(t)$$

defines a regular linear functional on $\mathcal{T}$ by the condition imposed on ρ. Since $m = \tau^*$ on $\mathfrak{K}$ and τ is $\mathfrak{K}$-tight by hypothesis, Proposition 14.2 and Theorem 14.11 prove that m is the only semiregular content on $\mathfrak{K}$ representing τ. By Lemma 14.5, m is $\mathfrak{G}$-regular. □

REMARK. *Berg, Christensen and Ressel* [1984], Theorem 2.2, p. 35, consider the special case of a point-separating min-stable cone $\mathcal{S}$, also stable w.r.t. positive differences, of continuous real-valued functions on a Hausdorff space X. The formula

$$\max(s,t) = s + (t - \min(s,t))^+$$

shows that $\mathcal{S}$ is a lattice cone. The finiteness of ρ^* on $\mathfrak{K}(X)$ together with the point-separation implies that $\mathfrak{K}(X)$ is Urysohn-separated by $\mathcal{T}$, using Example 14.9.4. In fact, if $x \neq y$ and $r(x) > r(y)$ for some $r \in \mathcal{S}$, choose $s \in \mathcal{S}$ with $s \geq 1_{\{x,y\}}$ and $\rho(s) < \infty$. By the Stonian condition, we may assume that $s \in \mathcal{S}^\nabla$

and $s(x) = s(y) = 1$. Then

$$t := r - \frac{1}{2}[r(x) + r(y)] \cdot s \in \mathcal{T}$$

satisfies $t(x) > 0$ and $t(y) < 0$. Finally, Proposition 14.8 shows that $\mathfrak{K}(X)$ is τ-separated by $\mathcal{T}$, i.e. ρ-separated by $\mathcal{S}$. The representing content is a Radon measure, since $\mathfrak{G}(\mathcal{S}) \subset \mathfrak{G}(X)$.

By the same argument, condition (i) of Lemma 8.5 is satisfied. Hence we could also have applied the Representation Theorem 8.4.

16.8 F. Topsoe's representation theorem for cones of positive functions without stability with respect to positive differences.

Following his joint work with Pollard, *Topsoe* [1976] considers again an increasing linear functional $\rho : \mathcal{S} \longrightarrow \mathbb{R}_+$ defined on a Stonian lattice cone $\mathcal{S} \subset \mathbb{R}_+^X$ of positive real-valued functions, and a lattice $\mathfrak{K}$ such that $\mathfrak{K}$ is ρ-separated by $\mathcal{S}$. Stability with respect to positive differences is no longer assumed, however the previous semicontinuity assumption $\mathfrak{G}(\mathcal{S}) \subset \mathfrak{G}(\mathfrak{K})$ is replaced by the requirement

$$\{s < \gamma\} \in \mathfrak{G}(\mathfrak{K})$$

for all $s \in \mathcal{S}$ and $\gamma > 0$.

Let $\mathcal{T} := -\mathcal{S}$ and $\tau : \mathcal{T} \longrightarrow \mathbb{R}$ be defined by

$$\tau(t) = -\rho(-t) = \rho_*(t) .$$

Then τ is an increasing linear functional on the Stonian lattice cone $\mathcal{T}$, and by the above condition, $\mathcal{T}$ consists of $\mathfrak{K}$-measurable, i.e. $\mathscr{E}(\mathfrak{K})$-measurable functions. Topsoe's assumption (t*) is slightly weaker than the semiregularity of τ , which we assume in what follows, as well as Topsoe's condition (E) which is equivalent to τ being $\mathfrak{K}$-tight. Together with Topsoe's version (7) of the truncation formula at $-\infty$, by Proposition 14.2 this yields that τ is $\mathscr{E}_-(\mathfrak{K})$-tight. Since obviously

$$\rho_* \leq \tau_\times ,$$

we infer that the upper functional ρ^* is $\mathscr{E}_-(\mathfrak{K})$-tight on $\mathcal{T}$. *Topsoe* [1976] proves in Lemma 4, p. 241, that the restriction m of ρ^* to $\mathfrak{K}$ is semiregular. By Lemma 14.10, m is a finite content and ρ_* coincides with m on $\mathscr{E}_-(\mathfrak{K})$, in particular ρ_* is linear on this cone. Proposition 7.4 shows that the restriction τ of ρ_* to $\mathcal{T}$ is represented by m , i.e.

$$\rho(s) = m_*(s) = m^\bullet(s)$$

holds for all $s \in \mathcal{S}$, since m is semiregular and s is positive.

16.9 An abstract version of Henry's extension theorem.

The theorem alluded to is Theorem 1 of *Henry* [1969], p. 239.

THEOREM. *Let τ be a regular linear functional on a min-stable function cone $\mathcal{T}$, and $\mathcal{S}$ be a min-stable function cone such that $\tau_* > -\infty$ on $\mathcal{S}$. If*

$$\tau(t) = \inf_{s \in \mathcal{S},\, s \geq t} \tau^*(s)$$

holds for all $t \in \mathcal{T}$, then there exists a regular linear functional μ on $\mathcal{S}$ representing τ. We even have

$$\tau(t) = \mu_*(t) = \mu^*(t)$$

for $t \in \mathcal{T}$.

By Theorem 2.2.(iii), τ_* is superlinear, whereas τ^* is sublinear on $\mathcal{S}$, with

$$-\infty < \tau_* \leq \tau^* \quad \text{on } \mathcal{S}.$$

In Theorem 17.3 of the appendix, we will prove that there exists an $\mathcal{S}$-minimal increasing linear functional κ on $\mathcal{S}$ with

$$\tau_* \leq \kappa \leq \tau^* .$$

The difference-regular linear extension μ to $\mathcal{S}$ of $\kappa_{|\mathcal{S}_-}$ represents τ. In fact, by Proposition 6.3, for $t \in \mathcal{T}$ we have

$$\mu^\bullet(t) \leq \mu^*(t) \leq \inf_{s \in \mathcal{S},\, s \geq t} \kappa(s) \leq \inf_{s \in \mathcal{S},\, s \geq t} \tau^*(s) = \tau(t),$$

hence

$$\tau(t) = \tau_*(t) = \sup_{u \in \mathcal{T},\, -u \leq t} -\tau(u) \leq \sup_{u \in \mathcal{T},\, -u \leq t} -\mu^*(u) =$$

$$= \sup_{u \in \mathcal{T},\, -u \leq t} \mu_*(-u) \leq \mu_*(t) \leq \mu_\bullet(t) \leq \mu^\bullet(t) \leq \mu^*(t)$$

and therefore

$$\tau(t) = \mu_*(t) = \mu_\bullet(t) = \mu^\bullet(t) = \mu^*(t) .$$

By Proposition 7.2, we have

$$\tau_* \le \mu_\bullet = \mu \le \kappa \le \tau^* \text{ on } \mathcal{S},$$

hence $\mu = \kappa$ by the $\mathcal{S}$-minimality of κ.

We claim that μ is regular. By Theorem 6.9.(ii), we only have to show that $\mu(s) = \mu_*(s)$ for $s \in \mathcal{S}_-$.

Suppose that $\mu_*(s) < \mu(s)$ and choose $\gamma \in]\mu_*(s),\mu(s)[$. We prove the existence of an increasing linear functional ξ on $\mathcal{S}$ with

$$\tau_* \le \xi \le \tau^* \text{ on } \mathcal{S}, \quad \xi \le \kappa \quad \text{and} \quad \xi(s) \le \gamma.$$

Then we have $\xi = \kappa$ on $\mathcal{S}$ by the $\mathcal{S}$-minimality of κ, hence

$$\mu(s) = \xi(s) \le \gamma$$

contradicting the choice of γ. Therefore $\mu(s) = \mu_*(s)$.

In Proposition 17.1 and Theorem 17.3 we will show that such a functional ξ exists iff

$$\tau_*(r) \le \tau^*(u) + \kappa(v) + \alpha \cdot \gamma$$

holds for all $r,u,v \in \mathcal{S}$ and $\alpha \in \mathbb{R}_+$ with

$$r \le u + v + \alpha \cdot s.$$

Since

$$\tau_*(r) \le \tau_*(u + v + \alpha \cdot s) \le \tau^*(u) + \tau_*(v + \alpha \cdot s),$$

we have to prove that

$$\tau_*(v + \alpha \cdot s) \le \kappa(v) + \alpha \cdot \gamma.$$

But for $t \in \mathcal{T}$ with $-t \le v + \alpha \cdot s$, we have

$$-\tau(t) \le \kappa(v) + \alpha \cdot \gamma.$$

In fact, by the above, μ represents τ, i.e.

$$-\tau(t) = -\mu^*(t) = \sup_{w \in \mathcal{S},\, -w \le -t} -\mu(w),$$

and for any such w we have $-(w + v) \le -(t + v) \le \alpha \cdot s$. This yields

$$-[\mu(w) + \kappa(v)] = -\mu(w + v) \le \mu_*(\alpha \cdot s) \le \alpha \cdot \gamma$$

and thus

$$-\mu(w) \le \kappa(v) + \alpha \cdot \gamma. \quad \square$$

COROLLARY. *Let $\mathfrak{K}$ and $\mathfrak{G}$, respectively $\mathfrak{L}$ and $\mathfrak{H}$, be compatible lattices of sets and suppose that l is an $\mathfrak{H}$-regular content on $\mathfrak{L}$ such that for any $K \in \mathfrak{K}$ there exists $H \in \mathfrak{H}$ with $K \subset H$ and $l_*(H) < \infty$,*

$$l(L) = \sup_{K\in\mathfrak{K},\, K\subset L} l_*(K)$$

for $L \in \mathfrak{L}$ *and*

$$l_*(H) = \inf_{G\in\mathfrak{G},\, G\supset H} l^*(G)$$

for $H \in \mathfrak{H}$ *. Then there exists a* $\mathfrak{G}$*-regular content* m *on* $\mathfrak{K}$ *representing* l *.*

Note that $l(1_H) = l_*(1_H)$ for $H \in \mathfrak{H}$, hence l^* is finite on $\mathfrak{K}$, i.e. $l_* > -\infty$ on $\mathcal{E}(\mathfrak{K},\mathfrak{G})$, and

$$l(t) = \inf_{s\in\mathcal{E}(\mathfrak{K},\mathfrak{G}),\, s\geq t} l^*(s)$$

for $t \in \mathcal{E}(\mathfrak{L},\mathfrak{H})$. □

We can now deduce the Theorem of Henry :

EXAMPLE. *Let* X *be a Hausdorff space and* $\mathfrak{A}$ *an algebra of subsets of* X *. Let* l *be a finite content on* $\mathfrak{A}$ *such that for any* $A \in \mathfrak{A}$ *and* $\varepsilon > 0$ *there exists* $K \in \mathfrak{K}(X)$ *with* $K \subset A$ *and* $l^*(A \setminus K) \leq \varepsilon$ *. Then there exists a Radon measure* m *on* X *with*

$$m_*(A) = l(A) = m^*(A)$$

for $A \in \mathfrak{A}$ *.*

In fact, l is $\mathfrak{A}$-regular, in particular bounded, and the condition yields

$$l(A) = l_*(A) \leq l^*(A \setminus K) + l_*(K) \leq l_*(K) + \varepsilon ,$$

hence

$$l(A) = \sup_{K\in\mathfrak{K}(X),\, K\subset A} l_*(K)$$

for $A \in \mathfrak{A}$. Therefore, by Proposition 2.7 we get

$$l(A) = l_*(A) = l(X) - l(\complement A) = l(X) - \sup_{K\in\mathfrak{K}(X),\, K\subset\complement A} l_*(K) =$$

$$= \inf_{K\in\mathfrak{K}(X),\, K\subset\complement A} l(X) - l_*(K) = \inf_{K\in\mathfrak{K}(X),\, K\subset\complement A} l^*(X \setminus K) \geq$$

$$\geq \inf_{G\in\mathfrak{G}(X),\, G\supset A} l^*(G) \geq l(A) .$$ □

§ 17 HAHN–BANACH–ANDENAES THEOREM FOR CONOIDS

To make the text self-contained, we will now prove a sandwich theorem for minimal increasing linear functionals on function cones. This has been used previously in 7.7, 7.8, 14.3, and 16.9. The following exposition is a refinement of analogous results in the context of semigroups, due to *Anger* [1975].

We assume more generally that $\mathcal{S}$ is an *ordered conoid*, i.e. a commutative semigroup in which a multiplication with scalars from $\mathbb{R}_+$ is defined such that the associative and distributive laws hold, and such that the order is compatible with the conoid structure. In particular, the order might be replaced by equality, or by the reversed order.

We adopt the definitions of § 1, replacing function cones by ordered conoids. In particular, linear functionals are always $\tilde{\mathbb{R}}$–valued, in contrast to sub– or superlinear functionals.

17.1 In the following, we will use the notation

$$\rho \leq \theta$$

for functionals ρ and θ, defined respectively on subsets $\mathcal{R}$ and $\mathcal{T}$ of $\mathcal{S}$, if this inequality holds on $\mathcal{R} \cap \mathcal{T}$. We say that ρ is *increasingly smaller* than θ if

$$\rho(r) \leq \theta(t) \quad \text{for all } r \in \mathcal{R} \text{ and } t \in \mathcal{T} \text{ with } r \leq t\,.$$

If $\mathcal{R} = \mathcal{T}$ and ρ or θ is increasing, this is equivalent to $\rho \leq \theta$.

For i in some index set I, let θ_i be a sublinear and ρ_i a superlinear functional, each defined on a subconoid of $\mathcal{S}$. We define increasing functionals $\bigwedge_{i\in I} \theta_i$ and $\bigvee_{i\in I} \rho_i$ on $\mathcal{S}$ by

$$\bigwedge_{i\in I} \theta_i(s) := \inf\left\{ \sum_{i\in J} \theta_i(t_i) : J \text{ finite} \subset I\,,\ t_i \in \{\theta_i < \infty\} \text{ with } s \leq \sum_{i\in J} t_i \right\}$$

and

$$\bigvee_{i\in I} \rho_i(s) := \sup\left\{ \sum_{i\in J} \rho_i(r_i) : J \text{ finite} \subset I\,,\ r_i \in \{\rho_i > -\infty\} \text{ with } \sum_{i\in J} r_i \leq s \right\}.$$

We obviously have $\bigwedge_{i\in I} \theta_i \le \theta_j$ and $\bigvee_{i\in I} \rho_i \ge \rho_j$ for all $j \in I$, as well as $\bigwedge_{i\in I} \theta_i(0) \le 0$ and $\bigvee_{i\in I} \rho_i(0) \ge 0$.

In the case of two functionals θ_1, θ_2, respectively ρ_1, ρ_2, we denote the corresponding functionals by $\theta_1 \wedge \theta_2$, respectively $\rho_1 \vee \rho_2$.

PROPOSITION. *There exists an increasing sublinear functional* θ *on* $\mathcal{S}$ *with*

$$\theta \le \theta_i \quad \textit{for all } i \in I$$

iff $\bigwedge_{i\in I} \theta_i(0) \ge 0$. *In this case,* $\bigwedge_{i\in I} \theta_i$ *is the largest such functional.*

For any increasing sublinear functional θ with $\theta \le \theta_i$ for all $i \in I$, $s \in \mathcal{S}$ and finitely many $t_i \in \{\theta_i < \infty\} \subset \{\theta < \infty\}$ with

$$s \le \sum t_i ,$$

we have

$$\theta(s) \le \theta(\sum t_i) \le \sum \theta_i(t_i) ,$$

hence

$$\theta(s) \le \bigwedge_{i\in I} \theta_i(s) .$$

The condition is therefore necessary.

Conversely, the set

$$\mathcal{T} := \{\bigwedge_{i\in I} \theta_i < \infty\}$$

and the functional $\bigwedge_{i\in I} \theta_i$ are positively homogeneous.

To prove that $\mathcal{T}$ is stable w.r.t. addition and $\bigwedge_{i\in I} \theta_i$ is subadditive on $\mathcal{T}$, let $s,s' \in \mathcal{T}$ and finitely many $t_i,t_i' \in \{\theta_i < \infty\}$ with

$$s \le \sum t_i \quad \text{and} \quad s' \le \sum t_i'$$

be given. Then

$$s + s' \le \sum (t_i + t_i') ,$$

hence

$$\bigwedge_{i\in I} \theta_i(s + s') \le \sum \theta_i(t_i + t_i') \le \sum \theta_i(t_i) + \sum \theta_i(t_i') < \infty .$$

This shows that $s + s' \in \mathcal{T}$ and

$$\bigwedge_{i\in I} \theta_i(s + s') \le \bigwedge_{i\in I} \theta_i(s) + \bigwedge_{i\in I} \theta_i(s') . \quad \square$$

COROLLARY. *There exists an increasing superlinear functional* ρ *on* $\mathcal{S}$ *with*

$$\rho \geq \rho_i \quad \textit{for all } i \in I$$

iff $\bigvee_{i \in I} \rho_i(0) \leq 0$. *In this case,* $\bigvee_{i \in I} \rho_i$ *is the smallest such functional.*

This follows immediately from the proposition, since $\bigvee_{i \in I} \rho_i = - \bigwedge_{i \in I} (-\rho_i)$ with respect to the reversed order on $\mathcal{S}$. □

REMARKS.

(1) Without loss of generality, we may assume that all θ_i and ρ_i are defined on $\mathcal{S}$. In fact, we may always extend θ_i by ∞ on $\mathcal{S} \setminus \{\theta_i < \infty\}$, and ρ_i by $-\infty$ on $\mathcal{S} \setminus \{\rho_i > -\infty\}$, preserving the sublinearity of θ_i and the superlinearity of ρ_i , without changing the functionals $\bigwedge_{i \in I} \theta_i$ and $\bigvee_{i \in I} \rho_i$.

(2) Let families (r_i), (t_i) in $\mathcal{S}$ and (β_i), (γ_i) in $\tilde{\mathbb{R}}$ be given. For an increasing linear functional μ on $\mathcal{S}$, the sandwich problem

$$\beta_i \leq \mu(r_i) \quad \text{and} \quad \mu(t_i) \leq \gamma_i \quad \text{for all } i \in I$$

is equivalent to

$$\bigvee_{i \in I} \rho_i \leq \mu \leq \bigwedge_{i \in I} \theta_i ,$$

considering the linear functionals

$$\rho_i : \alpha \cdot r_i \longmapsto \alpha \cdot \beta_i \quad \text{and} \quad \theta_i : \alpha \cdot t_i \longmapsto \alpha \cdot \gamma_i ,$$

defined respectively on the subconoids $\mathbb{R}_+ \cdot r_i$ and $\mathbb{R}_+ \cdot t_i$.

17.2 **PROPOSITION.** *Let* θ *be a sublinear and* ρ *a superlinear functional. The functional* $\theta \frown \rho$ *, defined on* $\mathcal{S}$ *by*

$$\theta \frown \rho\,(s) := \inf \left\{ \theta(t) - \rho(r) : t \in \{\theta < \infty\} \, , \, r \in \{\rho > -\infty\} \, , \, s + r \leq t \right\} ,$$

is increasing and satisfies

$$\theta \frown \rho \leq \theta \, .$$

It is sublinear iff ρ *is increasingly smaller than* θ .

The first part is obvious, in particular $\theta \frown \rho\,(0) \leq 0$. The condition is equivalent to $\theta \frown \rho\,(0) \geq 0$, hence to $\theta \frown \rho$ being positively homogeneous.

To prove that $\mathcal{U} := \{\theta\ulcorner\rho < \infty\}$ is stable w.r.t. addition and $\theta\ulcorner\rho$ is sub-additive on $\mathcal{U}$, let $s,s' \in \mathcal{U}$, $r,r' \in \{\rho > -\infty\}$ and $t,t' \in \{\theta < \infty\}$ with

$$s + r \leq t \quad \text{and} \quad s' + r' \leq t'$$

be given. Then

$$(s + s') + (r + r') \leq t + t' ,$$

hence $s + s' \in \mathcal{U}$ and

$$\theta\ulcorner\rho\,(s + s') \leq \theta(t + t') - \rho(r + r') \leq [\theta(t) - \rho(r)] + [\theta(t') - \rho(r')] ,$$

thus

$$\theta\ulcorner\rho\,(s + s') \leq \theta\ulcorner\rho\,(s) + \theta\ulcorner\rho\,(s') . \quad \square$$

COROLLARY 1. *The functional* $\rho\llcorner\theta$ *, defined on* $\mathcal{S}$ *by*

$$\rho\llcorner\theta\,(s) := \sup\left\{\rho(r) - \theta(t) : r \in \{\rho > -\infty\} , t \in \{\theta < \infty\} , r \leq s + t\right\} ,$$

is increasing, satisfies

$$\rho\llcorner\theta \geq \rho ,$$

and $\{\rho\llcorner\theta > -\infty\}$ *is the set of all* $s \in \mathcal{S}$ *for which there exist* $r \in \{\rho > -\infty\}$ *and* $t \in \{\theta < \infty\}$ *with* $r \leq s + t$ *.*

It is superlinear iff ρ *is increasingly smaller than* θ *.*

This follows from the proposition, since $\rho\llcorner\theta = -(-\rho)\ulcorner(-\theta)$ with respect to the reversed order on $\mathcal{S}$. $\square$

COROLLARY 2. *An increasing linear functional* μ *on* $\mathcal{S}$ *satisfies*

$$\rho \leq \mu \leq \theta \quad \textit{iff} \quad \rho\llcorner\theta \leq \mu \leq \theta\ulcorner\rho .$$

We have

$$\rho\llcorner\theta \leq \theta\ulcorner\rho$$

iff ρ *is increasingly smaller than* θ *.*

For the first part, we only have to show that the condition is necessary. Let $s \in \mathcal{S}$, $r,r' \in \{\rho > -\infty\}$ and $t,t' \in \{\theta < \infty\} \subset \{\mu < \infty\}$ with

$$r' \leq s + t' \quad \text{and} \quad s + r \leq t$$

be given. Then

$$\rho(r') - \theta(t') \leq \mu(r') - \mu(t') \leq \mu(s) \leq$$

$$\leq \mu(t) - \mu(r) \leq \theta(t) - \rho(r)$$

and therefore

$$\rho \llcorner \theta\,(s) \le \mu(s) \le \theta \ulcorner \rho\,(s)\,.$$

For the second part, it remains to show that the condition implies $\rho \llcorner \theta\,(s) \le \theta \ulcorner \rho\,(s)$ for all $s \in \mathcal{S}$. For r,r' and t,t' as above, we have

$$r + r' \le r + s + t' \le t + t'$$

and hence

$$-\infty < \rho(r) + \rho(r') \le \rho(r + r') \le \theta(t + t') \le \theta(t) + \theta(t') < \infty\,.$$

This gives

$$\rho(r') - \theta(t') \le \theta(t) - \rho(r)$$

and therefore

$$\rho \llcorner \theta\,(s) \le \theta \ulcorner \rho\,(s)\,. \quad \square$$

REMARK. Let θ be a sublinear functional on a function cone $\mathcal{S} \subset \tilde{\mathbb{R}}^X$. Then

$$\theta^* = \theta \ulcorner 0 \quad \text{and} \quad \theta_* = 0 \llcorner \theta$$

on $\tilde{\mathbb{R}}^X$, where 0 is defined on $\{0\}$.

If θ is an increasing linear functional and τ denotes its restriction to $\{\theta < \infty\}_-$, then

$$\theta^\times = \tau \ulcorner \tau \quad \text{and} \quad \theta_\times = \tau \llcorner \tau$$

on $\tilde{\mathbb{R}}^X$.

17.3 THEOREM. *Let $\theta,\rho : \mathcal{S} \longrightarrow \tilde{\mathbb{R}}$ be increasing functionals on $\mathcal{S}$, respectively sublinear and superlinear, with*

$$\rho \le \theta\,.$$

If $\mathcal{K}$ is a subcone of $\mathcal{S}$, then the set $\Xi(\theta,\rho)$ of all increasing sublinear functionals ξ on $\mathcal{S}$ with

$$\rho \le \xi \le \theta$$

has a linear $\mathcal{K}$-minimal element μ, i.e. every $\xi \in \Xi(\theta,\rho)$ with $\xi \le \mu$ on $\mathcal{K}$ coincides with μ on $\mathcal{K}$.

We introduce a "lexicographic" order relation $\sqsubseteq$ on $\Xi := \Xi(\theta,\rho)$ with "initial" set $\mathcal{K}$ by

$$\xi \sqsubseteq \eta \quad \text{iff} \quad \xi_{|\mathcal{K}} < \eta_{|\mathcal{K}}\,, \text{ or } \xi_{|\mathcal{K}} = \eta_{|\mathcal{K}} \text{ and } \xi \le \eta\,.$$

Note that $\xi \le \eta \Rightarrow \xi \sqsubseteq \eta \Rightarrow \xi_{|\mathcal{K}} \le \eta_{|\mathcal{K}}$.

By Zorn's lemma, Ξ contains a maximal chain (ξ_i) . Since $(\xi_{i|\mathcal{K}})$ is a chain,

$$\kappa : k \longmapsto \inf \xi_i(k)$$

is an increasing sublinear functional on $\mathcal{K}$ with $\theta\wedge\kappa \in \Xi$ by Proposition 17.1. In fact, for $i \in I$, $s \in \mathcal{S}$, $t \in \{\theta < \infty\}$ and $k \in \{\kappa < \infty\}$ with $s \leq t + k$ we have

$$\rho(s) \leq \xi_i(s) \leq \xi_i(t) + \xi_i(k) \leq \theta(t) + \xi_i(k) ,$$

hence

$$\rho(s) \leq \theta(t) + \kappa(k)$$

and therefore

$$\rho(s) \leq \theta\wedge\kappa\,(s) .$$

We either have $\kappa < \xi_{i|\mathcal{K}}$ and hence $\theta\wedge\kappa_{|\mathcal{K}} < \xi_{i|\mathcal{K}}$, or $\kappa = \xi_{i|\mathcal{K}}$ and hence $\xi_i \leq \theta\wedge\kappa$ by Proposition 17.1. This gives $\theta\wedge\kappa \sqsubseteq \xi_i$, respectively $\xi_i \sqsubseteq \theta\wedge\kappa$. Thus, $\theta\wedge\kappa$ belongs to the chain, and the set J of all $i \in I$ with $\xi_i \sqsubseteq \theta\wedge\kappa$ is non-empty. Since $\xi_{j|\mathcal{K}} \leq \kappa$ for $j \in J$, we have $\xi_{j|\mathcal{K}} = \kappa$ which proves that $(\xi_j)_{j\in J}$ is a chain w.r.t. the pointwise order. Therefore,

$$\mu : s \longmapsto \inf_{j\in J}\ \xi_j(s)$$

is a sublinear functional on $\mathcal{S}$ which belongs to Ξ and coincides with κ on $\mathcal{K}$. Since $\mu \sqsubseteq \xi_i$ for all $i \in I$, the functional μ is a minimal element of Ξ . In particular, if $\xi \in \Xi$ and $\xi \leq \mu$ on $\mathcal{K}$, then $\xi = \mu$ on $\mathcal{K}$. In fact, if we had $\xi_{|\mathcal{K}} < \mu_{|\mathcal{K}}$, then $\xi \sqsubseteq \mu$, hence $\xi = \mu$, contradicting the assumption.

It therefore suffices to prove that μ is linear, i.e. that

$$\mu(s) + \mu(t) \leq \mu(s + t)$$

for all $s,t \in \mathcal{S}$ with $\mu(s + t) < \infty$.

By Corollary 17.2.2, we have

$$\rho \leq \rho\llcorner\mu \leq \mu\ulcorner\rho \leq \mu \leq \theta .$$

Therefore, $\mu\ulcorner\rho \in \Xi$ which, by the minimality of μ and $\mu\ulcorner\rho \sqsubseteq \mu$, implies

$$\mu = \mu\ulcorner\rho .$$

For $\gamma \in \mathbb{R}$ with $\gamma < \mu(s)$, consider the linear functional

$$\tau : \alpha\cdot s \longmapsto \alpha\cdot\gamma$$

on $\mathbb{R}_+\cdot s$. Let $\alpha \in \mathbb{R}_+$, $r,u \in \mathcal{S}$ with $\mu(u) < \infty$ and $\alpha\cdot s + r \leq u$. Then

$$\tau(\alpha\cdot s) = \alpha\cdot\gamma \leq \mu(\alpha\cdot s) = \mu\ulcorner\rho\,(\alpha\cdot s) \leq \mu(u) - \rho(r) ,$$

hence $\rho(r) \le \mu(u) - \tau(\alpha \cdot s)$ and therefore $\rho(r) \le \mu \llcorner \tau\, (r)$. This proves that

$$\rho \le \mu \llcorner \tau \le \mu \le \theta .$$

Proposition 17.2 shows that $\mu \llcorner \tau \in \Xi$, and hence $\mu = \mu \llcorner \tau$ by the minimality of μ. Thus

$$\mu(t) \le \mu(s+t) - \tau(s) = \mu(s+t) - \gamma ,$$

so

$$\gamma + \mu(t) \le \mu(s+t)$$

holds for any $\gamma < \mu(s)$. Therefore,

$$\mu(s) + \mu(t) \le \mu(s+t)$$

which finishes the proof that μ is linear. □

In the proof of Theorem 7.7 we have used the following

ADDENDUM. *The subset* $\Xi' := \Xi(\theta,\rho,\mathrm{P})$ *of all* $\xi \in \Xi := \Xi(\theta,\rho)$ *having a property* (P) *has a linear* $\mathcal{K}$*-minimal element if the following hold*:

(i) $\theta \in \Xi'$.

(ii) *If* (ξ_j) *is a chain in* Ξ' *w.r.t. the pointwise order, then* $\inf \xi_j \in \Xi'$.

(iii) *For any* $\xi \in \Xi'$ *and any superlinear functional* ζ *defined on a subcone of* $\mathcal{S}$ *with* $\xi \llcorner \zeta \in \Xi$, *we have* $\xi \llcorner \zeta \in \Xi'$.

One can modify the proof of the theorem suitably, noting first that Ξ' contains a maximal chain (ξ_i) w.r.t. $\sqsubseteq$ by (i). Since $(\xi_j)_{j \in J}$ is a chain w.r.t. the pointwise order, the functional $\mu := \inf_{j \in J} \xi_j$ belongs to Ξ' by (ii). Finally, (iii) shows that $\mu \llcorner \rho, \mu \llcorner \tau \in \Xi'$. □

COROLLARY. *For* $s \in \mathcal{S}$ *and* $\gamma \in \tilde{\mathbb{R}}$ *there exists an increasing linear functional* μ *on* $\mathcal{S}$ *with*

$$\rho \le \mu \le \theta \quad \textit{and} \quad \mu(s) \le \gamma \ , \ \textit{respectively} \quad \mu(s) \ge \gamma ,$$

iff

$$\gamma \ge \rho \llcorner \theta\, (s) \ , \ \textit{respectively} \quad \gamma \le \theta \llcorner \rho\, (s) .$$

In particular, there exists an increasing linear functional μ *on* $\mathcal{S}$ *with*

$$\rho \le \mu \le \theta \quad \textit{and} \quad \mu(s) = \rho \llcorner \theta\, (s) \ , \ \textit{respectively} \quad \mu(s) = \theta \llcorner \rho\, (s) .$$

The functionals $\rho \llcorner \theta$ *and* $\theta \llcorner \rho$ *are the lower, respectively upper envelope of all increasing linear functionals* μ *with* $\rho \le \mu \le \theta$.

Furthermore, if $\theta \llcorner\!\!\!- \rho\ (s) < \infty$ *, then there exists an increasing linear functional* μ *on* $\mathcal{S}$ *with*

$$\rho \le \mu \le \theta \quad \textit{and} \quad \mu(s) = \gamma$$

iff

$$\gamma \in [\rho \llcorner \theta\ (s),\ \theta \ulcorner \rho\ (s)]\ .$$

Let τ denote the linear functional defined on $\mathbb{R}_+ \cdot s$ by $\tau(s) = \gamma$. By the theorem, the existence of μ is equivalent to

$$\rho \le \theta \wedge \tau \ , \ \text{respectively} \quad \rho \vee \tau \le \theta \ ,$$

which in turn is equivalent to the respective condition for γ.

In the last case, there exist increasing linear functionals μ_1 and μ_2 with $\rho \le \mu_i \le \theta$ on $\mathcal{S}$ and

$$-\infty < \mu_1(s) \le \gamma \le \mu_2(s) \le \theta \ulcorner \rho\ (s) < \infty\ .$$

Therefore, there exists $\alpha \in [0,1]$ with

$$\alpha \cdot \mu_1(s) + (1-\alpha) \cdot \mu_2(s) = \gamma\ .$$

Then $\mu = \alpha \cdot \mu_1 + (1-\alpha) \cdot \mu_2$ has the desired properties. □

Remark. We say that $s \in \mathcal{S}$ satisfies the *cancellation law* in $\mathcal{S}$ if $r + s \le t + s$ implies $r \le t$ for all $r,t \in \mathcal{S}$.

In this case, one can omit the restriction $\theta \ulcorner \rho\ (s) < \infty$ in the last part of the corollary.

Indeed, one has to prove that $\rho \vee \tau \le \theta \wedge \tau$. Let $\alpha,\beta \in \mathbb{R}_+$ with $\beta\gamma < \infty$, $t \in \{\theta < \infty\}$ and $r \in \{\rho > -\infty\}$ such that

$$r + \alpha s \le t + \beta s$$

If $\alpha \le \beta$, then $r \le (\beta - \alpha) \cdot s + t$ by the cancellation law, hence

$$\rho(r) \le \theta(t) + (\beta - \alpha) \cdot \gamma\ ,$$

since $\gamma \ge \rho \llcorner \theta\ (s)$. The case $\alpha \ge \beta$ is similar. □

17.4 Theorem. *Let* $\mathcal{K}$ *,* $\mathcal{R}$ *and* $\mathcal{T}$ *be subconoids of* $\mathcal{S}$ *. Suppose that the functionals* $\theta : \mathcal{S} \longrightarrow \tilde{\mathbb{R}}$ *,* $\rho : \mathcal{R} \longrightarrow \tilde{\mathbb{R}}$ *and* $\tau : \mathcal{T} \longrightarrow \tilde{\mathbb{R}}$ *are respectively sublinear, superlinear and linear, such that*

$$\rho(r) + \tau(t) \leq \theta(s) + \tau(t')$$

holds for all $s \in \mathcal{S}, r \in \mathcal{R}$ *and* $t,t' \in \mathcal{T}$ *with* $r + t \leq s + t'$.

If

$$(\rho \vee \tau) \llcorner (\theta \wedge \tau) > -\infty \quad on\ \mathcal{S},$$

e.g. $\mathcal{R} + \mathcal{T}$ *is coinitial in* $\mathcal{S}$ *, then there exists a* $\mathcal{K}$*-minimal increasing linear extension* ξ *of* τ *to* $\mathcal{S}$ *with*

$$\rho \leq \xi \leq \theta .$$

Note that $\theta \wedge \tau$ and $\rho \vee \tau$ are increasing and sublinear, respectively superlinear by 17.1. The assumption is equivalent to $\rho \vee \tau \leq \theta \wedge \tau$, hence by Corollary 17.2.2 to

$$(\rho \vee \tau) \llcorner (\theta \wedge \tau) \leq (\theta \wedge \tau) \ulcorner (\rho \vee \tau) .$$

Therefore, $(\rho \vee \tau) \llcorner (\theta \wedge \tau)$ is superlinear with values in $\tilde{\mathbb{R}}$, and $(\theta \wedge \tau) \ulcorner (\rho \vee \tau)$ is sublinear. The result now follows by application of Theorem 17.3 to these increasing functionals. □

REMARK. By 17.1, the assumption is also necessary.

INDEX OF SYMBOLS

SUBJECT INDEX

Abbreviations: Corollary, Definition, Example, Lemma, Proposition, Remark, Theorem

REFERENCES

Adamski, W.: *Extensions of tight set functions with applications in topological measure theory.* Trans. Am. Math. Soc. 283 (1984), 353–368.

Alexandroff, A.D.: *Additive set-functions in abstract spaces.* Mat. Sb. 50 (1940), 307–348, 51 (1941), 563–628, 55 (1943), 169–238.

Alfsen, E.M.: *On a general theory of integration based on order.* Math. Scand. 6 (1958), 67–79.

– *Order theoretic foundations of integration.* Math. Ann. 149 (1963), 419–461.

Andenaes, P.R.: *Hahn–Banach extensions which are maximal on a given cone.* Math. Ann. 188 (1970), 90–96.

Anger, B.: *Minimale Fortsetzungen additiver Funktionale.* Sitzungsber. Bayer. Akad. Wiss. Math.-Naturwiss. Kl. (1975), 23–42.

Anger, B., Portenier, C.: *Radon integrals and Riesz representation.* To appear in: Proc. Conf. Measure Theory, Oberwolfach 1990. Rend. Circ. Mat. Palermo, Suppl. (1991).

Arens, R.: *Representation of moments by integrals.* Ill. J. Math. 7 (1963), 609–614.

Bachman, G., Sultan, A.: *On regular extensions of measures.* Pac. J. Math. 86 (1980), 389–395.

Bassan, B.: *Comparison of two approaches to integration on separated topological spaces.* J. Math. Anal. Appl. 112 (1985), 391–395.

Bauer, H.: *Über die Beziehungen einer abstrakten Theorie des Riemann-Integrals zur Theorie Radonscher Maße.* Math. Z. 65 (1956), 448–482.

– *Maß- und Integrationstheorie.* De Gruyter, Berlin–New York (1990).

Berg, C., Christensen, C.P.R., Ressel, P.: *Harmonic analysis on semigroups.* Springer, New York (1984).

Bhaskara Rao, K.P.S., Bhaskara Rao, M.: *Theory of charges, a study of finitely additive measures.* Academic Press, London (1983).

Bourbaki, N.: *Intégration, Chap. V.* Hermann, Paris (1956, 1967).

– *Intégration, Chap. IX.* Hermann, Paris (1969).

– *Topologie générale, Chap. V–X.* Hermann, Paris (1974).

Choksi, J.R.: *On compact contents.* J. Lond. Math. Soc. 33 (1958), 387–398.

Choquet, G.: *Theory of capacities.* Ann. Inst. Fourier 5 (1955), 131–295.

– *Le problème des moments.* Sém. Choquet, 1re année (1962), Exp. n°. 4.

Daniell, P.J.: *A general form of integral.* Ann. Math. 19 (1917/18), 279–294.

– *Further properties of the general integral.* Ann. Math. 21 (1919/20), 203–220.

Darboux, M.G.: *Mémoire sur les fonctions discontinues.* Ann. Sci. Ec. Norm. Sup. 4 (1875), 57–112.

Dunford, N., Schwartz, J.T.: *Linear Operators Part I.* Interscience Publishers, New York (1958).

Fremlin, D.H.: *Topological measure spaces : two counter-examples.* Math. Proc. Camb. Phil. Soc. 78 (1975), 95–106.

Goldstine, H.H.: *Linear functionals and integrals in abstract spaces.* Bull. Am. Math. Soc. 47 (1941), 615–620.

Günzler, H.: *Linear functionals which are integrals.* Rend. Sem. Mat. Fis. Milano 43 (1973), 167–176.

– *Integral representations with prescribed lattices.* Rend. Sem. Mat. Fis. Milano 45 (1975), 107–168.

Haupt, O., Pauc, C.: *Mesures et topologies adaptées. Espaces mesurés topologiques.* C. R. Acad. Sci. Paris 230 (1950), 711–712.

Heiden, U. an der: *On the representation of linear functionals by finitely additive set functions.* Arch. Math. 30 (1978), 210–214.

Henry, J.P.: *Prolongements de mesures de Radon.* Ann. Inst. Fourier 19.1 (1969), 237–247.

Ionescu Tulcea, A. and C.: *Topics in the theory of lifting.* Springer, Berlin-Heidelberg-New York (1969).

Kelley, J.L., Srinivasan, T.P.: *Pre-measures on lattices of sets.* Math. Ann. 190 (1971), 233–241.

– *Measure and integral - a new gambit.* Measure Theory, Oberwolfach 1983, Lect. Notes Math. 1089 (1984), 120–126.

Kelley, J.L., Nayak, M.K., Srinivasan, T.P.: *Pre-measures on lattices of sets II.* Proc. Symp. Vector and Operator valued Measures and Applications, Alta, Utah 1972, 155–164. D.H. Tucker, H.B. Maynard eds. Academic Press, New York (1973).

Kisynski, J.: *Remark on strongly additive set functions.* Fund. Math. 63 (1968 a), 327–332.

– *On the generation of tight measures.* Studia Math. 30 (1968 b), 141–151.

König, H.: *On the basic extension theorem in measure theory.* Math. Z. 190 (1985), 83–94.

Lebesgue, H.: *Intégrale, Longueur, Aire.* Ann. di Mat. 7 (1902), 231–359.

LeCam, L.: *Convergence in distribution of stochastic processes.* Univ. Calif. Publ. Statistics 2 (1957), 207–236.

Lembcke, J.: *Konservative Abbildungen und Fortsetzung regulärer Maße.* Z. Wahrsch. verw. Geb. 15 (1970), 57–96.

– *Reguläre Maße mit einer gegebenen Familie von Bildmaßen.* Sitzungsber. 1976 Bayer. Akad. Wiss. Math.-Naturwiss. Kl. (1977), 61–115.

Loomis, L.H.: *Linear functionals and content.* Am. J. Math. 76 (1954), 168–182.

Luther, N.Y. : *A decomposition of measures.* Can. J. Math. 20 (1968), 953–959.

Markoff, A.: *On mean values and exterior densities.* Mat. Sb. 46 (1938), 165–191.

Pettis, B.J.: *On the extension of measures.* Ann. Math. 54 (1951), 186–197.

Pollard, D., Topsoe, F.: *A unified approach to Riesz type representation theorems.* Studia Math. 54 (1975), 173–190.

Portenier, C.: *Caractérisation de certains espaces de Riesz.* Sém. Choquet, 10e année (1970/71), Exp. n°. 6.

– *Formes linéaires positives et mesures.* Sém. Choquet, 10e année (1970/71), Comm. n°. 6.

– *Le théorème de Prokhorov-Sazonov comme théorème de représentation.* To appear in: Proc. Conf. Measure Theory, Oberwolfach 1990. Rend. Circ. Mat. Palermo (Suppl. 1991).

Radon, J.: *Theorie und Anwendungen der absolut additiven Mengenfunktionen.* Sitzungsber. Österr. Akad. Wiss. Math.-Naturwiss. Kl. 122 (1913), 1295–1438.

Riemann, B.: *Über die Darstellbarkeit einer Funktion durch eine trigonometrische Reihe* (1854). In: Gesammelte mathematische Werke. Teubner, Leipzig (1876), 213–250.

Riesz, F.: *Sur les opérations fonctionnelles linéaire.* C. R. Acad. Sci. Paris 149 (1909), 974–977.

– *Sur certains systèmes singuliers d'équations intégrales.* Ann. Sci. Ec. Norm. Sup. 28 (1911), 33–62.

Schwartz, L.: *Mesures de Radon sur des espaces topologiques arbitraires.* Cours de 3ème cycle, Institut H. Poincaré, Paris (1964/65).

Schwartz, L.: *Radon measures on arbitrary topological spaces and cylindrical measures.* Oxford University Press, London (1973).

Smiley, M.F.: *An extension of metric distributive lattices with an application in general analysis.* Trans. Am. Math. Soc. 56 (1944), 435–447.

Srinivasan, T.P.: *On extension of measures.* J. Indian Math. Soc. 19 (1955), 31–60.

Stone, M.H.: *Notes on integration I–IV.* Proc. Nat. Acad. Sci. USA 34 (1948), 336–342, 447–455, 483–490, 35 (1949), 50–58.

Topsoe, F.: *Topology and measure.* Lect. Notes Math. 133 (1970).

– *Compactness in spaces of measures.* Studia Math. 36 (1970), 195–212.

– *On construction of measures.* Proc. Conf. Topology and Measure I, Zinnowitz 1974, 343–381. Ernst-Moritz-Arndt-Universität, Greifswald (1978).

– *Further results on integral representations.* Studia Math. 55 (1976), 239–245.

– *Radon measures, some basic constructions.* Measure Theory and Applications, Sherbrooke 1982, Lect. Notes Math. 1033 (1983), 303–311.

Varadarajan, V.S.: *Measures on topological spaces.* Am. Math. Soc. Transl. 48 (1965), 161–228.

Young, W.H.: *On a new method in the theory of integration.* Proc. Lond. Math. Soc. 9 (1911), 15–50.

Progress in Mathematics

Edited by:

J. Oesterlé
Département de Mathématiques
Université de Paris VI
4, Place Jussieu
75230 Paris Cedex 05, France

A. Weinstein
Department of Mathematics
University of California
Berkeley, CA 94720
U.S.A.

Progress in Mathematics is a series of books intended for professional mathematicians and scientists, encompassing all areas of pure mathematics. This distinguished series, which began in 1979, includes authored monographs and edited collections of papers on important research developments as well as expositions of particular subject areas.

We encourage preparation of manuscripts in some form of TeX for delivery in camera-ready copy which leads to rapid publication, or in electronic form for interfacing with laser printers or typesetters.

Proposals should be sent directly to the editors or to: Birkhäuser Boston, 675 Massachusetts Avenue, Cambridge, MA 02139, U. S. A.

A complete list of titles in this series is available from the publisher.

53 LAURENT. Théorie de la Deuxième Microlocalisation dans le Domaine Complexe.
54 VERDIER/LE POTIER. Module des Fibres Stables sur les Courbes Algébriques: Notes de l'Ecole Normale Supérieure, Printemps, 1983
55 EICHLER/ZAGIER. The Theory of Jacobi Forms
56 SHIFFMAN /SOMMESE. Vanishing Theorems on Complex Manifolds
57 RIESEL. Prime Numbers and Computer Methods for Factorization
58 HELFFER/NOURRIGAT. Hypoellipticité Maximale pour des Opérateurs Polynomes de Champs de Vecteurs
59 GOLDSTEIN. Séminaire de Théorie des Nombres, Paris 1983–84
60 PROCESI. Geometry Today: Giornate Di Geometria, Roma. 1984
61 BALLMANN/GROMOV/SCHROEDER. Manifolds of Nonpositive Curvature
62 GUILLOU/MARIN. A la Recherche de la Topologie Perdue
63 GOLDSTEIN. Séminaire de Théorie des Nombres, Paris 1984–85
64 MYUNG. Malcev-Admissible Algebras
65 GRUBB. Functional Calculus of Pseudo-Differential Boundary Problems
66 CASSOU-NOGUES/TAYLOR. Elliptic Functions and Rings and Integers
67 HOWE. Discrete Groups in Geometry and Analysis: Papers in Honor of G.D. Mostow on His Sixtieth Birthday
68 ROBERT. Autour de L'Approximation Semi-Classique
69 FARAUT/HARZALLAH. Deux Cours d'Analyse Harmonique
70 ADOLPHSON/CONREY/GHOSH/YAGER. Analytic Number Theory and Diophantine Problems: Proceedings of a Conference at Oklahoma State University

71 GOLDSTEIN. Séminaire de Théorie des Nombres, Paris 1985–86

72 VAISMAN. Symplectic Geometry and Secondary Characteristic Classes

73 MOLINO. Riemannian Foliations

74 HENKIN/LEITERER. Andreotti-Grauert Theory by Integral Formulas

75 GOLDSTEIN. Séminaire de Théorie des Nombres, Paris 1986–87

76 COSSEC/DOLGACHEV. Enriques Surfaces I

77 REYSSAT. Quelques Aspects des Surfaces de Riemann

78 BORHO /BRYLINSKI/MACPHERSON. Nilpotent Orbits, Primitive Ideals, and Characteristic Classes

79 MCKENZIE/VALERIOTE. The Structure of Decidable Locally Finite Varieties

80 KRAFT/PETRIE/SCHWARZ. Topological Methods in Algebraic Transformation Groups

81 GOLDSTEIN. Séminaire de Théorie des Nombres, Paris 1987–88

82 DUFLO/PEDERSEN/VERGNE. The Orbit Method in Representation Theory: Proceedings of a Conference held in Copenhagen, August to September 1988

83 GHYS/DE LA HARPE. Sur les Groupes Hyperboliques d'après Mikhael Gromov

84 ARAKI/KADISON. Mappings of Operator Algebras: Proceedings of the Japan-U.S. Joint Seminar, University of Pennsylvania, Philadelphia, Pennsylvania, 1988

85 BERNDT/DIAMOND/HALBERSTAM/ HILDEBRAND. Analytic Number Theory: Proceedings of a Conference in Honor of Paul T. Bateman

86 CARTIER/ILLUSIE/KATZ/LAUMON/ MANIN/RIBET. The Grothendieck Festschrift: A Collection of Articles Written in Honor of the 60th Birthday of Alexander Grothendieck. Vol. I

87 CARTIER/ILLUSIE/KATZ/LAUMON/ MANIN/RIBET. The Grothendieck Festschrift: A Collection of Articles Written in Honor of the 60th Birthday of Alexander Grothendieck. Volume II

88 CARTIER/ILLUSIE/KATZ/LAUMON/ MANIN/RIBET. The Grothendieck Festschrift: A Collection of Articles Written in Honor of the 60th Birthday of Alexander Grothendieck. Volume III

89 VAN DER GEER/OORT / STEENBRINK. Arithmetic Algebraic Geometry

90 SRINIVAS. Algebraic K-Theory

91 GOLDSTEIN. Séminaire de Théorie des Nombres, Paris 1988–89

92 CONNES/DUFLO/JOSEPH/RENTSCHLER. Operator Algebras, Unitary Representations, Envelopping Algebras, and Invariant Theory. A Collection of Articles in Honor of the 65th Birthday of Jacques Dixmier

93 AUDIN. The Topology of Torus Actions on Symplectic Manifolds

94 MORA/TRAVERSO (eds.) Effective Methods in Algebraic Geometry

95 MICHLER/RINGEL (eds.) Represenation Theory of Finite Groups and Finite Dimensional Algebras

96 MALGRANGE. Equations Différentielles à Coefficients Polynomiaux

97 MUMFORD/NORI/NORMAN. Tata Lectures on Theta III

98 GODBILLON. Feuilletages, Etudes géométriques

99 DONATO /DUVAL/ELHADAD/TUYNMAN. Symplectic Geometry and Mathematical Physics. A Collection of Articles in Honor of J.-M. Souriau

100 TAYLOR. Pseudodifferential Operators and Nonlinear PDE

101 BARKER/SALLY. Harmonic Analysis on Reductive Groups

102 DAVID. Séminaire de Théorie des Nombres, Paris 1989-90

103 ANGER /PORTENIER. Radon Integrals